Electric Circuits - Analysis and Design

Electrical and Electronic Engineering Design Series
Electric Circuits Analysis and Design

Electronic Circuit Design   with Bipolar and MOS Transistors

CMOS Circuit Design Analog, Digital, IC Layout

Digital Design Logic, Memory, Computers

Analog Filter Design

Error Correction Code Design

Computer Science Design Series
Programming with MFC & Visual C++

Mathematics
Arithmetic – Integers, Fractions, Decimals

Algebra – A Clear Presentation

Electric Circuits

Analysis and Design

Nicholas L. Pappas. Ph.D.

Copyright © 2013 by Nicholas L. Pappas, Ph.D.

All Rights Reserved Worldwide.

Except as permitted under the Copyright Act of 1976, no part of this book may be reproduced in whole or in part in any manner. Not in any form or by any electronic or mechanical means, nor stored in a retrieval system, or, transmitted, in any form or by any means, electronic, mechanical, photocopying, recording, or otherwise, without the express permission of Nicholas L. Pappas, Ph.D.

ISBN-13: 978-1494273385   ISBN-10: 1494273381

A Message about this Text: The subject is essentially endless. The purpose here is to say enough about the subject so that you, the reader, have a running start when you apply this knowledge to your work.

Algebraic skill is an essential prerequisite. Knowledge of some elementary calculus is also a prerequisite.[1]

We believe important benefits accrue by doing the problems carefully, by reconstructing the Spice programs and running them to reproduce the text figures, by deriving the text equations, and and the very important effort doing the experiments. These efforts in effect provide "startup" work experience.

Once you have some work experience we are confident that you will be able to expand your know how with reasonable effort.

A Message from the Author: I have worked continuously in the electronics industry since 1950 except for 11 semesters teaching at San Jose State University (Professor and Chair Computer Engineering 1988-1993). There I discovered my talent for teaching such as it may be. After War2 I attended Lehigh University, and then transferred to Stanford where I earned the MS degree and, while working at HP in the early 1950's, the Ph.D. EE degree. (Somehow I did not get the word and formally apply for the BS degree.) Hardware design has been my principal activity. I learned enough about assembly language, Forth, C and C++ to design the software I needed for my projects. My current activity is designing integrated circuits.

---

[1] Thompson and Gardner, *Calculus Made Easy*, ISBN 0312 185 480 (pbk)
   Morris Kline, *Calculus,*   ISBN 0486 404 536 (pbk)

Preface

This entry level text is for anyone who wants to know how to analyze and, in time, design electric circuits. The prerequisites are algebra and the elementary calculus.

Once you have made your way through this text you will be able to do a node or mesh analysis of any linear circuit, while understanding resistors R, capacitors C, inductors L, transformers & mutual inductance M, as well as independent and dependent sources of current and voltage. Transistors create electronic circuits.[1] The results of an analysis are usually what are referred to as a transfer function(s) of a signal input to one or more outputs. With the transfer function in hand you then can produce a frequency response or transient response (having learned about the Laplace Transform). And, that you can write Spice programs to do the number crunching for you by calculating and plotting the results. Here are the details.

The text starts with the brilliant experiments that discovered electricity, which revealed that electricity is charge q at rest and in motion. Charge q is made available by free electrons each of which have a charge q_E attached. This activity led to the concepts of current, voltage and power.

Then we show that there are two basic classes of laws (1) connection constraints, which are Kirchhoff's' laws showing how currents and voltages in any circuit relate to each other and (2) voltage-current *vi* constraints for resistors, capacitors and inductors showing how current relates to voltage in each component.

We show how to analyze circuits for currents and voltages that are assembled from resistors only, because only basic algebra is required which allows us to focus on the methods.

We explain capacitors C and inductors L as we derive their simple differential equation voltage-current constraints, which escalate the math required to the elementary calculus.

Transformers based on mutual inductance M are explained as we derive their equivalent circuits and frequency response.

---

[1] N. L. Pappas *Electronic Circuit Design – with Bipolar and MOS Transistors*

Electric Circuits - Analysis and Design

Two general analysis methods, node and mesh, are presented so that you can analyze any circuit. The node method is based on Kirchhoff's current connection constraint law, and the mesh method is based on Kirchhoff's voltage connection constraint law.

A circuit is in the steady state after any transients die out. Since the impedances of capacitors C and inductors L vary with frequency any circuit's properties with L and C components are frequency dependent. In other words in the steady state a circuit with L and, or C has a frequency response. Frequency responses are important, because many circuit design goals are a specified over a range of frequencies.

Transient response shows how a circuit responds to signals. We show how to use the Laplace Transform Method to find any circuit's transient response (and frequency response). The Laplace transform is a terrific tool for solving ordinary and partial differential equations. The important related ideas of real and complex frequencies, and the complex frequency plane are discussed.

We explain the phasor idea so that you know about it. Then you can set it aside, because the idea is a special case of the Laplace transform. We chose not to burden you with special cases in our texts.

One mesh RL and RC circuits are widely used. They are worth knowing about. We apply the straightforward Laplace Transform method to analyze elementary RC and RL circuits emphasizing transfer functions, time constant, impedance, and stored energy.

Hendrick Bode invented a widely adopted method for making graphical displays of the magnitude and phase of electric circuit functions over frequency ranges. We show how to use the Bode graphical method that converts abstract ratios of polynomials encountered in electric circuit mathematics into clear visual displays.

The reactance chart is a graphical display of the impedance magnitude of R, L, and C components over a frequency range. The log-log scales span many decades of magnitude and frequency on one page. We show how to use this eminently practical time-saving chart for making estimates and selecting "in the ballpark" values for components appropriate to the problem at hand.

Preface

We show how to write Spice programs that illustrate direct voltage and current (DC) analysis, alternating voltage and current (AC) analysis, and transient (TRAN) analysis. There are many Spice features for special applications that we do not present. Nevertheless this is a running start. Spice does the nitty gritty numerical calculations and data plotting for you. Spice is used in most chapters to calculate results and plot data. Spice has an important role in the modern design process.

We give the simulation program Spice a significant role in the text.

Many useful ideas and important topics are found in the Appendix.

We include useful *experiments* that give you real world experience.

What is an *analog* electric circuit? Well, it is an assembly of electrical parts connected by wires, which may be in discrete or integrated circuit form. The good news is that the number of *different basic parts* is less than ten. So we need to learn the properties of these parts, and how they can relate to each other. We learn how they can be designed into a circuit.

We start with their names - *resistor, capacitor, inductor, voltage source, current source, wire, bipolar junction transistor, MOS transistor, and diode*. We use most of the parts in this text's circuit analysis's and designs. Transistors are presented in our text *Electronic Circuit Design - with Bipolar and MOS Transistors*

The good news is that a mathematical theory for analysis and synthesis is available. The mathematics takes several forms. There is the traditional form of written mathematics. There is Spice, a software form, using mathematics behind the scenes to evaluate circuit performance. And, there are Bode diagrams and Reactance charts that are graphical forms that convert electric circuit mathematics into comprehensible displays. We use all of these forms in the text.

> Our blog *npappasee.blogspot.com* may offer you additional information. Take a look.

> We would appreciate receiving your comments and views on this text at *npappasz@yahoo.com*.

Electric Circuits - Analysis and Design

1 Circuit Theory Fundamentals To start we go back to the early 1800's. when a number of very smart people ran a series of brilliant experiments whose results produced the basis for the theory of electricity. They showed that electricity is charge q at rest and in motion. In the beginning Michael Faraday and others discovered that charge creates a force field, the *Electric Field*. This discovery led to the concepts of *current, voltage, and electric power*. Later, as the concept of a circuit evolved, Gustav Kirchhoff published his *current and voltage connection constraint* laws about circuits, while Georg Ohm published his *voltage-current constraint* law (*vi constraint law*) relating current, voltage, and resistance R.

2 Circuits with Resistors Starting with resistors-only circuits that only require algebra, we show how elementary circuit analysis solves for currents and voltages in circuits where resistors are connected in series, in parallel, or in a series-parallel combination. The analysis methods are based on Kirchhoff's laws that produce equations that the sum of voltages around a mesh equals zero or the sum of currents at a node equal zero. And, Ohm's law that replaces voltages or currents in the equations via v=iR or i=v/R expressions. (The general analysis methods for circuits with all types of components are fully developed in Chapter 5.)

A DC Voltmeter Circuit, DC Ammeter Circuit, Ohmmeter Circuit, and AC Voltmeter Circuit are designed.

3 RLC Circuit Components   Three passive electric circuit components, resistors R, capacitors C, and inductors L are explained. (Passive means there are no internal sources of energy.) The R, L, C components' properties are defined by laws relating the voltage difference between a component's two terminals and the current flowing through the component. The generic term for these laws is *vi constraint*. Ohm's law is the vi constraint for R. We derive the vi constraints for L and C, and discuss their transient and steady state responses. The mathematics escalates here, because the L and C vi constraints are differential equations.

4. Transformers and Mutual Inductance A transformer is an assembly of two or more windings (coils of wire) wound on a structure known as a core. The principal transformer application is the production of different sinusoidal voltage levels that are determined by the turns ratio of the windings. The mutual inductance coefficient of coupling k (0 to 1) is a measure of what fraction of the magnetic field produced by one winding links with another winding. If a core material has high magnetic permeability then k is, in effect, equal to one. The ideal transformer is defined, and its properties are examined. Practical transformer equivalent circuits are derived. Practical transformers low and high frequency responses show how transformer properties affect transformer bandwidth. A full wave rectifier circuit design is presented.

5 Circuit Analysis The laws we need to solve circuit problems are Kirchhoff's *connection constraint* laws and the *vi constraint laws for R, L, and C*. In what follows we focus on learning how to analyze circuits in the easiest way we know how. We get ready by discussing the circuit analysis process. Then we present the two general methods of circuit analysis: the *node method* and the *mesh method*.

The node method is based on Kirchhoff's *current connection constraint law* that sets up the circuit analysis equations so that *the algebraic sum of all currents entering and leaving one node is zero*. We show how to deal with a voltage source in a branch, which is not a current we can add to the sum. This is important: *this is done without specifying the types of parts*.

The mesh method is based on Kirchhoff's *voltage connection constraint law* that sets up the circuit analysis equations so that *the algebraic sum of the voltage differences around every closed circuit contour is zero*. We show how to deal with a current in a branch, which is not a voltage we can add to the sum.

In either method specific parts are placed in the circuit by invoking the vi constraints for R, L, and C. For example a voltage is replaced by *Ldi/dt*, or *iR*, and a current by *Cdv/dt*.

The infinite number of electric circuits are readily analyzed in this way.

We illustrate these ideas by analyzing two circuits with the same topology - a resistor only Attenuator Circuit, and an LC Low Pass Filter Circuit.

6 Frequency Response A circuit is either in a *transient state* or, after any transients die out, the *steady state*. Connecting a sinewave source to a circuit input forces the input voltage to jump to some non-zero value. The jump creates transients within the circuit as signal energy propagates into the circuit. Soon the transients die out and only constant amplitude sinewaves, *which do not generate transients*, appear at every circuit node. This is the steady state.

Electric Circuits - Analysis and Design

Know that impedance z is the general term for resistance to current flow by any type of component. The general *steady state* expression is z=v/i. We show that the impedances of capacitors C and inductors L are $z_C=1/pC$ and $z_L=pL$ where $p=j\omega=j2\pi f$.

Since the impedances of capacitors C and inductors L vary with frequency the circuit properties are frequency dependent. In other words *in the steady state* a circuit with L and or C has a *frequency response*. For example vary the frequency of a *constant amplitude* sinewave input voltage to produce an output voltage whose amplitude varies as a function of frequency.

Frequency responses are important, because many circuit design goals are a specified over a range of frequencies. The frequency responses of four circuits are analyzed so you will know how to find frequency responses. They are a resistor R, an RC low pass filter, an RC phase equalizer, and an RLC series resonant circuit.

We show how to use the Spice simulation program to calculate a frequency response so that you do not have to spend time calculating numerical solutions. We include frequency response plots for various circuits, which relate performance to schematics.

7 Transient Response Transient responses of electric circuits require solutions of ordinary and partial differential equations. This is why we use *The Laplace Transform Method*. The Laplace transform is a terrific tool for solving ordinary and partial differential equations. The Laplace transform transforms the time domain differential equations of electric circuits into frequency domain algebraic equations (the algebraic equations are used to produce frequency responses). The algebraic equations are manipulated to solve for the variables of interest. Then the manipulated equations are inverse transformed back to the time domain solution of the original problem represented by the differential equations.

The transient response equations of RL, RC, and RLC circuits are developed, and Spice programs plot the frequency and transient responses. The important ideas of real and complex frequencies, and the complex frequency plane are discussed.

8 The Steady State and the ω Phasor Method. We explain the phasor idea so that you know about it. Then you can set it aside, because the idea is a special case of the Laplace transform. We chose not to burden you with special cases.

Preface

We show that traditional AC analysis methods producing solutions to the differential equations of electric circuits are tedious at best, have limited capability, and are definitely not efficient. The process solving the equations is simplified when sin x and cos x functions are replaced by e^{jx} that is then used as a transform. We refer to this process as the ω phasor method, because one could argue it is a special case of the Laplace Transform "p phasor method." The ω phasor method focuses on the steady-state terms.

9 One Mesh RL and RC Circuits One mesh RL and RC circuits are widely used. They are worth knowing about. We apply the straightforward Laplace Transform method to analyze elementary RC and RL circuits emphasizing *transfer functions, time constant, impedance, and stored energy*. The steady state frequency response, and the transient response to steps, pulses, and ramps are presented for the series and parallel circuit combinations of one R and one C, and of one R and one L.

10 The Bode Method Hendrick Bode invented a widely adopted method for making graphical displays of the magnitude and phase of electric circuit functions over frequency ranges. The Bode method is an important, very useful, graphical method that converts abstract ratios of polynomials encountered in electric circuit mathematics into clear visual displays.

We use the method in our work. In the majority of applications response plots accurate to within 1% or 2% are sufficient. The Bode method avoids laborious calculations of the points to be plotted. You only draw straight lines when you use the Bode method.

11 Reactance Chart Method The reactance chart is a graphical display of the impedance magnitude of R, L, and C components over a frequency range. The log-log scales span many decades of magnitude and frequency *on one page*. This eminently practical chart allows you to make estimates and to select "in the ballpark" values for components appropriate to the problem at hand. The visual display provided by reactance charts saves a great deal of effort. Several examples of impedance and transfer function plots illustrate how the chart is used.

12 How to write AC, DC, and TRAN Spice Programs We show how to write Spice programs that illustrate direct voltage and current (DC), alternating voltage and current (AC), and transient (TRAN) analysis. There are many Spice features for special applications that we do not present. Nevertheless this is a running start.

Electric Circuits - Analysis and Design

Spice is used to plot performance of circuits analyzed in the text: DC response, AC frequency response, and TRAN transient response. You see results immediately.

We recognize that you do not have to crunch numbers today, because Spice is the modern way to crunch numbers. That is why we include Spice in the text, and do not delegate it to "labs."

> Spice has an important role in the modern design process.

Appendix Many useful ideas and important topics are found here.
A1 Partial Fraction Expansions
A2 Complex Numbers
A3 Oliver Heaviside's Method
A4 Easy Method for Evaluating Laplace Transforms
A5 Topology Applied to Electric Circuits
A6 Delta-Wye Transformation
A7 Equivalent circuits
A8 Signal Sources
A9 Impedance, Admittance, and Immittance

Answers to Most of the Problems are included

Ten Experiments *Each experiment has several "sub" experiments.*
The Solderless Breadboard and the Power Supply
Experiment 1 Circuit with Resistors
Experiment 2 Kirchhoff's Laws
Experiment 3 Bridged-T Attenuator
Experiment 4 AC Voltmeter
Experiment 5 Capacitance
Experiment 6 Inductance
Experiment 7 Transformers
Experiment 8 Circuit Analysis
Experiment 9 Frequency Response
Experiment 10 Transient Response
Solved

> We will use multiple copies of various figures in order to have a figure on the same page as the text referring to that figure.

Electric Circuits - Analysis and Design

Contents

1 Circuit Theory Fundamentals 1
1.1 Electrostatic Field Theory 1
Charge 4
Fields of Force 5
1.2 Current, Voltage, and Power 6
1.2.1 Current 6
1.2.2 Voltage 8
1.2.3 Power 10
1.3 Kirchhoff's Laws 11
1.3.1 Kirchhoff's Current Law (KCL) 11
1.3.2 Kirchhoff's Voltage Law (KVL) 12
The Forest and the Trees 13

2 Circuits with Resistors 14
2.1 Resistors in Series 14
2.2 Resistors in Parallel 16
2.3 Resistors in Series-Parallel 18
2.4 Voltage Division 20
2.5 Current Division 20
2.6 Multimeter Design 21
2.6.1 Display Meter Circuit 21
2.6.2 DC Voltmeter Circuit 22
2.6.3 DC Ammeter Circuit 23
2.6.4 Ohmmeter Circuit 24
2.6.5 AC Voltmeter Circuit 25

3 RLC Circuit Components 27
3.1 Resistor R 27
3.2 Capacitor C 28
3.2.1 Theoretical Basis of the Capacitor 28
3.2.2 Capacitor in a circuit 32
3.2.3 Capacitors in series 33
3.2.4 Capacitors in parallel 34
3.2.5 Capacitor Impedance, Transient and Steady State Responses 35
3.3 Inductor L 36
3.3.1 Theoretical Basis of the Inductor 37
3.3.2 Inductor in a circuit 38
3.3.3 Inductors in series 40
3.3.4 Inductors in parallel 40
3.3.5 Inductor Impedance, Transient and Steady State Responses 41

4 Transformers and Mutual Inductance ... 42
4.1 Mutual Inductance $M = k\sqrt{L_1 L_2}$ ... 43
4.1.1 Self and Mutual Inductance vi constraints ... 44
4.1.2 Induced Voltage and the Dot Convention ... 45
4.1.3 Coefficient of Coupling k ... 46
Total Inductance of Coupled Coils ... 47
4.2 Ideal Transformers ... 48
4.3 Transformer Equivalent Circuit ... 49
4.4 Transformer Frequency Response ... 50
4.4.1 Low Frequency Circuit Analysis ... 51
4.4.2 High Frequency Circuit Analysis ... 53
4.4.3 High and Low Frequency Circuit Analysis ... 56
4.5 Full Wave Rectifier ... 57

5 Circuit Analysis ... 59
5.1 Circuit Analysis Process ... 59
Here is Why We Use Laplace ... 63
5.2 Node Method ... 64
5.2.1 Circuit Node Equations ... 64
5.2.2 Branch with Voltage Source ... 66
5.3 Maxwell's Mesh Method ... 69
5.3.1 Circuit Mesh Equations ... 69
5.3.2 Branch with a Current Source ... 71
5.4 Attenuator Circuit Equations ... 74
5.5 Low Pass Filter Circuit Equations ... 75
Cramer's Rule ... 76
5.6 Solution of Mesh Equations, Attenuator ... 77
5.7 Solution of Node Equations, Low Pass Filter ... 78
5.8 Practice Analyzing Circuits ... 81
Complex Numbers ... 85
Sinusoidal Current and Voltage ... 86

6 Frequency Response ... 87
6.1 Resistor Circuit ... 88
6.2 RC Low Pass Filter ... 90
6.3 RC Phase Equalizer ... 95
6.4 RLC Series Resonant Circuit ... 100

7 Transient Response 105
7.1 The Laplace Transform 106
7.2 Transforms Simplify Functions 107
7.3 Transforms Simplify Operations.. 109
7.4 RL Circuit Transient Response.. 111
Laplace Transforms of vi Constraints 113
The Exponential Function 114
7.5 RC Circuit Transient Response.. 115
7.6 RLC Circuit Transient Response 118
7.7 Transient State and Steady-State 121
7.8 Real and Complex Frequencies 123
7.9 The Complex Frequency Plane.. 125

8 The Steady State and the ω Phasor Method............................ 127
8.1 Traditional AC Analysis 128
Sinusoidal alternating current and voltage 130
8.2 ω Phasor Transforms.. 131
8.3 RLC Steady-State Impedances 132
8.4 Phasor AC analysis... 133
Examples... 135
Summary 8 140

9 One Mesh RL and RC Circuits... 141
9.1 RL Steady State 142
9.2 RL Transient State.. 145
9.2.1 Steps u(t)... 145
9.2.2 Pulses u(t)–u(t–T) 148
9.2.3 Ramps t/T×u(t) 150
9.3 RC Steady State.. 152
9.4 RC Transient State 155
9.4.1 Steps u(t)... 155
9.4.2 Pulses u(t)–u(t–T) 158
9.4.3 Ramps t/T×u(t) 160
9.5 Series RC Differentiator... 162
9.6 Square Wave Fidelity in an RC Circuit 164

10 The Bode Method 166
Logarithms 167
10.1 Corner Frequency 168
10.2 Factor K 170

10.3 Factors p and 1/p 171
10.4 Factors (p+a) and 1/(p+a) 173
10.5 Factors (p+a+jb) and (p+a−jb) 176
Table 1001 Magnitude and phase of 20Log $(p+\omega_0)$ 178
Table 1002 Magnitude and phase of 20Log $(1+\lambda/Q+\lambda^2)$ 179
10.6 Bode Plots 180
Problems 10 182

11 Reactance Chart Method 184
11.1 Impedance Plots - $|Z(\omega)|$ 185
    11.1.1 RC in series 185
    11.1.2 RC in parallel 186
    11.1.3 RL in series 187
    11.1.4 RL in parallel 188
    11.1.5 RLC in series 189
11.2 Transfer Function Plots - $|T(\omega)|$ 190
    11.2.1 Series C, Shunt R High Pass Filter 190
    11.2.2 RC phase compensation network 190

12 How to write AC, DC, and TRAN Spice Programs 192
12.1 DC Spice program Fig4011.ckt 194
12.2 AC Spice program Fig4021.ckt 196
12.3 TRAN Spice program Fig5051.ckt 198

Appendix
A1 Partial Fraction Expansions 200
A2 Complex Numbers 203
A3 Oliver Heaviside's Method 206
A4 Easy Method for Evaluating Laplace Transforms 210
A5 Topology Applied to Electric Circuits 212
A6 Delta-Wye Transformation 217
A7 Equivalent circuits 219
A8 Signal Sources 226
A9 Impedance, Admittance, and Immittance 228

Answers to Most of the Problems 233

Ten Experiments 253

Index 303

The SI System of Units

SI Prefixes

Prefix	Multiplier	Symbol
tera	10^{12}	T
giga	10^{9}	G
mega	10^{6}	M
kilo	10^{3}	K
milli	10^{-3}	m
micro	10^{-6}	μ
nano	10^{-9}	n
pico	10^{-12}	p
femto	10^{-15}	f
atto	10^{-18}	a

Basic Units

Quantity	Name	Symbol
Length	meter	m
Mass	kilogram	kg
Time	second	s
Current	ampere	A
Temperature	Kelvin	K
Luminous Intensity	candela	cd

Derived Units

Quantity	Name	Formula	Symbol
Acceleration	meter per sec per sec	m/s²	a
Velocity	meter per sec	m/s	v
Force	newton	kg×m/s	N
Pressure (stress)	pascal	N/m²	Pa
Density	kg per cubic meter	kg/m³	r
Energy or work	joule	N×m	J
Power	watt	J/s	W
Charge	coulomb	A×s	C
Potential	volt	W/A = J/C	V
Resistance	ohm	V/A	Ω
Capacitance	farad	C/V	F
Magnetic flux	weber	V×s	Wb
Inductance	henry	Wb/A	H

Greek Alphabet

A	α	alpha	a[1]
B	β	beta	b
Γ	γ	gamma	g
Δ	δ	delta	d
E	ε	epsilon	e
Z	ζ	zeta	z
H	η	eta	h
Θ	θ	theta	q
I	ι	iota	i
K	κ	kappa	k
Λ	λ	lambda	l
M	μ	mu	m
N	ν	nu	n
Ξ	ξ	xsi	x
O	o	omicron	o
Π	π	pi	p
P	ρ	rho	r
Σ	σ	sigma	s
T	τ	tau	t
Y	υ	upsilon	u
Φ	φ	phi	f
X	χ	chi	c
Ψ	ψ	psi	y
Ω	ω	omega	w

---

[1] equivalent computer keyboard English letter keys

1 Circuit Theory Fundamentals

Electricity is about charge q. Electricity is charge q. In the beginning Faraday and others discovered that charge creates a force field, the Electric Field. They followed up this discovery by creating a series of brilliant experiments that uncovered the concepts of current, voltage, and electric power. Later, as the concept of a circuit evolved, Kirchhoff published his current and voltage laws about connections, while Ohm published his law relating current, voltage, and resistance.

1.1 Electrostatic Field Theory

Experiment 1 Rub a piece of silk or other material on pieces of copper that are suspended by insulated threads in space. Faraday, and others showed us this action places *charges at rest* on the copper piece surfaces that we now know are free electrons with negative charge. Move a particle that has a very small charge on it through the region surrounding the charged copper pieces to discover *the charged particle experiences a force **F** proportional to the charge q on the particle*. The force is everywhere. The force's magnitude and direction vary with position. We say the charged copper pieces have created a *force field*. This force field, created by charges at rest, is referred to as the electrostatic force field, or electric field **E**. The electric field **E** is defined by

(1) $\quad F = qE \quad\quad newtons = coulombs \times volts/meter$

Experiment 2 Move a particle that has a very small charge on it through an electric field in a closed path so that the particle returns to the starting point. You discover *that zero total work is done by the particle, or on the particle*. This makes sense when you realize that we have a perpetual motion machine if work > 0 is done on each round trip. And, so

(2) $\quad Work = integral\ of\ force \times distance = \oint F \cdot ds = 0$

(3) $\quad Work = q\oint E \cdot ds = 0$

(4) $\quad \oint E \cdot ds = 0$

Experiment 3 We learn that like charges repel each other, and unlike charges attract each other (+ + or − − repel, while + − attract).

Electric Circuits - Analysis and Design

> The theory of electricity is derived from the results of experiments. We do not know why the results are correct. We know they are correct, because the theory derived from these experiments produces designs that work.

Direction of the Electric Field By definition a *positive* charge Q creates electric field **E** that points *away* from the charge Q. Force **F**=q^+**E** on a positively charged particle q^+ moves the particle in the direction of the electric field. (**F** and **E** are vectors.) Thus positively charged particle q^+ in the **E** field is repelled, and moves away from Q so that + + repel. On the other hand, a negatively charged particle q^- reverses the direction of the force, because **F**= q^-**E**=–|q|**E**. Therefore the negatively charged particle q^- moves towards the positive charge Q so that + – attract.

On the other hand a *negative* charge Q creates electric field **E** that points *towards* the charge Q. In that field the force **F** on a negatively charged particle q^- repels the particle, because **F**=q^-**E**=–|q|**E** we have – – repel. Whereas the force attracts a positively charged particle q^+ and – + attract.

Emphasis The electric field, as defined, points from + charges towards – charges. And, a negative charge moving around in space makes the direction of the force *on it* the opposite of the electric field direction.

The Potential An electric field **E** exerts a force on a charged particle that imparts motion to the charged particle. Clearly work is done on the charge. Since work equals force times distance, the work required to bring the charge to position p from far away where the force is zero is

(5) $$\text{Work} = \int_{-\infty}^{p} F \cdot ds = \int_{-\infty}^{p} (qE) \cdot ds = q \int_{-\infty}^{p} E \cdot ds = qV \quad \Rightarrow \quad V = \int_{-\infty}^{p} E \cdot ds$$

This means charge q at position p in the field stores energy. This energy is equal to the work that moved q to position p from far away where the force is zero. Usually far away is referred to as minus infinity.

Integration is required, because **E** varies with position. However, the constant V is a *number* produced by the integration. V is referred to as the *potential* at a point in space that is easier to work with than electric field vectors. An alias for potential is *voltage*.

Emphasis: Work is charge times voltage: W=qV
            (joules=coulombs×volts).

1 Circuit Theory Fundamentals

Potential Difference Assume some electric field exists. Then bringing a test charge q from minus infinity to positions p_1 or p_2 in the field requires work qv_1 or qv_2 respectively where the potential at p_1 is v_1 and the potential at p_2 is v_2.

If W_{21} is the work required to move a charge q from position p_1 to position p_2, then work in terms of potential is $q(v_2-v_1)$.

(6) $\quad qv_2 = qv_1 + W_{21} \quad \Rightarrow \quad W_{21} = q(v_2 - v_1)$

On the other hand work is force times distance.

(7) $\quad W_{21} = F \cdot dx = qE \cdot dx$

Merging (6) and (7) shows (9) that *the voltage decreases in the direction of positive E when $v_1 > v_2$*.

(8) $\quad qE \cdot dx = q(v_2 - v_1) = -q(v_1 - v_2)$

(9) $\quad \dfrac{v_1 - v_2}{dx} = -E \quad \Rightarrow \quad \dfrac{\partial v}{\partial x} = -E$

Electric field Units

(10) $\quad E = \dfrac{F}{q} = \dfrac{newtons}{coulomb} = \dfrac{1}{coulomb} \cdot \dfrac{joules}{meter} = \dfrac{joules}{coulomb} \cdot \dfrac{1}{meter} = \dfrac{volts}{meter}$

Transition from field theory to circuit theory The voltages of points in space occupied by an electric field become the voltages of circuit points (nodes). Current flow of charge moving in space, or in some material, becomes current flowing in wires and components from node to node. How much current depends on the voltage current constraints. The transition from field to circuit is possible when the circuit dimensions are less than the *wavelength* λ of a sinewave source signal. The wavelength of a wave is the distance the wave travels in one period. For example a 300MHz sinewave has a 1 meter wavelength. A circuit with dimensions less than say 1/10 meter can be represented by wires and components for a 300MHz signal, that does not vary very 'much' as it travels the 1/10 meter through the circuit. Lower frequency signals vary even less.

$velocity = \dfrac{distance}{time} \quad \Rightarrow \quad c = \dfrac{\lambda}{T} = \lambda f \quad \Rightarrow \quad \lambda = \dfrac{c}{f} = \dfrac{3 \cdot 10^8 \, m/sec}{300 MHz} = 1 \, meter$

Electric Circuits - Analysis and Design

Charge

Everything in the world is some assembly of atoms of the elements. For example, a copper wire is an assembly of atoms. The structure of any atom includes electrons, some of which may be stripped from the atoms under various conditions. For example, thermal agitation frees electrons from atomic bonds to create an electron "gas" in metals. And so we can say the atoms of the elements are the source of electrons. Why do we care about electrons? We care, because each electron has a charge $-q_E$ permanently attached to it. We care about charge, because *charge at rest and charge in motion* are the basis of all electrical phenomena.

The smallest unit of electricity is the permanently attached charge $-q_E$ on the electron. One or more electrons are part of the structure of the atoms of all chemical elements. We are in awe of the fact there are no omissions in the number of electrons in the atoms of chemical elements that range from 1, 2 ,3, 4, 5, 6, 7, ... , 117, 118. The number of electrons is what differentiates the elements. See a Periodic Table of the Elements.

Atoms are electrically neutral, because their structure includes a center of an equal number of protons carrying positive charge q_E. Protons are about 1800 times heavier than electrons. This is one reason why electrons, not protons, are stripped from atoms by electrical forces.

Robert A. Millikan accurately measured this charge q_E in 1909 as $4.774 \pm 0.005 \times 10^{-10}$ electrostatic units (1.5913×10^{-19} coulombs). His great achievement was his experimental proof of the atomic nature of the electron. Millikan demonstrated experimentally that any charge q is some *integral multiple n* of q_E ($q = n \times q_E$). He also verified experimentally Einstein's photoelectric equation, and made the first direct photoelectric determination of Planck's constant *h*. He was awarded the Nobel Prize for Physics in 1923 "for his works on the uniform electric charge and the photoelectric effect." Standard value of the charge is now 1.60209E−19.

One coulomb = n electron charges : $1 = n \times q_E$

$$q_E = 0.160209 \times 10^{-18} = 1.60209 \times 10^{-19} \frac{coulombs}{electron}$$

$$n = \frac{1}{q_E} = 6.241847 \times 10^{18} = 6.242 \times 10^{18} \frac{electrons}{coulomb}$$

1 Circuit Theory Fundamentals

Fields of Force

Gravitational field Consider the idea of a *field*. A small object, such as the earth's moon, is attracted by the gravitational forces of two large objects, such as the sun and the earth. The gravitational attraction is some force vector **F**. Newton showed that the force vector **F** can be written as a mass m times the *gravitational* vector **G** so that **F** = m**G**. **G** is interpreted as a vector field created by the sun and the earth that is independent of the mass m. **G** exists whether or not mass m is there for **G** to act on it. This vector we call **G** is an example of a field. If we place a mass m in the field **G** the mass m will experience force **F** = m**G**. This is how we know **G** is there. The mass m does not modify the field **G**.

Work in a gravitational field The field applies a force on a mass m that is moved from point a to point b. The velocity is v, and s is the mass position at time t so that v=ds/dt

the kinetic energy, the work $W = \frac{1}{2}mv^2$ and $\frac{dW}{dt} = mv\frac{dv}{dt}$

force $F = ma = m\frac{dv}{dt}$ so that $\frac{dW}{dt} = vm\frac{dv}{dt} = vF = Fv = F\frac{ds}{dt}$

consequently $dW = Fds$ so that $W = \int_a^b dW = \int_a^b F ds$

Electric field If we replace the sun and earth with + and - charge on two bodies in space an *electric field vector* **E** is created. This was confirmed by Charles Augustin de Coulomb (1746-1806), when he *demonstrated in a series of experiments* that a free charge q, placed at any point in the electric field, experiences a force vector **F** = q**E**. The negligible charge q does not modify the field **E**. The existence of the force means that work is required to move the charge q from point a to point b. This is why potential V *per unit charge* at point b is defined as $V_B = V_A + W/q$.

$W = \int_a^b F ds = \int_a^b qE ds = q\int_a^b E ds \Rightarrow \frac{W}{q} = \int_a^b E ds$

Since $V_b = V_a + \frac{W}{q} \Rightarrow V_{ba} = V_b - V_a = \frac{W}{q} = \int_a^b E ds$

Electric Circuits - Analysis and Design

1.2 Current, Voltage, and Power

When two charged conductors at positions p_2 and p_1 with voltages $v_2 > v_1$ are connected by a conducting wire the electric field in the wire will drive electrons (– charge) from the conductor at p_1 to the conductor at p_2. The electron flow continues until the charge distribution is at the same potential on the two conductors and wire. This happens, because like charges repel until the mutual forces equalize. Then the potentials are equal, because the electrons are now at rest on "one" piece of metal.

On the other hand if some means replenished the charges on the conductors as they were carried away by conduction through the wire, then the electron flow would remain steady, and the conductors would remain at potentials v_2 and v_1. *The electron flow rate is referred to as a current.*

1.2.1 Current

Electric currents are electrons in motion with a net flow in some direction.

> *Electric current is the charge passing through a cross section in some unit of time* (Figure 101).

Current i delivers charge *dq* in time *dt* when it flows through a wire or circuit component.

(11) $\quad i = \dfrac{dq}{dt} \quad units: 1\ Ampere = 1 \dfrac{Coulomb\ of\ charge}{second} \quad i.e.\ 1A = 1\dfrac{C}{s}$

Figure 101 Moving charge defines a current

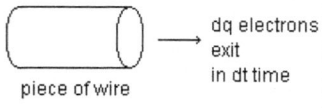

piece of wire → dq electrons exit in dt time

One ampere means 6.242×10^{18} electrons per second are flowing through a "wire." The very large number of electrons flowing is the reason why the "granularity" of charge flowing is not an issue. In practice the current is considered to be continuous when mathematical methods are applied to electric circuits.

1 Circuit Theory Fundamentals

In different electric circuits all magnitudes of current are found. At one extreme ion detectors in effect count electrons one by one. Currents in integrated circuits can range from 10^{-12} amperes (6.242×10^6 electrons per second flowing) to many amperes. In many circuits typically one finds microampere (10^{-6} amperes), and milliampere (10^{-3} amperes) currents. Ten amperes flow when a 1.8V power supply delivers 18 watts. On the other hand, hundreds of amperes can flow in large electric motors, and power systems.

Note: The *electron rest mass* is 9.109×10^{-31} kg. A coulomb of electrons has a mass of 57×10^{-13} kg. Any force produces large accelerations. No wonder electric circuits can be super fast performers. The 'electricity' that flows in objects is *charge transported* by electrons. An object is a wire, a trace on a printed circuit board, the channel of a transistor in a VLSI chip, and so forth.

Experiment 4 Ohm's Law Georg Ohm connected voltage to current for one type of component, the *resistor*. We have seen that when current flows in a conductor, electricity (charge) is in motion, and consequently there is a potential difference along the conductor. After many experiments Georg Ohm produced what is now known as Ohm's Law. Later others produced a theoretical equation supporting Ohm's Law.

Ohm's Law The potential difference v between any two points of a linear conductor in which a current i is flowing is equal to the resistance R to current flow times the current: $v=iR$. We refer to the equation $v=iR$ as a voltage-current constraint. Ohm's law is one example of a *vi constraint*. Ohm's law is the *vi constraint* for any linear resistor R that is one type of electrical component.

Note: The Battery The voltaic cell (alias battery) employs a chemical reaction to produce a steady flow of electrons in a circuit. Hence the battery maintains a potential difference v. This is the means that replenishes the charges on conductors connected by a wire as they are carried away by conduction through the wire.

Problem 101 Current i delivers charge dq in time dt. A 2mA current flows. How many coulombs/second flow?

Electric Circuits - Analysis and Design

1.2.2 Voltage

Experiment 2 related work to charge. Force field **F** created by an electric field **E** requires that work must be done to move any charge against the force. If the electric field **E** has uniform intensity the work done is the product of the constant force q**E** and the distance the charge moves. If the field is not uniform, the field varies from point to point, and the integral of force times distance is required.

In order to simplify computations the concept of potential v, or electrical level, was introduced. This is analogous to potential energy in mechanics. In mechanics a body moved from one level to another level has work done upon it. The body's potential energy is changed.

(12a) $\quad work \ w = \int force \times distance = \int_{x}^{y} qE \times ds = qv_{xy}$

(12b) $\quad v_{xy} = \int_{x}^{y} E \, ds$

Consequently every point in an electric field has some value of voltage v that is a scalar magnitude and not a vector. The voltage does not depend on the amount of charge.

> The key observation is that the **difference** of voltage between any two points x and y represents the work done in moving charge from point x to point y. The integral of the electric field **E** from x to y equals the voltage difference from x to y.

Again: The work w expended while moving a charge q from node x to node y is proportional to the voltage v where v is the potential difference between the two nodes (w=qv).

Voltage produces the electromotive force that moves electrons through a circuit. If $v_S(t)$ is a 3 volt battery, then the potential difference from the + terminal to the − terminal is $v^+ - v^- = 3$ volts. This potential difference exists, because chemical action in the battery, in effect, produces + (ions) and − charge (electrons) at the terminals. In turn the + and − charges create an *electric field of force* that moves electrons inside circuit components attached to the battery.

1 Circuit Theory Fundamentals

The battery's 3 volt potential is converted into work when a circuit is attached so that an electric field appears throughout the attached circuit. By throughout the circuit we mean inside the circuit's wires and components so that electrons with negative charge flow from the − battery terminal through the circuit's wires and components and return to the + battery terminal. Electrons flow towards the positive potential, because the terminal's + charge attracts the − electron charge..

A voltage exists at each component terminal or circuit node. Energy and current are related to the voltage difference from one terminal to another terminal of a component or components that are connected between any two nodes. This is why the *voltage (potential) difference of any two nodes is what matters*. Think about the positive and negative nodes on any battery such as the 3 volt battery.

The easiest way to define circuit voltages is to select one circuit node as the reference node, and to define all circuit voltages v_J as the voltage difference between node j and the reference node, i.e. $v_J - v_{REF}$. If $v_{REF}=0$ then v_J is understood to be the voltage difference between node j and ground (0 volts). This means you select one node as the reference with value zero volts, and label the remaining nodes v_1, v_2, ... v_{N-1} when there are N circuit nodes. Then a component connected to nodes with voltages v_J and v_K has $v_J - v_K$ volts across it.

One must keep in mind that a voltage is always measured between two nodes, and that ground is a node.

Voltages in some applications are as small as 10^{-9} volts. Voltages in digital integrated circuits are in the 0.5 to 5 volt range. Ten volts is the order of magnitude of voltage used in typical analog electric circuits. When low power vacuum tubes are used the order of magnitude jumps to a hundred volts. Some high power transmission lines transfer energy at the million volt level.

In many electric circuits a reasonable approximation is that the electric field exists only *inside* the circuit components. Such components are referred to as lumped parameter components. *The points representing connections between components are called nodes* that are visible to the outside world. Each node has some value of potential whose unit is the volt.

Electric Circuits - Analysis and Design

Voltage is energy/charge. Dimensional analysis confirms this.

voltage v = electric field E × distance d

$$v = E \times d = \frac{F \times d}{q} = \frac{newton \times meter}{columb} = \frac{joules/meter \times meter}{coulomb} = \frac{joule}{coulomb}$$

Definition: A voltage difference of one volt exists between two nodes of a circuit when one joule of energy is required to move one coulomb of charge through the circuit from one node to the other node.

(13) $\quad v = \dfrac{dw}{dq} \quad\quad units: 1\,volt = 1\dfrac{joule}{coulomb} \quad i.e. \quad 1V = 1\dfrac{J}{C}$

1.2.3 Power

Power is an important parameter, because all components are rated to dissipate less than some number of watts of power. Power ratings are attended to in order to avoid burning up components. Power is the rate of energy expended.

(14) $\quad p = \dfrac{dw}{dt} = \dfrac{dw}{dq} \times \dfrac{dq}{dt} = v \times i$

Definition of the watt If one joule of energy w transfers one coulomb of charge per second, then one watt of power p is expended.

(15) $\quad 1\,watt = 1\dfrac{joule}{second} = 1\dfrac{joule}{coulomb} \times 1\dfrac{coulomb}{second} = 1\,volt \times 1\,ampere$

The watt unit of power is named after James Watt (1736-1819).

Problem 102 The work w expended while moving a charge q from node x to node y is proportional to the voltage v. *Voltage is the change in energy w per charge q.* What is the voltage when 5 joules of energy push 0.2 coulombs of charge through a wire?

---

1 Circuit Theory Fundamentals

1.3 Kirchhoff's Laws

In 1848 G. Kirchhoff (1827-1887) published his two laws. The KVL law constrains voltages, and the KCL law constrains currents, in an electric circuit. The laws are *connection constraints*, and they are discussed in upcoming paragraphs. Connection constraints impose conditions on voltages or currents in a circuit that depend on how the components are wired into the circuit. Connection constraints are independent of the specific types of components used. This independence becomes clearer when you read about the topology of circuits (Appendix 5).

1.3.1 Kirchhoff's Current Law (KCL)

The algebraic sum of all currents entering and leaving one node is zero.

Figure 102 KCL

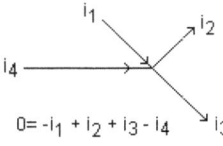

This is proved by the following counterexample. Suppose two unequal currents enter a node and two unequal currents leave it (Figure 102). If the sum of the currents entering the node is greater than the sum of the currents leaving it, then the sum of the currents cannot be zero. A non-zero sum means more charge enters than leaves the node, because input dq exceeds output dq in the same dt. So, charge must accumulate at the node and be destroyed. On the other hand, if more current leaves a node than enters it, then the sum of the currents is again not zero. This means the node must create the excess charge leaving the node. This is why the first law (KCL) is a consequence of the principle of conservation of electric charge that states that charge cannot be created or destroyed. Currents flowing in the reference circuit (Figure 100) are constrained by the first law (KCL) as shown below.

Figure 100 Reference Circuit

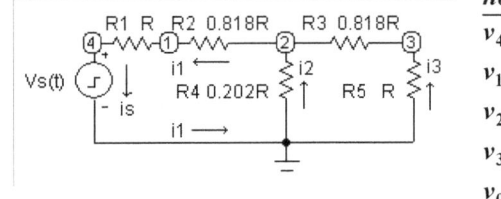

node	KCL
v_4	$i_S = i_1$
v_1	$i_1 = i_1$
v_2	$i_1 = i_2 + i_3$
v_3	$i_3 = i_3$
v_0	$i_1 = i_S$ and $i_2 + i_3 = i_1$

11

Electric Circuits - Analysis and Design

1.3.2 Kirchhoff's Voltage Law (KVL)

The algebraic sum of the voltage differences around every closed circuit contour is zero.

Figure 103 KVL

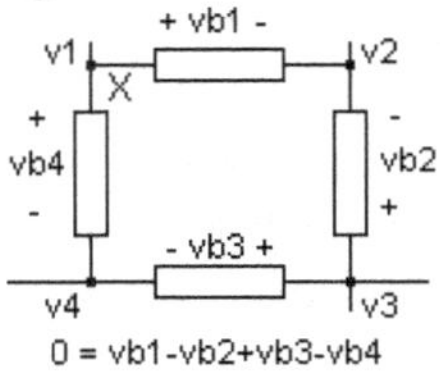

This is proved by the following counterexample. Suppose a charge starts at node X (Figure 103) and flows through components connected to different nodes as it goes around a path that returns it to node X. The charge experiences branch voltages v_{b1}, v_{b2}, v_{b3}, and v_{b4}. If the charge's potential is different at the start and end points, then the sum of the voltages around the path is not zero. This means node X has two potentials, two energy levels, which is impossible. This is why the second law (KVL) is a consequence of the principle of conservation of energy that states that energy cannot be created or destroyed.

Voltages in the reference circuit (Figure 100b) are constrained by the second law as shown here.

v_{bk} branch voltages

$v_{b1} = v_4 - v_1$

$v_{b2} = v_1 - v_2$

$v_{b3} = v_2 - v_3$

$v_{b4} = v_2 - 0$

$v_{b5} = v_3 - 0$

$v_{bS} = v_4 - 0 = v_S$

Figure 100b Reference Circuit Branch Voltages

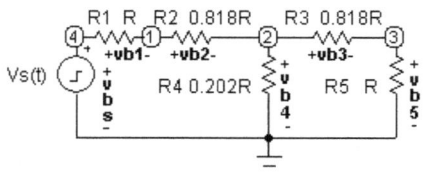

Emphasis: Kirchhoff's laws describe how currents and voltages are constrained by the circuit connections. These connection-constraints do not depend upon the types of devices wired into the circuit.

1 Circuit Theory Fundamentals

The Forest and the Trees

The infinite number of trees in the electric circuits forest precludes memorization as an effective learning technique. Knowing and understanding puts you in the powerful position of being able to deal with any electric circuit. What is there to know and understand?

Analysis of any circuit has two basic phases that are two very different subjects. The subjects are electrical engineering (formulation of equations) and mathematics (solving the equations).

Formulation of equations requires knowing about Kirchhoff's circuit connection constraints KVL and KCL, and voltage-current constraints for circuit components such as Ohm's law for resistors $v = iR$.

Every circuit component type such as an inductor L (or a transistor) has a vi constraint.

A real voltage source v_S has two parts: an ideal v_S in series with a source Z_S. *Analysis is simplified* if you think of Z_S as part of the circuit now driven by the ideal v_S. The voltage of an ideal v_S does not change as it delivers whatever current the circuit requires. Treat a real current source i_S in the same way.

Kirchhoff's laws are connection constraints for v (KVL) and i (KCL). The connection constraints on various circuit v and i are derived from how the circuit components are wired together. The KCL and KVL equations do not need to know what types of components are in the circuit. The vi constraints provide this information. In the end, formulation of equations for analysis boils down to the following (and everything else is just so many details):

Define node voltages, or mesh currents.

Derive KVL or KCL connection constraint equations from the circuit topology.

Use vi constraints to insert specific circuit components into the circuit.

Then mathematics takes over.

Electric Circuits - Analysis and Design

2 Circuits with Resistors

We show how elementary circuit analysis based on Kirchhoff's *connection constraint laws*, and Ohm's *voltage-current constraint law* for resistors solves for currents and voltages in circuits where resistors are connected in series, in parallel, or in a series-parallel combination. Then a DC volt meter, a DC ammeter, an ohm meter, and an AC volt meter are designed. The circuit analysis methods used here are fully developed in Chapter 5.

In any circuit real electrons transport *negative* charge. In many books electrons are assumed to transport positive charge. Either convention will do. Select one convention and use it consistently to avoid writing of a multitude of minus signs. The absolute values of results are not affected. There is no need to make an issue about using positive or negative charge. *The choice does not matter.* Here is our choice for *branch* currents, where electrons transport *negative* charge. The world defines *conductance* g to equal 1/R, the reciprocal of the resistance.

Figure 201

(19) $v_2 - v_1 = iR$

(20) $i = \dfrac{(v_2 - v_1)}{R} = g(v_2 - v_1) \qquad v_2 > v_1, g = 1/R$

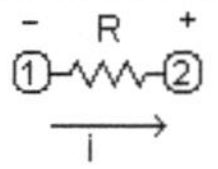

2.1 Resistors in Series

Resistors are connected in series when the same current flows through them (Figure 202).

In other words one terminal of a device with resistance R_1 is connected, via a wire, to one terminal of another device with resistance R_5, and so forth (Figure 202). All series combinations are possible. Wire components in series in any sequence. The wire is assumed to have zero resistance that in practice is correct. Use KVL to solve for current (Figure 202).

(21a) $v_1 = v_{12} + v_{20} = (iR_1) + (iR_5)$
(21b) $v_1 = i(R_1 + R_5)$
(21c) $i = \dfrac{v_1}{R_1 + R_5}$

Use KCL to write that the sum of the currents at node 2 equals zero.

Figure 202

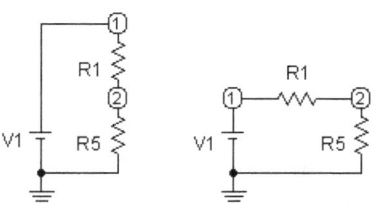

(23a) $i_1 + i_5 = 0$
(23b) $g_1(v_2 - v_1) + g_5(v_2 - 0) = 0$
(23c) $(g_1 + g_5)v_2 - g_1 v_1 = 0$
(23d) $\dfrac{v_2}{v_1} = \dfrac{g_1}{g_1 + g_5} \times \dfrac{R_1 R_5}{R_1 R_5} = \dfrac{R_5}{R_5 + R_1}$

An important point of view is the transmission of a signal from node 1 to node 2. In this circuit (Figure 202) the signal is battery voltage v_1. Continuing from equation 21c we get the *transfer function* v_2/v_1.

(22a) $v_2 = iR_5 = \dfrac{v_1}{R_1 + R_5} R_5$ (22b) $\dfrac{v_2}{v_1} = \dfrac{R_5}{R_1 + R_5}$

Problem 201 The value of v_2 depends on the load R. Show that $v_2 = v_S$, $v_2 = 9v_S/10$, $v_2 = v_S/2$, $v_2 = v_S/10$, $v_2 = 0v_S$, (Figures p201a, b, c, d. e)

Figure p201 Constant voltage sources with various loads

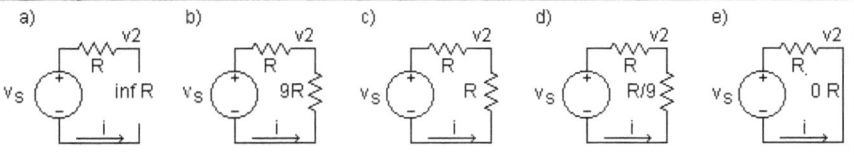

Problem 202 Let v_S=3V, R_1=1KΩ and R_2=2KΩ (Figure p202. When the switch is closed find branch voltages v_{b1} and v_{b2} across R_1 and R_2 (1V, 2V). Find the current through and the voltage drop across the switch (1mA, 0V). Next, open the switch. Find branch voltages v_{b1} and v_{b2} across R_1 and R_2 (0V, 0V) and the current through and the voltage drop across the switch (0mA, 3V).

Figure p202

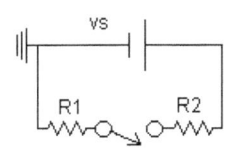

Problem 203 Show that v_2=1.5V, power=1.5mW (Figure p203a).

Problem 204 Show that v_2=2.25V, p_2=2.25mW, power R_1=1.5mW, and power R_2=0.75mW (Figure p203b).

Figure p203

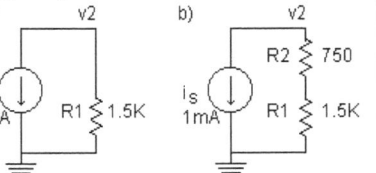

Electric Circuits - Analysis and Design

2.2 Resistors in Parallel

Resistors are connected in parallel when the same voltage difference is across their terminals (Figure 203).

Figure 203

The source current i_S flows from node 1 (Figure 203), down through the battery, and out the negative terminal. The current does not flow into the ground, because there is no return connection from ground to the battery + terminal at node 1. The current i_S flows into the wire connected to the lower terminals of the resistors, through resistors R_1 and R_2 as i_1 and i_2, and back to node 1. Currents i_1 and i_2 flow into node 1, and i_{source} flows away from node 1 down through the battery. The sum of the currents at a node equal zero (Section 1.3.1 KCL).

(24a) $\quad 0 = -i_S + i_1 + i_2$

(24b) $\quad i_S = i_1 + i_2$

The two resistor branch currents are each equal to a gv term.

(25a) $\quad i_1 = g_1(v_1 - 0) = g_1 v_1 \qquad i_2 = g_2(v_1 - 0) = g_2 v_1$

(25b) $\quad i_S = i_1 + i_2 = g_1 v_1 + g_2 v_1 = (g_1 + g_2)v_1 = \left(\dfrac{1}{R_1} + \dfrac{1}{R_2}\right)v_1$

(25c) $\quad \dfrac{v_1}{i_S} = \dfrac{1}{g_1 + g_2} = \dfrac{1}{g_1 + g_2} \times \dfrac{R_1 R_2}{R_1 R_2} = \dfrac{R_1 R_2}{R_2 + R_1}$

The battery "sees" a resistor equivalent to the two resistors in parallel. The equivalent value is the product over the sum of R_1 and R_2 (25c). *This is one "rule" you will use almost every day.*

Problem 205 The value of i depends on the load R. Show that i=0, i=i_S/10, i=i_S/2, i=9i_S/10, i=i_S, (Figures p205a,b,c,d.e).

Figure p205 Constant current sources with various loads

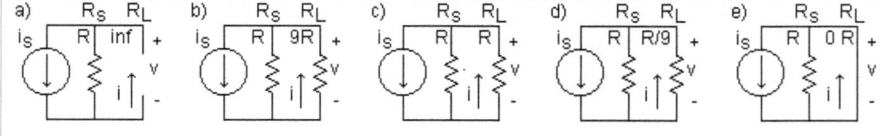

2 Circuits with Resistors

Problem 206 Ohm's law applies here: I=v/R. Show that i_{B1}=1mA, i_{B2}=2mA, i_S=3mA (Figure p206d),.

Problem 207 Show that the equivalent resistor load is 500Ω (Figure p206d) by calculating v_S/i_S. Then use equation 25c to find the load.

Figure p206 Two resistors in parallel

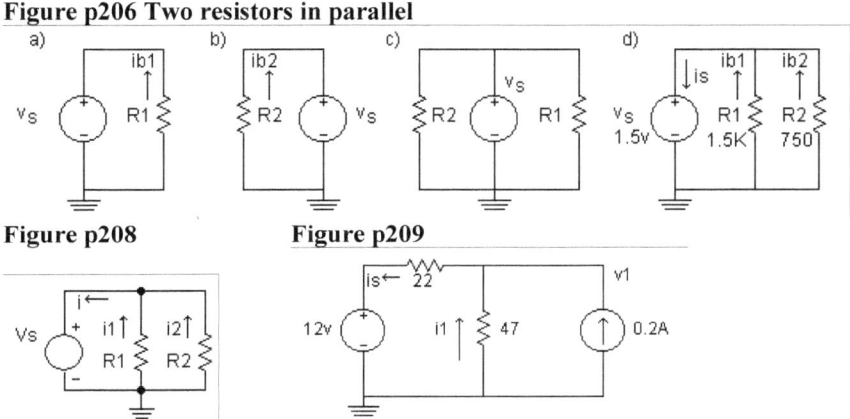

Figure p208   **Figure p209**

Problem 208 Refer to Figure p208. Let v_S=3V, R_1=1KΩ and R_2=2KΩ. Find i_1 and I_2 (3mA, 1.5mA). Write an expression for the ratio i_1/i_2 (= R_2/R_1).

Problem 209 Refer to Figure p209. Think about how the currents are related. Solve for i_S, i_1 and v_1.

Problem 210 Refer to Figure p210. Think about the voltage drop across the 10K resistor. Solve for v_2.

Problem 211 Refer to Figure p211. Think about the voltage drop across the 100Ω resistor. How does i_S relate to 1.2A? Solve for i_S and v_L.

Figure p210   **Figure p211**

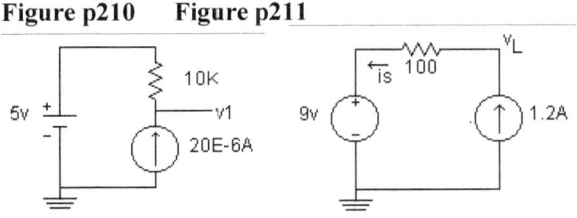

17

Electric Circuits - Analysis and Design

2.3 Resistors in Series-Parallel

R_2 and R_3 are in parallel (Figure 204). Right? And, R_1 is in series with the parallel pair. Immediately we can write that the battery "sees" a resistor equivalent to

Figure 204

(26) $R = R_1 + \dfrac{R_2 R_3}{R_2 + R_3}$

Node 1 voltage equals battery voltage v_1, and we want to find a value, or expression for v_2. We use Kirchhoff's current law KCL (Section 1.3.1). The sum of the node 2 currents equals zero (27a). We replace currents with Ohm's vi constraint (27b), collect terms (27c), and solve for v_2 (27d).

(27a) $i_1 + i_2 + i_3 = 0$

(27b) $g_1(v_2 - v_1) + g_2(v_2 - 0) + g_3(v_2 - 0) = 0$

(27c) $(g_1 + g_2 + g_3)v_2 - g_1 v_1 = 0$

(27d) $\dfrac{v_2}{v_1} = \dfrac{g_1}{g_1 + g_2 + g_3} = \dfrac{g_1}{g_1 + g_2 + g_3} \times \dfrac{R_1 R_2 R_3}{R_1 R_2 R_3} = \dfrac{R_2 R_3}{R_2 R_3 + R_1 R_3 + R_1 R_2}$

Problem 212 Reference Figure p212. Show that i_{B2}=1.64mA, i_{B3}=1.10mA. Does their sum equal i_1? Show that source power=32.88mW, R_1 power = 24.77 mW, R_2 power=4.85mW, R_3 power = 3.26 mW.

Figure p212

Problem 213 Reference Figure p213. When the switch is open find the value of R_1 when the ratio $v_2/v_S = 2/3$. ($R_1 = 1500$ ohms) Given this value of R_1, close the switch and find the value of R_2 so that $v_2/v_S = 1/3$ ($R_2 = 500$ ohms).

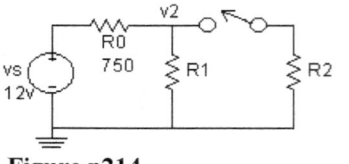

Figure p213

Problem 214 Reference Figure p214. Switch open: find v_2, i_1 and i_2 (80V, 8mA, 4mA). Switch closed: find v_2, i_1 and I_2 (40V, 4mA, 8mA).

Figure p214

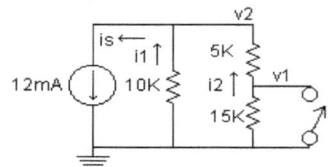

2 Circuits with Resistors

Problem 215 How do series and parallel resistors combine? Reference Figure p215. Let $R_1 = 10K\Omega$, $R_2 = 5K\Omega$, $R_3 = 4K\Omega$, $R_4 = 12K\Omega$. Find equivalent R of each circuit in $K\Omega$. Results as fractions are satisfactory (15K, 10/3 K, 40/9 K).

Figure p215

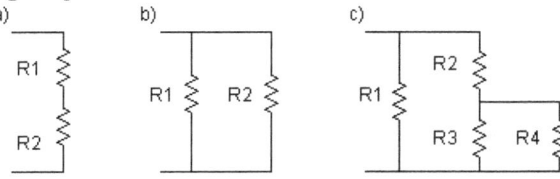

Problem 216. Reference Figure p216. Solve for i_2, v_2 and i_1, v_1. (5.82mA, 3.44mA, 41.85V, 9.28V)

Problem 217 Reference Figure p217 Solve for i_1, i_2, i_3, v_1, and v_2 (72mA, 6mA, 9mA, 72V, 27V)

Figure p216

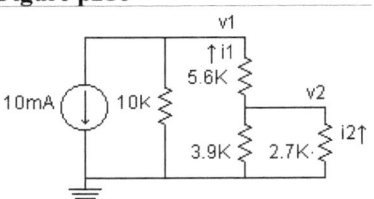

Figure p217

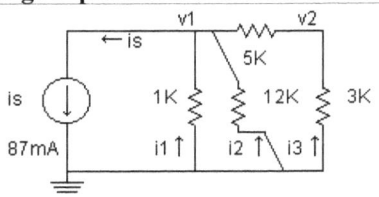

T Network: A precise output/input ratio is one application (Figure p220).

Problem 218 Show that $v_1/v_S = 1/2$, $v_3/v_S = 1/4$

Figure p218 T Network

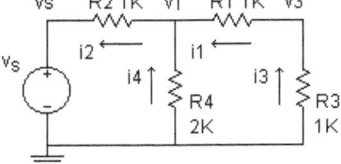

Problem 219 Formal analysis that the input to output *transfer function* is as shown below. Use this "formula" to show that $v_3/v_S = 1/4$ for the values in Figure p218. Hint convert all G to 1/R.

$$\frac{v_3}{v_S} = \frac{1}{(G_4 R_2 G_3 R_1) + (G_4 R_2 + G_3 R_2 + G_3 R_1) + 1}$$

Electric Circuits - Analysis and Design

2.4 Voltage Division

Two resistors in series implement the *voltage divider idea* (Figure 205).

Battery voltage v_S is 'divided' into resistor voltages v_1 and v_2. The same current i flows in the two resistors. Using Ohm's law we get $i=v_S/R$ where $R=R_1+R_2$ in this circuit.

Figure 205

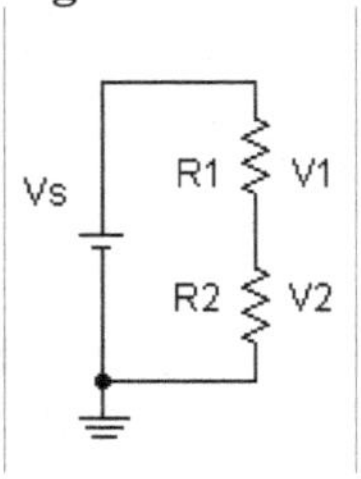

$$(28a) \quad v_s = v_1 + v_2 = iR_1 + iR_2 = i(R_1 + R_2) \Rightarrow i = \frac{v_s}{R_1 + R_2}$$

$$(28b) \quad \frac{v_1}{v_s} = \frac{iR_1}{i(R_1 + R_2)} = \frac{R_1}{R_1 + R_2}$$

$$(28c) \quad \frac{v_2}{v_s} = \frac{iR_2}{i(R_1 + R_2)} = \frac{R_2}{R_1 + R_2}$$

Note: A voltage divider can have as many R in series as you need.

2.5 Current Division

Two resistors in parallel implement the *current divider idea* (Figure 206).

Battery current i_S is 'divided' into resistor currents i_1 and i_2. The same voltage v_S is across the two parallel resistors.

$$(29a) \quad i_s = i_1 + i_2 = g_1 v_S + g_2 v_S = (g_1 + g_2) v_S$$

Figure 206

$$(29b) \quad \frac{i_1}{i_s} = \frac{g_1 v_S}{(g_1 + g_2) v_S} = \frac{g_1}{g_1 + g_2} = \frac{R_2}{R_2 + R_1}$$

$$(29c) \quad \frac{i_2}{i_s} = \frac{g_2 v_S}{(g_1 + g_2) v_S} = \frac{g_2}{g_1 + g_2} = \frac{R_1}{R_2 + R_1}$$

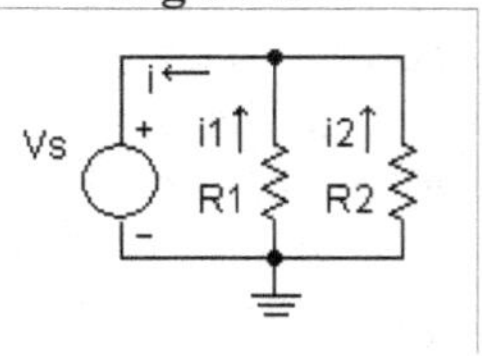

Note: A current divider can have as many R in parallel as you need.

2 Circuits with Resistors

2.6 Circuit Design of an Analog Multimeter

An analog multimeter measures volts, ohms, and amperes. The meter's analog display of the measured value is implemented by an analog panel meter, which converts a current to a 0° to 90° rotation of a pointer. This is why all measurements are converted to a current.

Scales behind the pointer display ranges of volts, ohms, and amperes

2.6.1 Display Meter Circuit

What standard meter movement can we buy or order custom made? We can buy a "50µA" full scale analog panel *meter* based on the D'Arsonval movement mechanism. The meter has about 1,800 ohms resistance and full scale meter current is 48µA. In order to accommodate manufacturing tolerances we specify the

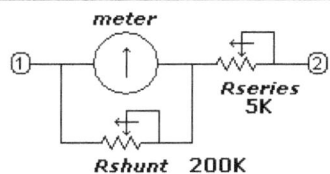

Figure 207
The Display Meter Circuit

display meter circuit "black box" (Figure 207) as a 50µA/5,000Ω meter that has a 50µA × 5KΩ =250 millivolt drop across the display meter.

We add potentiometers R_{series} and R_{shunt} to allow for adjustment to the 50µA/5,000Ω specification when using any manufactured meter with parameters 48±1.5 µA full scale, and 1.8K±100 ohms resistance.

The same voltage difference v_{meter} is across the meter movement and R_{shunt}. The full scale current flowing from node 1 to node 2 (Figure 207) is specified as 50µA. Therefore we want $50 - i_{meter}$ flowing in R_{shunt}. Now from 48±1.5 µA full scale we can calculate max and min values of R_{shunt}. The calculations (equation 30) lead us to select a standard 200K value for the R_{shunt} "pot." Having set R_{shunt}, R_{series} needs a slight adjustment.

(30) $v_{meter} = i_{meter} R_{meter} = i_{shunt} R_{shunt}$

(31) $R_{shunt} = \dfrac{i_{meter}}{i_{shunt}} R_{meter} = \dfrac{i_{meter}}{50 - i_{meter}} 1.8K$ can ignore $\pm 100\Omega$

(32) $R_{max\_shunt} = \dfrac{49.5}{50 - 49.5} 1.8K = 178.2K$ $R_{min\_shunt} = \dfrac{46.5}{50 - 46.5} 1.8K = 23.9K$

R_{series} max is set by R_{shunt} min. The $R_{meter}\|R_{shunt} = 1.674K$ min is satisfied by R_{series} 3.326K max. Select a 5K R_{series} potentiometer.

2.6.2 DC Voltmeter Circuit

Any node in any circuit can be represented by an equivalent circuit that is a voltage source V_S and a resistor R_S in series (A7 Equivalent Circuits). Any load R_L on the node creates a voltage divider that reduces the measured node voltage. This is why an ideal voltmeter has an infinite input resistance. However we can only design a "real world" voltmeter with resistors as follows.

The 50μA display meter circuit current requires that the voltmeter input resistance has to be 20KΩ *per volt* of the selected scale.

(33) $1 volt = 50\mu A \times 20 K\Omega \Rightarrow 20K\Omega$ *input resistance per scale volt*

A resistor converts voltage to current. Using the display meter means we want the maximum 50μA to flow at *any* full scale voltage. For example the 2.5V scale has a 50K input resistance so that 50μA flows when the voltage being measured is 2.5 volts.

(34) $R_{2.5V} = \dfrac{V_{2.5V}}{I_{meter}} = \dfrac{2.5V}{50\mu A} = 50K\Omega$

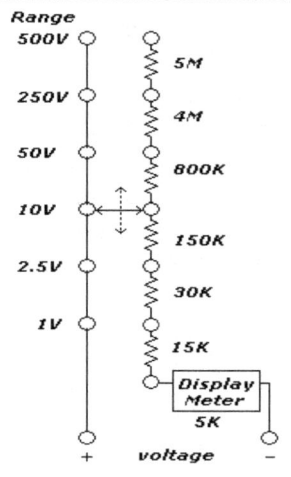

Figure 208 DC Voltmeter Circuit

The resistance in the circuit for each full scale is calculated, while accounting for the 5K display meter resistance. For example

(35) $1 volt\ scale\ R_{1V} = 1V \times \dfrac{20K}{V} = 20K$

(36) $2.5 volt\ scale\ R_{2.5V} = 2.5V \times \dfrac{20K}{V} = 50K$

(37) $10 volt\ scale\ R_{10V} = 10V \times \dfrac{20K}{V} = 200K$

And so forth. As a practical matter resistors are switched (inserted) in series (Figure 208). For example

(38) $50 volt\ scale$

$R_{50V} = 1000K = 800K + 150K + 30K + 15K + 5K$

2.6.3 DC Ammeter Circuit

A (shunt) resistor in parallel with the display meter defines full scale current.

$$(39a) \quad v_{meter} = i_{meter} R_{meter} = i_{shunt} R_{shunt}$$

$$(39b) \quad R_{shunt} = \frac{v_{meter}}{i_{shunt}} = \frac{250mV}{i_{scale} - 50\mu A}$$

$$(40a) \quad R_{500mA} = \frac{250mV}{i_{500mA} - 50\mu A} = \frac{250mV}{499.95mA} = 0.50005\Omega = 0.5\Omega$$

$$(40b) \quad R_{100mA} = \frac{250mV}{i_{100mA} - 50\mu A} = \frac{250mV}{99.95mA} = 2.50125\Omega = 2 + 0.5$$

$$(40c) \quad R_{10mA} = \frac{250mV}{i_{10mA} - 50\mu A} = \frac{250mV}{9.95mA} = 25.12563\Omega = 25 = 22.5 + 2 + 0.5$$

$$(40d) \quad R_{1mA} = \frac{250mV}{i_{1mA} - 50\mu A} = \frac{250mV}{0.95mA} = 263.158\Omega = 238\Omega + 25\Omega$$

Figure 209 Ammeter Circuit

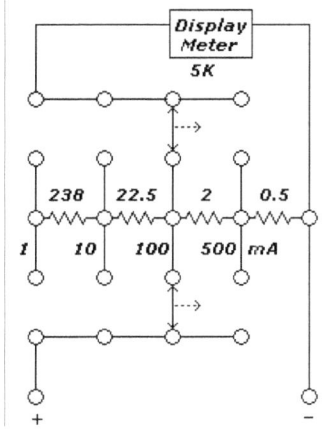

A series string of resistors is desirable, because total resistance and cost are reduced.

Again, as a practical matter resistors are switched in and out of the circuit. (Figure 209).

2.6.4 Ohmmeter Circuit

A conventional specification for a resistance measurement is that when the display pointer is at half scale (and the meter current is 25µA) an unknown resistor R_X equals a known resistor R_K (Figure 210). The battery, R_X, and R_K are in series. R_K is the ohmmeter's *equivalent* resistance.

The 10K meter circuit is in parallel with R_1, R_2, or R_3, and R_J is in series. Consequently the display pointer is at full scale (the meter current is 50µA) when the unknown resistor $R_X=0$ ohms (+ is shorted to –).

The "half scale" known resistors R_K are 12Ω, 1200Ω, and 120KΩ for scales R×1, R×100, R×10000 ohms, in some commercial ohmmeters. We will use those values and scales. If the battery voltage is 1.5V, then the "half scale" currents that flow in R_X when $R_X = R_K$ are as follows.

(41) $I_{R \times 1} = \dfrac{1.5V}{2 \cdot 12} = 62.5mA \quad I_{R \times 10^2} = \dfrac{1.5V}{2 \cdot 1200} = 0.625mA \quad I_{R \times 10^4} = \dfrac{1.5V}{2 \cdot 120K} = 6.25\mu A$

The 6.25µA is too small, because 25µA has to flow in the display meter. This means a higher voltage is required. The next higher standard battery voltage used in commercial ohmmeters is 9V. This increases the 6.25µA to 37.5µA. Current flows through the resistor R_X to be measured, which is in series with the known resistor R_K in parallel with the meter circuit. Internal batteries produce the current through the circuit. A 1.5V battery is used in the Rx1 and Rx100 scales. A 9V battery is used in the Rx10000 scale (Figure 210).

Figure 210 A Straightforward Ohmmeter Circuit

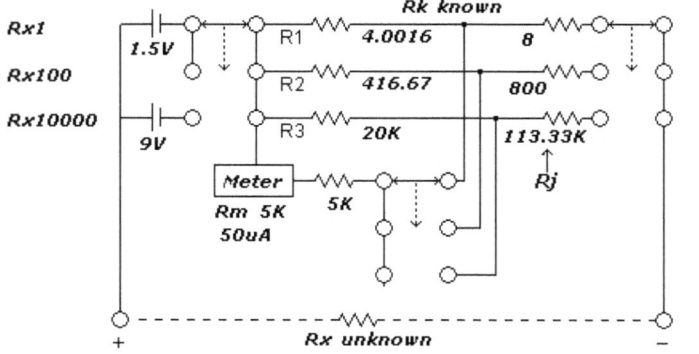

2 Circuits with Resistors

The meter circuit The meter voltage drop is 0.125 volts when $I_M=25\mu A$, and the meter resistance is 5K. If a series "calibrating" potentiometer *on all scales* also drops 0.125V, then its value is also 5K.

Rx10,000 scale The current through R_X is 37.5µA (was 6.25 equation 41), which means 12.5µA must flow in a resistor R_3 parallel to the meter. The voltage drop is 250mV, the current is 12.5µA, and so $R_3=20K$. The 20K in parallel with 10K (meter circuit 5K plus 5K in series) equals 6.67K. A 113.33K resistor makes the known resistor equal 120K as desired.

Rx100 scale The current through R_X is 625µA (equation 41), which means 600µA must flow in a resistor R_2 parallel to the meter. The voltage drop is 250mV, the current is 600µA, and so $R_2=416.67\Omega$. The 416.67Ω in parallel with 10K equals 400Ω. An 800Ω resistor makes the known resistor equal 1200Ω as desired.

Rx1 scale The current through R_X is 62500µA (equation 41), which means 62475µA must flow in a resistor R_1 parallel to the meter. The voltage drop is 250mV, the current is 62475µA, and so $R_2=4.0016\Omega$. The 4.0016Ω in parallel with 10K equals 4Ω. An 8Ω resistor makes the known resistor equal 12Ω as desired.

2.6.5 AC Voltmeter Circuit

An AC voltage waveform is a sinewave with maximum value V_{max}. The V_{rms} value, root mean square, or effective heating value of the voltage is $V_{max}/\sqrt{2}$. For example your house line voltage is 120VAC V_{rms} value, and it has a $V_{max}=120\sqrt{2}=170V$.

(43) $\quad v = V_{max} \sin \omega t = V_{rms} \sqrt{2} \sin \omega t$

A full wave bridge rectifier circuit (Figure 211) converts AC voltage to a DC current displayed by the meter by forcing the current to flow through the meter movement in the *same* direction in each half cycle.

When node 1 is positive with respect to node 2, diode D_1 conducts and diode D_2 is off. The current $k \times i_{meter}$ flowing through node 1 has two components: the meter current, and the "parallel" 3.9KΩ current (Figure 210). The meter current due to a "positive half sinewave" of AC voltage flows to the right through R_9, the meter, and the series 3.9KΩ resistor.

When node 1 is negative with respect to node 2, diode D_2 conducts and diode D_1 is off. The meter current due to a "negative half sinewave" of AC voltage flows through node 2, left up through the 3.9KΩ resistor, then to the right through R_9, the meter, and D_2.

The sequence of "half sinewaves" of current is *averaged* by the meter acting as a *mechanical* low pass filter. Average is 0.9 times rms.

(44) $V_{average} = \frac{2}{\pi} V_{max} = \frac{2\sqrt{2}}{\pi} V_{rms} = 0.9 V_{rms}$

(45a) If $R_{input} = 5K\Omega / ac\ volt$, then $R_{2.5v\_input} = 2.5 \times 5K = 12.5K$ and

(45b) $i_{in} = 2.5v / 12.5K = 200\mu A$

(45c) Full scale meter current i_m is 50µA DC, or 50/0.9µA AC

(45d) ∴ 3.9K shunt current i_{sh} is $200 - 50/0.9 = 130/0.9 \mu A$

(45e) check $i_m + i_{sh} = 50/0.9 + 130/0.9 = 180/0.9 = 200\mu A$

(45f) $R_{meter\ branch} = \frac{i_{sh}}{i_m} 3.9K = \frac{130}{0.9} \times \frac{0.9}{50} 3.9K = 2.6 \times 3.9 = 10.14K$

(45g) And so $R_9 = 10.14 - R_m - 3.9 = 10.14 - 5 - 3.9 = 1.24K$
Select a 2.5K potentiometer for R_9.

(45h) The bridge circuit resistance $R_{br} = \frac{10.14 \times 3.9}{10.14 + 3.9} = 2.217K$

(45i) And so $R_8 = 12.5 - 6.8 - R_{br} = 12.5 - 6.8 - 2.217 = 3.5K$
Select a 5K potentiometer for R_8.

Figure 211 AC Voltmeter

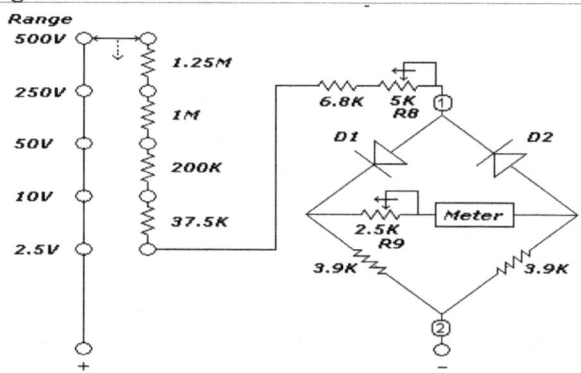

3 Circuit Components

The three passive electric circuit components, resistors R, capacitors C, and inductors L are explained. The R, L, C components' properties are defined by laws relating the voltage difference between a component's two terminals and the current flowing through the component. The generic term for these laws is *vi constraint*. Ohm's law is the R *vi* constraint. We derive the *vi* constraints for L and C, the L and C impedances, and discuss their transient and steady state responses.

Note: Linear electric circuits are assemblies of various combinations of R, L, C components, *dependent* voltage and current sources connected by *zero* impedance wires, plus *external* voltage and current sources energizing and driving the circuit. Any linear electric circuit you draw will have zero, one, or more copies of these components.

3.1 Resistor R

A resistor R is a passive two-terminal component representing an energy sink. The value of a linear resistor R is invariant with time, and not influenced by the current flowing through it, or the voltage difference between its two terminals. The jargon for "voltage difference" is "the voltage drop across the resistor." An irreversible conversion of electric energy into heat takes place in a resistor. The SI unit of resistance is the ohm. This unit is named after Georg Simon Ohm (1787-1854). In practice values of resistance range from about 10^{-3} ohms (1 milliohm) to 10^9 ohms (1000 megohms). The SI unit of conductance is the Siemens that replaced the mho (ohm spelled backwards). We prefer to use mho.

Ohm's law is the vi constraint for R. Georg Ohm showed *by experiments* that the voltage drop v across a linear resistor R is proportional to the current i flowing in the resistor. You will find Ohm's law written in at least three equivalent ways emphasizing voltage v, current i, and resistance R.

(1) *Forms of Ohm's Law* $v = iR$ $i = \dfrac{v}{R}$ $R = \dfrac{v}{i}$

Problem 301 A 2mA current flows through a 470 ohm resistor R. What is V_R? Power$_R$?

Electric Circuits - Analysis and Design

3.2 Capacitor C

A capacitor (C) is a passive element that stores energy in the form of stored charge. The value of C depends only on the shapes, sizes and serparation of the metal parts storing the charge and the dielectric constant of the material insulating the metal parts. Therefore C is invariant with time, and is not influenced by the charge stored in it. or the voltage differential (voltage drop) across it.

3.2.1 Theoretical Basis of the Capacitor

Charles Augustin de Coulomb (1746-1806) used his force law to show that charge Q on a capacitor C is proportional to voltage V (Q=CV), and that capacitance is only a function of the structure.

In Chapter 1 we discussed *experiments* leading to the conclusion that (1a) moving a charge q through a voltage difference V required work W=qV. (1b) the *experimentally determined* electric field E arising from the charges ±Q is proportional to the charge Q so that E=kQ. (1c) that voltage is the integral of the electric field. We merge the three equations to define capacitor C, and to calculate the capacitor *vi* constraint.

(1a) $W = qV$ (1b) $E = kQ$ (1c) $V_{ba} = \int_a^b E ds$

(2a) $W = qV = q \int E ds = q \int kQ ds = qQ \int k ds = qQ \dfrac{1}{C}$ by definition

(2b) $qV = qQ \dfrac{1}{C}$ $\Rightarrow$ $Q = CV$

From Q = CV the differential equation relating a capacitor's voltage and current follows immediately.

$$\boxed{(3) \ dq = C \, dv \quad \Rightarrow \quad \dfrac{dq}{dt} = C \dfrac{dv}{dt} \quad \Rightarrow \quad i = C \dfrac{dv}{dt}}$$

This is the capacitor vi constraint.

3. Circuit Components

The capacitor voltage is found by transposing the equation for current, and using a dummy variable x in the integral of the current to get the charge q. This returns us to V=Q/C.

(4) $\dfrac{dv}{dt} = \dfrac{1}{C}i$ so that $\displaystyle\int_0^t \dfrac{dv(x)}{dx}dx = \dfrac{1}{C}\int_0^t i(x)dx \;\Rightarrow\; v(t)-v(0) = \dfrac{Q(t)}{C}$

Energy in a capacitor equals the integral of the power at time t from far back in time to present time t.

(5) $w_C(t) = \displaystyle\int_{-\infty}^{t} v(x)i(x)dx = \int_{-\infty}^{t} v\left(C\dfrac{dv}{dx}\right)dx = C\int_{-\infty}^{t} v\,dv$

$w_C(t) = \dfrac{1}{2}Cv^2(t)\;\;where\;v(-\infty)=0$

Parallel plate capacitor Consider two flat parallel plates separated by distance d. Charge on a plate distributes itself uniformly across the plate, because like charges repel. Assume charge –Q is on plate a and charge +Q is on plate b (each with area A). Immerse the plates in a dielectric with dielectric constant ε. The two charge distributions create a uniform and constant electric field E from plate to plate. The *experimentally determined* electric field E arising from the charges ±Q is proportional to the charge density Q/A divided by the dielectric constant ε.

$$F = qE = q\dfrac{Q/A}{\varepsilon} = \dfrac{qQ}{A\varepsilon}$$

Work done on a charge moved from point a to point b is

$W = \displaystyle\int_a^b F\,ds = \int_a^b \dfrac{qQ}{A\varepsilon}ds = qQ\dfrac{1}{A\varepsilon}\int_a^b ds = qQ\dfrac{1}{A\varepsilon}(b-a)$

$W = qQ\dfrac{(b-a)}{A\varepsilon} = qQ\dfrac{1}{C}\;\;where\;\;\dfrac{1}{C} = \dfrac{(b-a)}{A\varepsilon}$

Observe that C is only a function of the dielectric constant ε, the area A, and the distance between plates b–a. The C equation depends only upon the geometry of its structure. This is true for any geometric structure.

Electric Circuits - Analysis and Design

Spice program Calculates the C Transient Response

```
Fig3021.ckt   Capacitor Transient Response

*PULSE( Vbase Vmax Tdelay Trise Tfall Twidth Tperiod )
V1 1 0 PULSE(0 1.8  100p   200p  200p  300p   1000p)

C1 1 0 100f ; 100fF
.TRAN 1e-011 1e-009 0
.TEMP 27
.PLOT TRAN V(1) 0,2
.PLOT TRAN I(V1) -2M,2M
.end
```

Figure 302

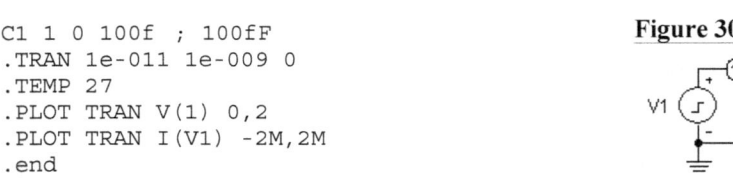

Figure 30211 Signal generator voltage and capacitor current

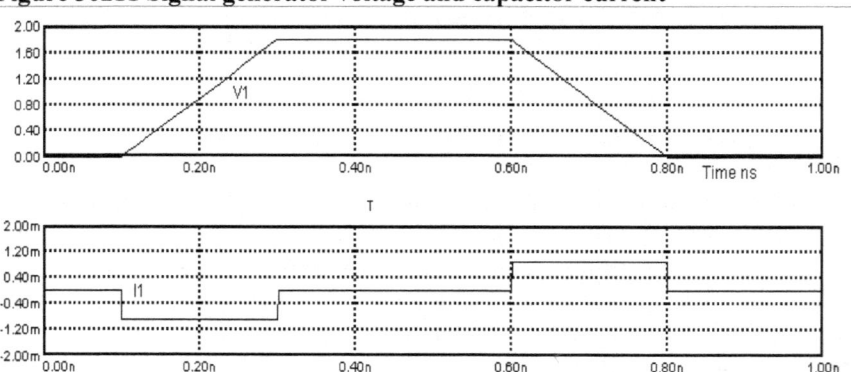

Capacitor current i = C dv/dt. Observe that when the voltage is constant at 0V and 1.8V the current is zero, because dv/dt = 0. When the voltage rises or falls with constant slope the current is a constant 0.9mA, because i = C dv/dt = 100fF × 1.8V/200ps = 0.9mA. Given the current we can calculate the capacitance.

$$C = i\frac{dt}{dv} = 0.9mA \frac{200ps}{1.8volts} = 0.9 \cdot 10^{-3} \frac{200 \cdot 10^{-12}}{1.8} = 100 fF$$

Problem 302 A 200pC charge passes through a 22 ohm resistor R in 5nS. What is V_R?

3. Circuit Components

The Spice program for the C circuit.

```
Fig3021.ckt  Capacitor Transient Response
*       PULSE(Vbase Vmax Tdelay Trise Tfall Twidth Tperiod )
V1 1 0 PULSE(0      1.8   100p   200p  200p  300p   1000p)

C1 1 0 100f ; 100fF
.TRAN 1e-011 1e-009 0
.PLOT TRAN V(1) 0,2
.PLOT TRAN I(V1) -2M,2M
.TEMP 27
.end
```

Figure 302

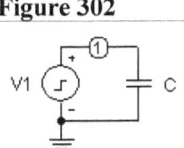

Write the program   (Spice is not case sensitive)

`Fig3021.ckt  Capacitor Transient Response`
  First line The first line of every program is assumed by Spice to be a title statement. The title statement can include any words.

`*PULSE( Vbase Vmax Tdelay Trise Tfall Twidth Tperiod )`
  The asterisk * defines a line as a comment.

`V1 1 0 PULSE(0 1.8 100p 200p 200p 300p 1000p)`
  Voltage source   Signal generator v_1 is connected to nodes 1 and 0. The v_1 signal is a pulse with 1.8 volt peak value (Figure 30211).

`C1 1 0 100f ; 100fF`
  Circuit components Capacitor C_1 is connected to nodes 1 and 0.

`.TRAN 1e-011 1e-009 0 1e-011`
  TRAN control statement Dot TRAN defines a time range from 0 to 1ns with points calculated every 0.01ns.

`.PLOT TRAN V(1) 0,2`
`.PLOT TRAN I(V1) -2m,2m`
  Plot the data dot PLOT TRAN executes the dot TRAN statement. Both lines are executed calculating v_1 and i from 0ns to 1ns

`.TEMP 27`
  Temperature Dot temp (.temp) defines temperature as 27 degrees C.

`.end`
  Last line The last line of every program is .end (dot end).

31

3.2.2 Capacitor in a circuit

Capacitors are two terminal devices. Two wires are connected from the capacitor's two insulated metal parts A and B to two terminals. Connect the two terminals of a capacitor to the two terminals of a battery of V_M volts. The connection forces the capacitor's voltage V_C to jump from zero to V_M volts. When the capacitor's voltage V_C jumps from zero to V_M volts electric field E is established inside the capacitor, where E=V/d and d is the average distance from A to B. At the same time electrons are removed from one terminal leaving it with a positive surface charge. The electrons flow through the battery to the other terminal creating a negative surface charge. The electron flow stops when equilibrium, $V_C=V_M$, is reached. Later, if you insert an ammeter in the circuit you will find that the ammeter reads zero current.

> An ammeter is a current measuring instrument.

When you remove the battery the capacitor stands alone, and a voltmeter that draws no current, measures V_M volts across the capacitor. No current is flowing, yet there is a voltage across the capacitor. The voltage exists, because there is stored charge q on metal part A and –q on metal part B maintaining the electric field that supports the voltage.

> That is why a charged capacitor is dangerous.

The electric field forces the electrons away from the surface of plate A leaving the required equivalent positive surface charge. No charge is created. The displaced electrons left the capacitor in the form of an current that decreases to zero. When the battery is disconnected the charge remains in place on the plates. We say the charge has been stored. This is possible because a conducting metal body has many free electrons in it. Since q=CV the value of q is CV_M.

The voltage across a capacitor is proportional to the charge, not the current. This is why Ohm's law does not apply to capacitors. Constant voltage implies constant charge that means the current i = dq/dt is zero. This means the capacitor's "resistance" to constant-voltage is infinite. Consequently a capacitor is an open circuit when the source is a (zero frequency) direct voltage such as a battery. This is not surprising when you realize a capacitor is, in effect, just two insulated metal plates.

3. Circuit Components

Figure 301 Capacitor models

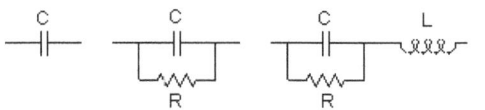

In practice a small *leakage* current flows through the insulator between the plates. A conductance G=1/R in parallel represents this effect (Figure 301). We refer to it as a parasitic resistance, because we intended to construct a perfect capacitor. The presence of the parallel R leads to the idea of a capacitor's quality Q. The Q of the capacitor is the ratio of stored energy to dissipated power.

Perfect capacitor $\qquad i = C\dfrac{dv}{dt}$

Imperfect capacitor $\qquad i = C\dfrac{dv}{dt} + gv \quad \rightarrow \quad vi = Cv\dfrac{dv}{dt} + gv^2$

For sinusoidal voltage $\qquad vi = Cv\omega v + gv^2$

(11) $\quad Q \equiv \dfrac{C\omega v^2}{gv^2} = \dfrac{\omega C}{g} = \omega CR$

Another practical attribute of any structure is the self-inductance L of its leads and plates (Figure 301). The SI unit of capacitance is the farad (F). This unit is named after Michael Faraday (1791-1867). In practice, one farad is a very large capacitor. In circuits that are on a semiconductor chip, capacitors measured in picofarads (1pF = 10^{-12}F), femtofarads (1fF = 10^{-15}F) and attofarads (1aF = 10^{-18} F) are common.

3.2.3 Capacitors in series

The key observation for capacitors in series is that the same current flows through each capacitor in the series string. The voltage across each capacitor is Q_J/C_J. The total voltage is the sum of the individual capacitor voltages that is the sum of the Q_J/C_J terms. Since each Q is the integral of the same current i, all of the Q's are equal to each other. This means Q factors out of the voltage expression so that the total voltage equals Q times the sum of the $1/C_J$ terms or Q/C_{Total}. If the C's have different values the individual capacitor voltages are not equal even though the charges Q are equal.

Electric Circuits - Analysis and Design

A calculation of the sum of the voltage drops in the series string shows that the sum of the reciprocal values of capacitors in series can be replaced by one capacitor whose reciprocal value equals the sum. For capacitors in series "add reciprocals."

$$v(t) = \frac{1}{C_1}\int_0^t i\,dx + \frac{1}{C_2}\int_0^t i\,dx + \cdots + \frac{1}{C_n}\int_0^t i\,dx$$

$$v(t) = \left(\frac{1}{C_1} + \frac{1}{C_2} + \cdots + \frac{1}{C_n}\right)\int_0^t i\,dx = \frac{1}{C}\int_0^t i\,dx = \frac{Q}{C}$$

(12) $$\frac{1}{C} = \frac{1}{C_1} + \frac{1}{C_2} + \cdots + \frac{1}{C_n}$$

3.2.4 Capacitors in parallel

The key observation for capacitors in parallel is that the same voltage is across each capacitor. The sum of the currents flowing in the capacitors shows that the sum of values of capacitors in parallel can be replaced by one capacitor whose value equals the sum. For capacitors in parallel "add the values."

$$i(t) = C_1\frac{dv}{dt} + C_2\frac{dv}{dt} + \cdots + C_n\frac{dv}{dt}$$

$$i(t) = (C_1 + C_2 + \cdots + C_n)\frac{dv}{dt} = C\frac{dv}{dt}$$

(13) $\therefore C = C_1 + C_2 + \cdots + C_n$

Problem 303 A 45nC charge produces 3V across capacitor C. What is C?

Problem 304 A 3.3V change takes place in 5nS on 500pF capacitor C. What is I_C?

Problem 305 A 100fF capacitor C experiences 1.8V change in 200pS. What is I_C?

Problem 306 Capacitor C stores 200fC at 1.8V. What is C?

Problem 307 If C2 = 10pF and I=(5×10^{-6}) sin($2\pi ft$) where frequency f = 100KHz, what is the expression for the voltage across the capacitor?.

---

3. Circuit Components

3.2.5 Capacitor Impedance, Transient and Steady State Responses.

Transient response Capacitor transient response to a specific voltage waveform is calculated in Spice program Fig3021.ckt page 30. The capacitor's vi constraint i=Cdv/dt determines the transient response to any voltage waveform (Chapter 7).

Steady state response Capacitor steady state response depends on the impedance of the capacitor, which, in turn, is dependent on the frequency f of the sinusoidal source driving the circuit.

$$v = V_m e^{j\omega t} \quad \Rightarrow \quad i = C\frac{dv}{dt} = j\omega C V_m e^{j\omega t} = I e^{j\omega t}$$

$$\text{where } I = j\omega C V_m \quad \Rightarrow \quad Z_C = \frac{V_m}{I} = \frac{1}{j\omega C} \quad \text{where } \omega = 2\pi f$$

A capacitor's impedance Z_C is 1/f frequency dependent. That means the behavior of circuits with capacitors is a function of frequency. The world says a circuit with capacitors has a *frequency response*. We are in the frequency domain and the circuit is in the steady state (Chapters 6, 8).

As frequency f decreases towards zero Z_C increases without limit.

As frequency f increases towards infinity Z_C decreases towards zero.

3.3 Inductor L

Hans Christian Oersted (1777-1851) discovered in 1820 that an electrical current in a wire created a magnetic field around that wire. This was the first evidence of a physical link connecting electricity and magnetism.

In 1831 Michael Faraday (1791-1867) moved a magnet (i.e. a magnetic field) across an electric circuit, and a voltage was induced in the circuit. Induction requires relative motion between the circuit and the magnet.

A magnet is not required, because a changing current flowing in a wire creates a changing magnetic field that passes by its own (source) wire inducing a voltage in that same wire. This voltage is referred to as the "back electromotive force" (back emf) because it *opposes* the source voltage that is forcing current flow in the wire.

> *A voltage will be induced in a wire whenever there is a change in any magnetic field linking it.*

This induced voltage, the back emf, is a voltage drop in the circuit with an effect analogous to an IR voltage drop that also opposes the source voltage. The circuit property that created the cycle of events culminating in an induced voltage opposing the voltage source is referred to as self inductance, or inductance L. Mutual inductance M is the circuit property that arises when a changing current in one wire induces a voltage in another wire.

> An inductor L is a structure that stores energy in the magnetic field created by the current flowing through its wire.

The two ends of a coil of wire or piece of metal are brought out from the inductor structure to two terminals. Examples of the structure are a piece of straight wire, a coil of wire wrapped on a cardboard tube (non-magnetic material), a coil of wire wrapped on an iron or ferrite core (non-linear magnetic material with variable magnetic permeability μ), or any piece of material a current can flow through.

The *value* of the inductance of a structure depends only on the geometry and material of that structure. This is why the value of L is invariant with time, and not influenced by the current it conducts, or the voltage drop across it. The value of L changes when the core material is non-linear.

3. Circuit Components

3.3.1 Theoretical Basis of the Inductor

Starting from the Law of Andre-Marie Ampere (1775-1836) we quantify Michael Faraday's discovery that voltage can be induced in an electric circuit by relative motion between the circuit and a magnetic field. The force F on a coil of N turns is proportional to the magnetic field B, the current i flowing through the coil, and the length y of the moving wire (Figure 303). Length y times differential distance ds equals differential area dA. Magnetic field B times differential area dA is the differential flux dφ (Figure 303).

Figure 303 Sliding bar crosses flux lines

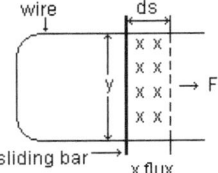

If current is i, magnetic field is B, force is F on wire length y, then on a coil with N turns Ampere's Law of Force is *F = NiBy*. The incremental mechanical work *dw* is force F times incremental distance *ds*. The incremental electrical work is power *vidt*. Equating mechanical work to electrical work produces Faraday's law after adding 10^{-8} for SI units..

$dw = Fds = NBiyds = NBi\, dArea = NiB\, dA = Ni\, dflux = Nid\varphi$

$dw = power\, dt = vidt$

$\therefore Ni\, d\varphi = vi\, dt$

(14) $\quad v = 10^{-8} N \dfrac{d\varphi}{dt} \quad$ Faraday's Law

Seeking a vi constraint we convert flux φ to Ni ampere-turns in Faraday's law of induction and derive the differential equation relating an inductor's voltage and current: For any geometric structure where k is a constant, A is the area the flux φ passes through, the flux density B leads to inductance.

$B = kNi$

$\varphi = BA = kNi\, A$

$\therefore v = 10^{-8} N \dfrac{d\varphi}{dt} = 10^{-8} N \cdot kNA \dfrac{di}{dt} = 10^{-8} kN^2 A \dfrac{di}{dt}$

(15) $\quad v = L \dfrac{di}{dt} \quad$ where $L = 10^{-8} kAN^2$

Electric Circuits - Analysis and Design

$$(16) \quad v = L\frac{di}{dt} \quad \text{This is the inductor vi constraint.}$$

The current is the integral of the inductor voltage.

$$(17a) \quad \frac{di}{dt} = \frac{1}{L}v \Rightarrow \int_0^t \frac{di(x)}{dx}dx = \frac{1}{L}\int_0^t v(x)dx$$

$$(17b) \quad i(t) - i(0) = \frac{1}{L}\int_0^t v(x)dx$$

Energy in an inductor equals the integral of the power.

$$(18) \quad w_L(t) = \int_{-\infty}^t vidx = \int_{-\infty}^t \left(L\frac{di}{dx}\right)idx = L\int_{-\infty}^t idi = \frac{1}{2}Li^2(t)$$

For example, when a wire is wound into a coil of N turns the N turns act as if they are N wires in parallel. This is why the total flux from the N wires is the merged flux from current i in N wires. In turn, N times the flux induces N times the voltage that makes the value of the inductance proportional to N^2.

The SI unit of inductance is the Henry (H). This unit is named after Joseph Henry (1797-1878). Values of inductance used in practice range from 10^{-9}H (1nH, 1 nanohenry) to about 100H (Henry). In practice, one Henry is a large inductor.

3.3.2 Inductor in a circuit

When the current flowing through an inductor is constant with time the magnetic flux is constant and not cutting across wires so that the voltage drop is zero. Common sense tells us this is correct. After all, an inductor is just a piece of wire, or metal, shaped in some way. The voltage across an inductor is proportional to the differential of the current, not the current. This is why Ohm's law does not apply to inductors. Furthermore, this implies the inductor's opposition to constant-current flow is zero. Constant-current is also known as direct current. A direct current does not vary with time so it can be considered as a zero-frequency sinusoid $I_M \cos 0t$ producing constant magnetic flux. As the rate of change of current increases the induced voltage increases.

3. Circuit Components

Spice program Calculates the L Transient Response

```
Fig3051.ckt   Inductor Transient Response

I1 1 0 PULSE(0 2m  100p 200p 200p 300p 1000p)

L1 1 0 100n ; 100nH
```

Figure 305

```
.TRAN 1e-011 1e-009 0
.TEMP 27
.PLOT TRAN I(I1) 0,2.5M
.PLOT TRAN V(1) -2,2
.end
```

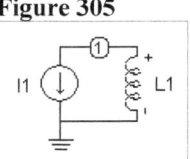

Figure 30511 Signal generator current and inductor voltage

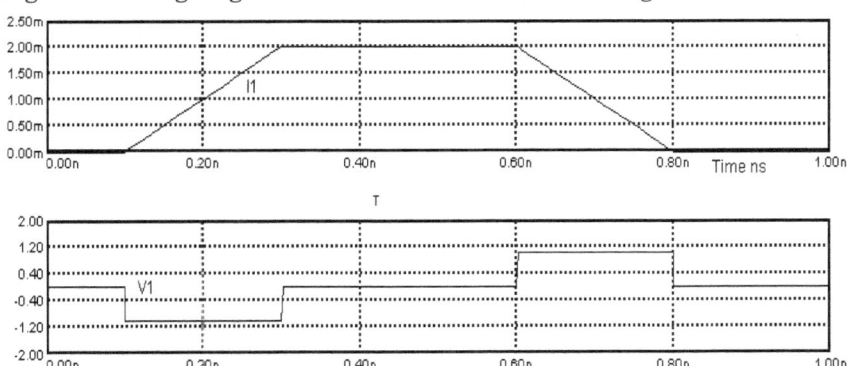

Inductor voltage v = L di/dt. The applied current starts at 0mA, and rises with constant slope di/dt=2mA/200ps. Then the current is a constant 2mA for 300ps, and finally falls to 0mA with slope dv/dt = −2mA/200ps.

Observe that when the current is constant at 0mA and 2mA the voltage is zero, because di/dt = 0. When the current rises or falls with constant slope the voltage is a constant ±1V (v=Ldi/dt=100nH×2mA/200ps=1V). Given the voltage we can calculate the inductance.

(31) $$L = v\frac{dt}{di} = 1v\frac{200\,ps}{2mA} = 1\frac{200 \cdot 10^{-12}}{2 \cdot 10^{-3}} = 100 \cdot 10^{-9} = 100nH$$

39

Electric Circuits - Analysis and Design

Figure 304 Inductor models

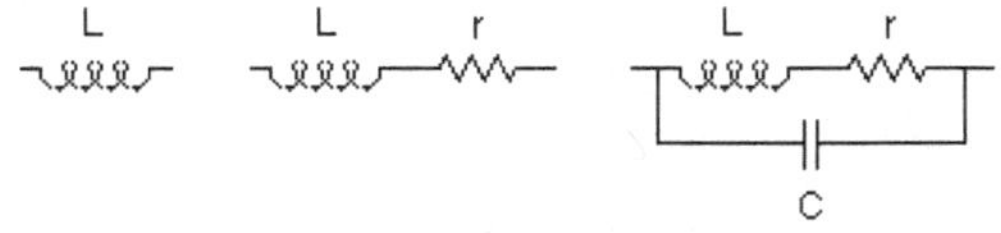

Consequently an inductor behaves like a short circuit in a (zero frequency) DC circuit. In practice the wire has a small non-zero resistance r (Figure 304). We refer to it as a parasitic resistance, because we intended to construct a perfect inductor. The presence of the series r leads to the idea of an inductor's quality; the Q of the coil. Another practical attribute of any structure is its self-capacitance C (Figure 304).

Perfect inductor $\qquad v = L\dfrac{di}{dt}$

Imperfect inductor $\qquad v = L\dfrac{di}{dt} + ri \quad \Rightarrow \quad vi = Li\dfrac{di}{dt} + ri^2$

For sinusoidal voltage

(19) $\quad vi = Li\omega i + ri^2 \quad \Rightarrow \quad Q = \dfrac{L\omega i^2}{ri^2} = \dfrac{\omega L}{r}$

3.3.3 Inductors in series

Inductors in series have the same current flowing through each inductor in the series string. Hence each inductor experiences the same di/dt. The induced voltage is proportional to L. Voltages in series add, and so do the inductance's. A calculation of the sum of the voltage drops in the series string shows that inductors in series can be replaced by one inductor whose value equals their sum.

$$v(t) = L_1\dfrac{di}{dt} + L_2\dfrac{di}{dt} + \cdots + L_n\dfrac{di}{dt}$$

$$v(t) = (L_1 + L_2 + \cdots + L_n)\dfrac{di}{dt} = L\dfrac{di}{dt}$$

(20) $\quad L = L_1 + L_2 + \cdots + L_n$

3.3.4 Inductors in parallel

Inductors in parallel have the same voltage across each inductor. A current entering a node to which inductors are connected in parallel divides among the inductors. The sum of the currents flowing in the inductors shows that the sum of reciprocal values of inductors in parallel can be replaced by one inductor.

/ 3. Circuit Components

$$i(t) = \frac{1}{L_1}\int_0^t v(x)dx + \frac{1}{L_2}\int_0^t v(x)dx + \cdots + \frac{1}{L_n}\int_0^t v(x)dx$$

$$i(t) = \left(\frac{1}{L_1} + \frac{1}{L_2} + \cdots + \frac{1}{L_n}\right)\int_0^t v(x)dx = \frac{1}{L}\int_0^t v(x)dx$$

(21) $\quad \dfrac{1}{L} = \dfrac{1}{L_1} + \dfrac{1}{L_2} + \cdots + \dfrac{1}{L_n}$

Problem 308 A 2mA current change produces a 1.8V change across 3.3µH inductor L. What is Δt?

Problem 309 A 5mA current passes through a 10mH inductor L in 2nS. What is V_L?

3.3.5 Inductor Impedance, Transient and Steady State Responses.

Transient response Inductor transient response to a specific current waveform is calculated in Spice program Fig3051.ckt. The inductor's vi constraint v=Ldi/dt determines the transient response to any voltage waveform (Chapter 7).

Steady state response Inductor steady state response depends on the impedance of the inductor, which, in turn, is dependent on the frequency f of the sinusoidal source driving the circuit.

$$i = I_m e^{j\omega t} \Rightarrow v = L\frac{di}{dt} = j\omega L I_m e^{j\omega t} = V e^{j\omega t} \quad \text{where } V = j\omega L I_m$$

$$Z_L = \frac{V}{I_m} = j\omega L \quad \text{where } \omega = 2\pi f$$

A inductor's impedance Z_L is f frequency dependent. That means the behavior of circuits with inductors is a function of frequency. The world says a circuit with inductors has a *frequency response*. We are in the frequency domain and the circuit is in the steady state (Chapters 6, 8).

As frequency f increases towards infinity Z_L increases without limit.

As frequency f decreases towards zero Z_L decreases towards zero

Electric Circuits - Analysis and Design

4 Transformers in Circuits

Transformers are found everywhere in electric power systems, because sinusoidal voltages of various values are required for practical reasons: 24V for low-voltage control systems, 120V for household power and light, 220V for large appliances, 2KV for local power distribution, 220KV for power transmission over longer distances, and so forth. Transformers are found in the power supply of almost every electronic product that is not battery operated.

Transformers are still found everywhere in telephone and communications systems. There are many applications in the telephone plant for transformers in filters, radio frequency amplifiers and so forth, where the emphasis is on signal processing. They are used to implement ingenious telephone circuits. Transformers in tuned circuits are found in every radio and television set. We need to understand these devices.

The basic transformer is an assembly of two or more windings (coils of insulated wire) on a core. Each winding has self inductance L and mutual inductance M with every other winding. Major properties of a transformer are conversion of voltage, and current to new levels with almost 100% efficiency when using high magnetic permeability cores. Consistent with conversion of voltage and current the associated impedance's are *transformed* by the square of the turns ratio (n^2) of the windings as a first approximation. Metallic connection between windings is not required.

Many transformers are essentially ideal transformers, because they have high magnetic permeability cores so that the coefficient of coupling k=1. They use mutual inductance to convert from one voltage to another voltage at any desired power level with greater than 99% efficiency.

The core material ranges from a cardboard tube to high permeability magnetic material. The windings are referred to as the primary and secondary windings. The only distinction between primary and secondary is that the energy source is usually connected to the primary. High efficiency transformer cores are made up from magnetic material of high permeability that can be $10,000\mu_o$ (μ_o is the permeability of free space). In transformers high permeability means all flux generated by one winding is coupled to the other windings (k=1). Then there are the transformers whose core permeability is a small multiple of μ_o, and k is less than one.

4 Transformers in Circuits

4.1 Mutual Inductance $M=k\sqrt{L_1 L_2}$

Hans Christian Oersted (1777-1851) discovered in 1820 that an electrical current in a wire created a magnetic field B around that wire. This was the first evidence of a physical link connecting electricity and magnetism.

In 1831 Michael Faraday (1791-1867) moved a magnet (i.e. a magnetic field) past an electric circuit, and voltage was induced in the circuit. Induction requires relative motion between the circuit and the magnet. A magnet is not required, because a changing current flowing in a wire creates a changing magnetic field that passes by its own (source) wire inducing a voltage in that same wire. This voltage is referred to as the "back electromotive force" (back emf), because it *opposes* the source voltage that is forcing current flow in the wire. This induced voltage, the back emf, is a voltage drop in the circuit with an effect analogous to an IR voltage drop that also opposes the source voltage.

> *A voltage will be induced in any wire whenever there is a change in any magnetic field linking it.*

The circuit property that created the cycle of events culminating in an induced voltage opposing the voltage source is referred to as self inductance, or inductance (L). Mutual inductance (M) is the circuit property that arises when a changing current in one wire induces a voltage in *another* wire.

A alternating current in a wire produces an alternating magnetic field that spreads out in space. The field induces a voltage in any wire (or any conductive material) the field passes across. A changing current in one wire that induces a changing voltage in a second wire extends the concept of inductance. This extension is referred to as mutual inductance (M) because it arises from a magnetic field *flux* coupling *both* wires. The difference is that the induced voltage is across one inductance (wire) while the changing current flows in another inductance (wire). Mutual inductance (M) has the same type of vi constraint as inductance. *M is represented in Spice by the coupling coefficient k* (Section 4.1.3 page 46).

There is no distinct mutual inductor you can touch. Mutual inductance is a property of an assembly of two or more inductors that may just be adjacent lengths of wire.

Electric Circuits - Analysis and Design

4.1.1 Self and Mutual Inductance vi constraints

Mutual inductance is defined as the *ratio* of voltage induced in one circuit to the rate of change of current in another circuit. Two network meshes have mutual inductance when there is an inductor in a branch common to both meshes (Figure 401a). Two unconnected network meshes have mutual inductance when there is *transformer* common to both meshes (Figure 401b). *Mutual* because both circuits produce the same types of terms (equations 3).

Figure 401 Mutual Inductance Examples

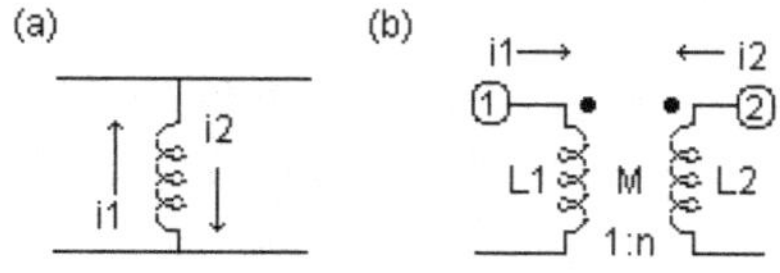

(a)   (b)

(1) $v = L\dfrac{di}{dt}$ This is the inductor vi constraint.

(2) $v_2 = \pm M_{21}\dfrac{di_1}{dt}$ This is the mutual inductance vi constraint.

Magnetic coupling to two or more inductors creates a voltage drop across an inductor in mesh 1 that has terms due to rates of change of current in other meshes.

(3a) $v_1 = L_{11}\dfrac{di_1}{dt} \pm M_{12}\dfrac{di_2}{dt} \pm M_{13}\dfrac{di_3}{dt} + \cdots$

(3b) $v_2 = \pm M_{21}\dfrac{di_1}{dt} + L_{22}\dfrac{di_2}{dt} \pm M_{23}\dfrac{di_3}{dt} + \cdots$

The time domain equation transforms to the p domain.
(4a) $v_1 = pL_{11}i_1 \pm pM_{12}i_2 \pm pM_{13}i_3 + \cdots$
(4b) $v_2 = \pm pM_{21}i_1 + pL_{22}i_2 \pm pM_{23}i_3 + \cdots$

Problem 401 Let L_1=100mH, L_2=200mH, k=0.93. What are the values of M, and turns ratio n? See 4.1.3.
Problem 402 Two coupled coils L_1 and L_2 have a turns ratio n=6, k=0.8, and M=3mH. What are the values of L_1 and L_2?
Problem 403 Two magnetically coupled windings have N_1=300 turns and N_2=900 turns. When I_1=50mA the fluxes produced are Φ_{11}=0.2 weber and Φ_{21}==0.6 weber. Find L_1, L_2, M, and k. Hint use equations 6 and 7.

4 Transformers in Circuits

4.1.2 Induced Voltage and the Dot Convention

A current in one winding on a core of permeability $\mu_r\mu_0 = 1{,}000\mu_0$ produces magnetic flux $\Phi = \Phi_{11} + \Phi_{21}$ (Figure 402). (μ_0 is the permeability of free space, and μ_r is the ratio of the permeability of the material relative to free space.) The free

Figure 402 Flux in Windings

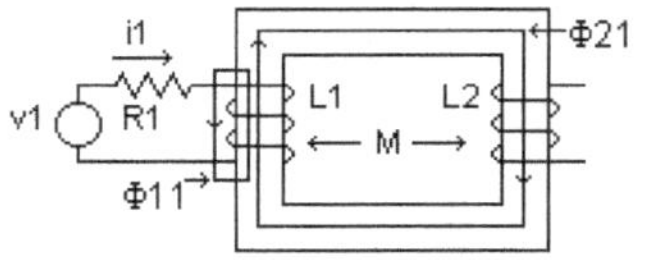

space surrounding the core has permeability μ_0. Thus flux in the free space is a 0.001 fraction of the total the flux in the core, and k=0.999 or 1 in practice. The Φ_1, Φ_2 flux directions are found by *the right hand rule* (Figures 403, 404). Wrap the fingers in the direction of the current, and the thumb points in the flux direction.

Figure 403 Fluxes subtract

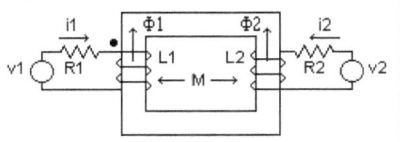

Figure 404 Fluxes add

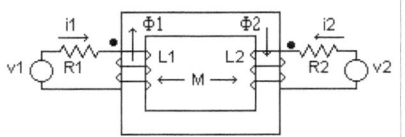

Formulation of equations with mutual inductance requires knowledge of the polarity of the induced voltage. The polarity of an induced voltage depends on how the coils are wound relative to each other. The dot convention that provides polarity information allows us to avoid the need to examine how the coupled coils are wound on a core. The *dot convention* is implemented by placing a dot on one end of one of the coupled coils (either end will do). Next, and this time it matters which end is marked, a dot is placed on the end of each of all other coupled coils (there may be more than one) whose polarity of the induced voltage is the same as the polarity at the first dot.

How to place the dots in practice We cannot see how the windings are wound, because transformers are encapsulated. One way to correctly place dots uses a two channel oscilloscope. Connect channel 1 input to node 1. Connect channel 2 input to

Figure 405 Adding Dots

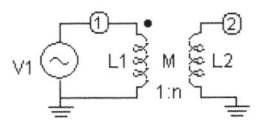

node 2. Connect a 1KHz sinewave generator to node 1. If the two signals are in phase place the dot at node 2. If the two signals are 180 degrees out of phase place the dot at the grounded node.

Electric Circuits - Analysis and Design

4.1.3 Coefficient of Coupling k

The coefficient of coupling k relates the mutual M and L_1, L_2 inductance of two coils so that we do not have to be concerned about flux and flux linkages. The coupling between two coils can range from zero to 100%. Coupling is zero when there are no flux linkages from $coil_1$ to $coil_2$ ($\Phi_{12} = 0$). Coupling is 100% when all of the flux generated by a current in one coil links the other coil ($\Phi_{11} = 0$). A proportionality constant (k) ranging from 0 to 1 is defined as the coefficient of coupling between two coils that is related to M, L_1, and L_2 (equation 6c).

The coupled fluxes dictate a relationship between mutual and self inductance's based on the numbers of turns. Mutual inductance M_{12} is proportional to $N_1 \times N_2$, because current in the turns of one coil creates flux that cuts the turns of a second coil. Self inductance is proportional to N^2, because current in the turns of a coil creates flux that cuts the same turns of the coil. Inductors L_1 and L_2 are proportional to N_1^2 and N_2^2.

(5) $v_1 = L_1 \dfrac{di}{dt} \;\Rightarrow\; L_1 = 10^{-8} N_1^2 \kappa = u_1 N_1^2$

(6a) let $L_1 = u_1 N_1^2$ and $L_2 = u_2 N_2^2$ and $M = M_{12} = M_{21} = u_M N_1 N_2$

(6b) $L_1 L_2 = u_1 u_2 N_1^2 N_2^2 = \dfrac{u_1 u_2}{u_M^2} M^2 = \dfrac{1}{k^2} M^2$

(6c) $M = k\sqrt{L_1 L_2}$ where $0 \le k \le 1$

> M is the actual mutual inductance, whereas $\sqrt{L_1 L_2}$ is the maximum possible value of mutual inductance.

Spice program Inductors L1 and L2 with coefficient of coupling K12 that can range from 0 to 1, are listed in a Spice program as follows.
```
L1  1 0 100m
L2  2 0 400m
K12 L1 L2 0.93
```

Problem 404 If L=22µH and v=5sin(2πft) where frequency f=10MHz, what is the expression for the current through the inductor?

Problem 405 Find an equation for M as a function of L_{plus} and L_{minus}.

4 Transformers in Circuits

Total Inductance of Coupled Coils

Current i_1 flows to its dot, and i_2 flows from its dot. *Therefore the sign of voltage induced by M is negative.* Note that $i_2 = -i_1$.

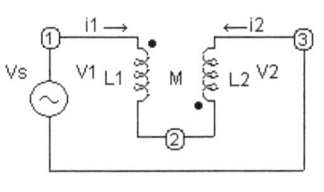

$$v_{12}(t) = +L_1 \frac{di_1}{dt} - M \frac{di_2}{dt} = +L_1 \frac{di_1}{dt} + M \frac{di_1}{dt}$$

$$v_{23}(t) = +M \frac{di_1}{dt} - L_2 \frac{di_2}{dt} = +M \frac{di_1}{dt} + L_2 \frac{di_1}{dt}$$

KVL for the loop in the circuit shows that the source voltage v_s equals the sum of v_{12} and v_{23}. The current i_2 is replaced by $-i_1$ to find the input impedance. In other words, the sign of induced voltages is positive here because i_1 flows "to" both dots. Note that there are *two* M terms.

$$v_S(t) = v_{12}(t) + v_{23}(t) = L_1 \frac{di_1}{dt} + M \frac{di_1}{dt} + M \frac{di_1}{dt} + L_2 \frac{di_1}{dt}$$

$$v_S(t) = (L_1 + L_2 + 2M) \frac{di_1}{dt}$$

$$L_{plus\_series} = L_1 + L_2 + 2M$$

If connections to the terminals of coil$_2$ are reversed then i_2 flows to its dot, and $i_2 = i_1$. This changes the sign of voltage induced by M to positive. However, the sign of induced voltages is negative here because i_1 now flows "to" one dot and "from" the other dot.

$$v_S(t) = v_{12}(t) + v_{23}(t) = L_1 \frac{di_1}{dt} - M \frac{di_1}{dt} - M \frac{di_1}{dt} + L_2 \frac{di_1}{dt}$$

$$v_S(t) = (L_1 + L_2 - 2M) \frac{di_1}{dt}$$

$$L_{minus\_series} = L_1 + L_2 - 2M$$

Total inductance for parallel inductors is

$$L_{minus\_parallel} = \frac{L_1 L_2 - M^2}{L_1 + L_2 + 2M} = \frac{L_1 L_2 (1 - k^2)}{L_1 + L_2 + 2k\sqrt{L_1 L_2}}$$

$$L_{plus\_parallel} = \frac{L_1 L_2 - M^2}{L_1 + L_2 - 2M}$$

Electric Circuits - Analysis and Design

4.2 Ideal Transformers

Consider a two winding transformer whose windings have N_1 and N_2 turns, and the turns ratio $n = N_2/N_1$. The basic properties of ideal transformers are
1. Voltage is scaled by turns ratio (n)
2. Current is scaled by turns ratio (1/n)
3. Impedance is scaled by n^2
4. $k=1$ and $M=\sqrt{L_1 L_2}$
5. No power is dissipated

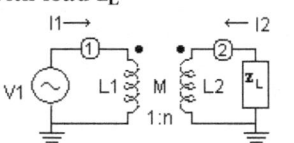

Figure 406 Ideal transformer with load z_L

If $z_L=\infty$, then $I_2=0$ (Figure 406)

$$v_2 = pMi_1 \quad\text{and}\quad i_1 = \frac{v_1}{pL_1} \Rightarrow v_2 = pM\frac{v_1}{pL_1}$$

(7) $\dfrac{v_2}{v_1} = \dfrac{pM}{pL_1} = \dfrac{M}{L_1} = \dfrac{\sqrt{L_1 L_2}}{L_1} = \dfrac{\sqrt{L_1 L_2}}{L_1} = \sqrt{\dfrac{L_2}{L_1}} = \dfrac{N_2}{N_1} = n$

If $z_L=0$, then KVL around the output mesh produces

(8) $0 = pL_2 i_2 - pMi_1 \Rightarrow \dfrac{i_2}{i_1} = \dfrac{pM}{pL_2} = \dfrac{N_1 N_2}{N_2^{\,2}} = \dfrac{N_1}{N_2} = \dfrac{1}{n}$

If $z_L \neq 0$ then $v_2 = i_2 z_L$.

(9) $z_{INPUT} = \dfrac{v_1}{i_1} = \dfrac{v_2}{n}\dfrac{1}{ni_2} = \dfrac{1}{n^2}\dfrac{v_2}{i_2} = \dfrac{z_L}{n^2}$

> The impedance z_L is transformed by the square of the turns ratio n. This is one reason why two or more windings coupled to each other are referred to as a transformer.

Problem 406 $k=0.8$ for all windings. $L_1=100$mH, $L_2=60$mH, $L_3=220$mH. Find the 3 M's and the total inductance L from terminals 1 to 2.

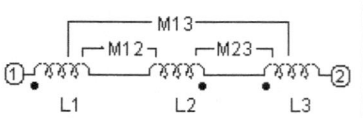

Problem 407 A transformer has 2 windings n_1, n_2 with turns in the ratio $n:1$. Resistor R_2 is connected to winding n_2. Find the equation for z_{INPUT} by equating P_1 power in to P_2 power out.

Problem 408 A transformer has 3 windings n_1, n_3, n_2 with turns ratios 1:3:2. Resistors R_3 and R_2 are connected to windings n_3, n_2 respectively. What is the input impedance z_{INPUT} at winding n_1? Hint use power.

4.3 Transformer Equivalent Circuit

Equations 10 represented a transformer with no resistance in the windings (Figure 407a). The process creating a two mesh (i_1, i_3) equivalent circuit inserts an ideal transformer (Figure 407b). Ideal equations $i_2=i_3/n$ and $v_2=nv_3$ are substituted into equations 10, and the tricks $-M/n+M/n=0$, $-M+M=0$ are used. The *total* inductance in mesh 1 is still L_1, and M/n is a *mutual inductance* common to both meshes. That is why we add and subtract M/n terms. We add and subtract M terms in mesh 2 to produce the T circuit (Figure 407b) whose mesh equations are

$$(10a) \quad v_1 = +L_1 \frac{di_1}{dt} + M \frac{di_2}{dt}$$

$$(10b) \quad v_2 = +M \frac{di_1}{dt} + L_2 \frac{di_2}{dt}$$

$$(11) \quad v_1 = \left(L_1 - \frac{M}{n} + \frac{M}{n}\right)\frac{di_1}{dt} - \frac{M}{n}\frac{di_3}{dt}$$

$$(12) \quad v_3 = -\frac{M}{n}\frac{di_1}{dt} + \left(\frac{L_2}{n^2} - \frac{M}{n} + \frac{M}{n}\right)\frac{di_3}{dt}$$

Figure 407 Transformer Equivalent Circuit

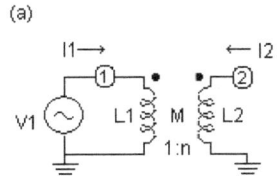
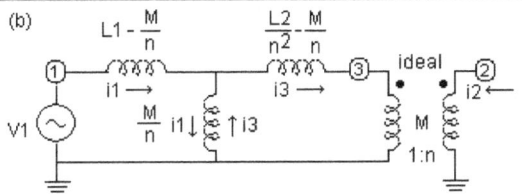

Replace M with $k\sqrt{L_1L_2}$ An alternative equivalent circuit replaces M with $k\sqrt{L_1L_2}$, and L_2 with n^2L_1. Then only primary side values appear in the equivalent circuit (Figure 408).

Figure 408 Another Equivalent Circuit

$$(13a) \quad \frac{M}{n} = \frac{k}{n}\sqrt{L_1L_2} = \frac{k}{n}\sqrt{L_1 n^2 L_1} = kL_1 \quad \text{where } L_2 = n^2 L_1$$

$$(13b) \quad L_1 - \frac{M}{n} = L_1 - kL_1 = (1-k)L_1$$

$$(13c) \quad \frac{L_2}{n^2} - \frac{M}{n} = L_1 - kL_1 = (1-k)L_1$$

Electric Circuits - Analysis and Design

4.4 Transformer Frequency Response

The ideal transformer transfer function T(p)=1. It has no flaws. However a practical transformer's T(p) has a low frequency zero arising from shunt inductance kL, and a high frequency pole arising from *leakage* inductance (1−k)L in series arising from the fact coupling coefficient k does not equal one. In addition the coils as well as the core have finite resistance and parasitic capacitance.

More realistic equivalent circuit We started by assuming there are two windings on a core. Each winding has a self inductance L_1 or L_2 that you can measure, while no current flows in the other winding. A T-equivalent circuit has two meshes (Figures 408, 409). The meshes are coupled together by the transformer mutual inductance M. The leakage inductances are in series. Additional components make the circuit more realistic. Each copper wire winding has finite resistance R_1, R_2. The core has ac losses represented by R_c.

Figure 409 Another version of a transformer equivalent circuit

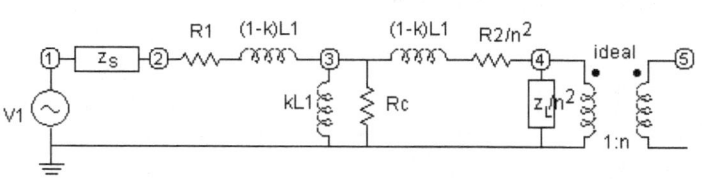

> ***Design*** The analysis in this section shows how transformer parameters interact with circuit parameters source resistance, load resistance, and capacitance. This information provides a basis for design.

Problems 408 and 409 show how $T_L(p)$ relates to Figures 41011, & 41012.

Problem 409 If R_C=100Kohm (Figure 409), z_S=1Kohm, R_1=200ohms show that the (Thevenin equivalent) value of R_x is 1186 ohms. If z_L=5Kohm, R_2=600ohms, n=2.5 show that the value of R_y is 896 ohms.

Problem 410 Reference Figure 409. If n=3 and z_L+R_2=900 ohms what is the value of z_L+R_2 reflected to the primary? If k=0.95 and L_1=50mH what are the L values in the T equivalent circuit?

50

4.4.1 Low Frequency Circuit Analysis

At low frequencies the impedance of the series inductors is negligible compared to the source impedance. Capacitors are not included, because their impedance is large and negligible compared to the impedance of kL_1. This simplifies the equivalent circuit (Figure 410). Assume $R_L \ll R_C$.

Figure 410 Low Frequency Model

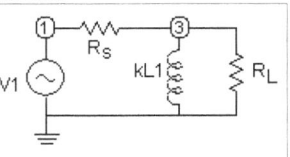

(14a) $\quad 0 = g_s(v_3 - v_1) + \left(g_L + \dfrac{1}{pkL_1}\right)v_3$

(14b) $\quad g_s v_1 = \left(g_s + g_L + \dfrac{1}{pkL_1}\right)v_3$

(14c) $\quad T_L(p) = \dfrac{v_3}{v_1} = \dfrac{g_s}{\left(g_s + g_L + \dfrac{1}{pkL_1}\right)} = \dfrac{R_L}{R_s + R_L} \cdot \dfrac{p}{\left(p + \dfrac{R_s \parallel R_L}{kL_1}\right)}$

$$T_L(p) = \dfrac{R_L}{R_s + R_L} \cdot \dfrac{p}{p + \omega_0}$$

If this transformer is in a 500 ohm line impedance system, then

(15a) $\quad$ let $R_s = R_L = 500\Omega$, $k = 0.99$, $L_1 = 1H$

(15a) $\quad f_0 = \dfrac{R_s \parallel R_L}{2\pi kL_1} = \dfrac{500 \parallel 500}{2\pi \cdot 0.99 \cdot 1} = 40.2 Hz$

Compare to Figures 41011 and 41012. Ascertaining the −3dB frequency in Figure 41011 is difficult, but Figure 41012 clearly shows the corresponding 45 degree phase at about 40Hz.

Spice Fig4101.ckt print statements produce numeric output of output v_3 magnitude and phase that allow you to ascertain the −3dB frequency.

Problem 411 Reference Figure 409. If n=3, k=0.95 and L_1=50mH what are the L values in the T equivalent circuit?

Problem 412 Reference problem 411. If k=0.99, and L_1=0.1H show that the value of f_1 is 820.5Hz Compare to Figure 41111 where the −3dB frequency is about 400Hz. Explain the difference in f_1.

Electric Circuits - Analysis and Design

Spice Program 4101

```
Fig4101.ckt transformer xmsn T

V1 1 0 AC 1 0        ; volts

* names/nodes/values
Rs 1 3 500
L1 3 0 990m          ; kL1, k=0.99
RL 3 0 500

*.PLOT AC VDB(3) -40,10
.AC DEC 200 1 1e+007
.PLOT AC VP(3) 0,100
.PRINT AC VDB(3)
.PRINT AC VP(3)

.TEMP 27
.end
```

Figure 41011 Transformer Low Frequency Transmission (dB)

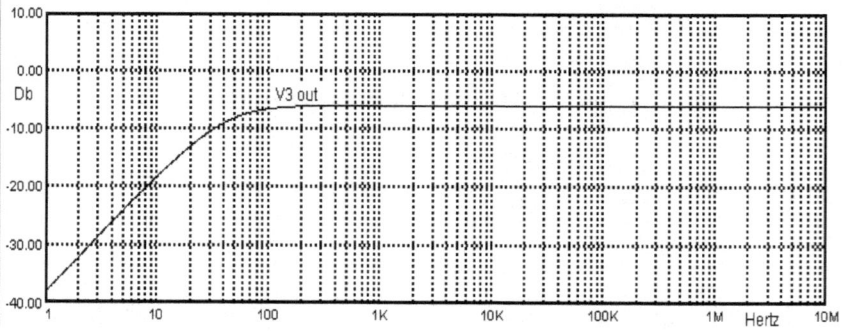

Figure 41012 Transformer Low Frequency Transmission (degrees)

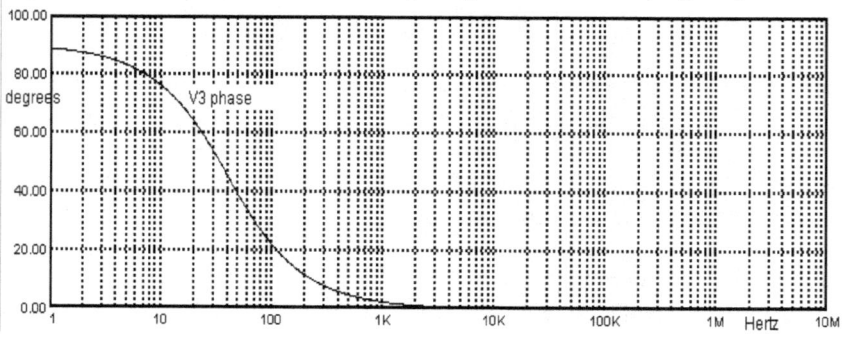

4 Transformers in Circuits

4.4.2 High Frequency Circuit Analysis

At high frequencies the shunt kL_1 and R_C are omitted, because their impedances are very large compared to the load, the source R and the series inductors' impedance. The windings have capacitance that is usually represented by one capacitor C_L *as part of* Z_L (Figure 411).

Figure 411 High Frequency Model

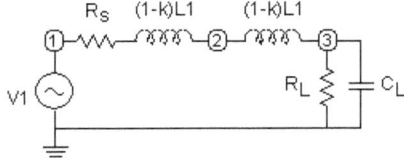

$z_L = R_L$ Spice Program 4111 shows the effect of the windings' leakage inductance $(1-k)L_1$ with C_L omitted.

(16a) $\quad v_1 = [R_s + 2(1-k)pL_1 + R_L]i_1$

(16b) $\quad i_1 = v_1 \dfrac{1}{R_s + R_L + 2(1-k)pL_1}$

(16c) $\quad v_3 = i_1 R_L = v_1 \dfrac{R_L}{R_s + R_L} \dfrac{1}{1 + \dfrac{2(1-k)L_1}{R_s + R_L} p}$

(17) $\quad T_H(p) = \dfrac{v_3}{v_1} = \dfrac{R_L}{R_s + R_L} \cdot \dfrac{1}{1 + \dfrac{p}{\omega_H}} \quad$ where $\quad \omega_H = \dfrac{R_s + R_L}{2(1-k)L_1}$

$z_L = 1/(G_L + pC_L)$ Spice Program 4112 shows the effect of the windings' leakage inductance $(1-k)L_1$ and load capacitance C_L.

Problem 413   If $R_s = R_L = 500$ ohms (Figure 411), $k = 0.99$, $L_1 = 1H$ show that $f_H = 7.96$KHz. Compare to Figure 41111. Explain any difference.

Problem 414   Find $T_H(p)$ when $z_L = 1/(G_L + pC_L)$. Hint replace R_L with $1/(G_L + pC_L)$.

Electric Circuits - Analysis and Design

Spice Program 4111

$z_L=R_L$
```
Fig4111.ckt   transformer xmsn T
V1 1 0 AC 1 0          ; volts
* names/nodes/values
R1 1 2 500
L1 2 3 20m             ;2(1-k)L1
R2 3 0 500
*.PLOT AC VDB(2) VDB(3) -60,10
.AC DEC 200 10 1e+008
.TEMP 27
.PLOT AC VP(2) VP(3) -200,100
.PRINT AC VDB(2) VDB(3)
.end
```

Observe the 6dB loss at low frequencies when $R_1=R_2$ ($R_S=R_L$).

Figure 41111 Transformer High Frequency Transmission (dB), C_L omitted

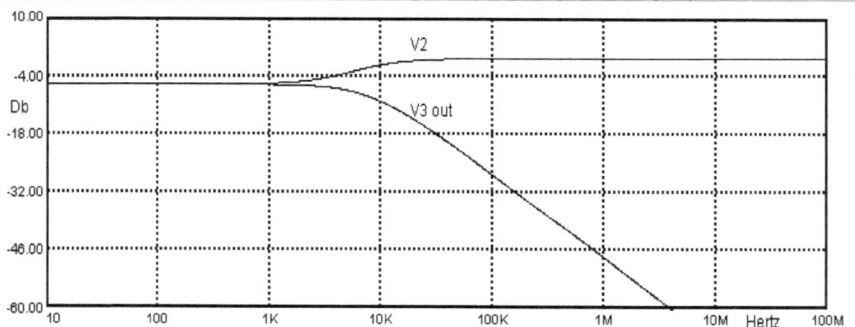

Figure 41112 Transformer High Frequency Transmission (degrees)

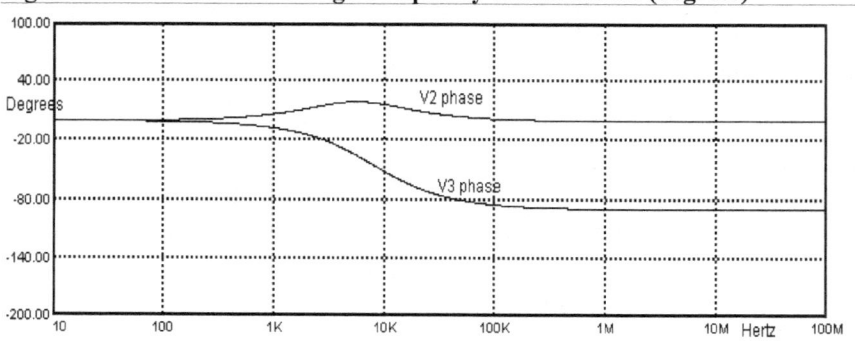

4 Transformers in Circuits

Spice Program 4112 $z_L=1/(G_L+pC_L)$

```
Fig4112.ckt    transformer xmsn T
V1 1 0 AC 1 0   ; volts
Rs 1 2 500
L1 2 3 20m
RL 3 0 500
CL 3 0 500pf

R11 1 12 500
L11 12 13 20m
R12 13 0 500
*.PLOT AC VDB(2) VDB(3) VDB(13) -60,10
.AC DEC 200 10 1e+008
.TEMP 27
.PLOT AC VP(2) VP(3) VP(13) -200,100
.PRINT AC VDB(2) VDB(3)
.end
```

Figure 41121 Transformer High Frequency Transmission (dB)

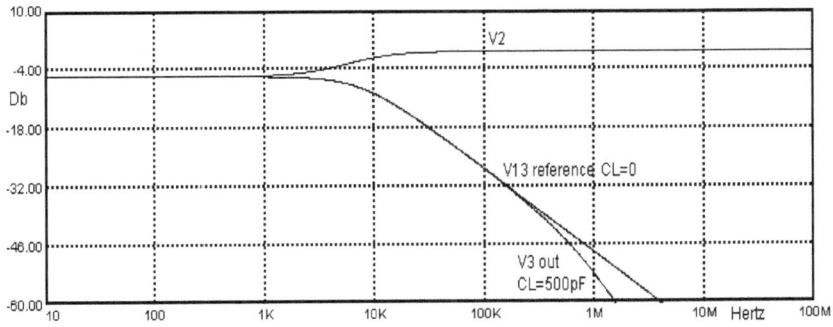

Figure 41122 Transformer High Frequency Transmission (degrees)

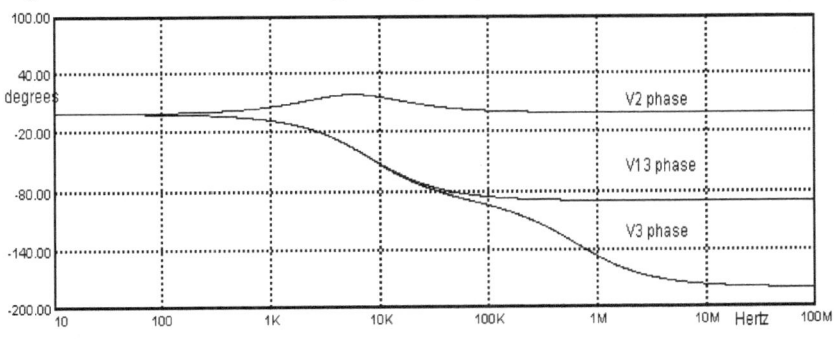

55

Electric Circuits - Analysis and Design

4.4.3 High and Low Frequency Circuit Analysis

Traditional analysis uses low frequency and high frequency equivalent circuits (Figures 410 and 411). The exercise is useful, because we learn which components contribute to high or low equivalents. This was the practice before Spice appeared. Spice allows you to simulate the whole circuit conveniently (Figure 40911).

Spice Program 4091
```
Fig4091.ckt    transformer xmsn T
V1 1 0 AC 1 0      ; volts
Rs 1 2 500
L1 2 0 1
L2 3 0 4
K12 L1 L2 0.99
RL 3 0 2000          ; 2000 here - 500 in prior plots
CL 3 0 500pf
*.PLOT AC VDB(2) VDB(3) -60,15
.AC DEC 200 1 1e+007
.TEMP 27
.PLOT AC VP(2) VP(3) -150,100
.end
```

Figure 40911 Transformer Transmission Function (magnitude dB)

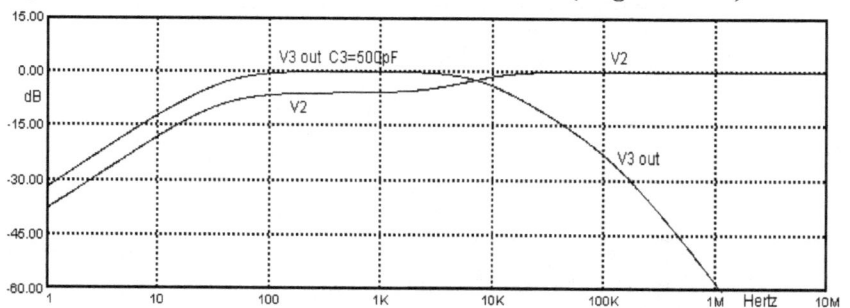

Figure 40912 Transformer Transmission Function (phase degrees)

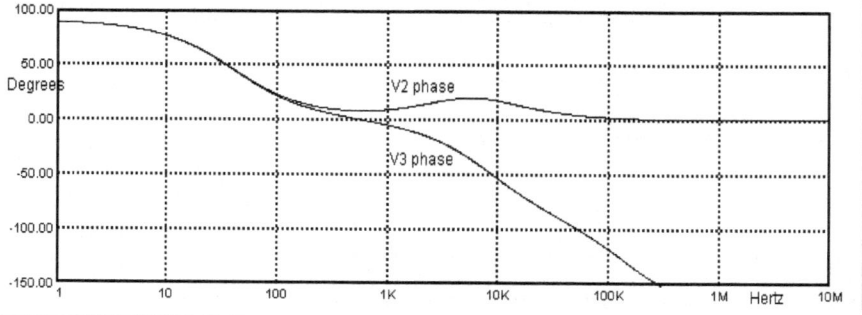

4 Transformers in Circuits

4.5 Full Wave Rectifier

The purpose of this circuit is to convert an AC voltage to a DC voltage.[2] Transformer action makes the voltages at nodes 3 and 4 follow the AC voltage difference from node 1 to node 2 (Figure 41211). The L_1/L_2 and L_1/L_3 ratios are 4/1 that makes the input/output turns ratios n equal to 2/1. This is why V_3 and V_4 are 1/2 the amplitude of V_1-V_2. Transformer action makes nodes 3 and 4 positive on alternate half cycles (Figure 41211). A positive node turns on the diode and current flows through load $R_1=10K$ ohms (assume C_1 is omitted). In this way alternate positive half cycles of V_3 and V_4 appear at node 5 (Figure 41212).

Fig 412 Full wave Rectifier

Add $C_1=1\mu F$ to the circuit. The $5R_1C_1$ time constant is 50ms that is why C_1 cannot discharge to zero when the diodes are off (Figure 41213). In this way a DC voltage is developed at node 5. The node 5 voltage is essentially constant when $C_1=10\mu F$.

An equation derived from Q=CV produces a minimum estimate for C_1.

$$(18a) \quad i = C\frac{dv}{dt} \quad or \quad C = \frac{Q}{V} \quad \rightarrow \quad C = \frac{idt}{dv} = \frac{I_{load}}{\Delta V \cdot 2f} = \frac{1}{\Delta V \cdot 2f} \frac{V_{load}}{R_{load}}$$

$$(18b) \quad C = \frac{1}{1V \cdot 2 \cdot 60} \frac{3V}{10K\Omega} = 2.5\mu F$$

Figure 41211 Transformer voltages

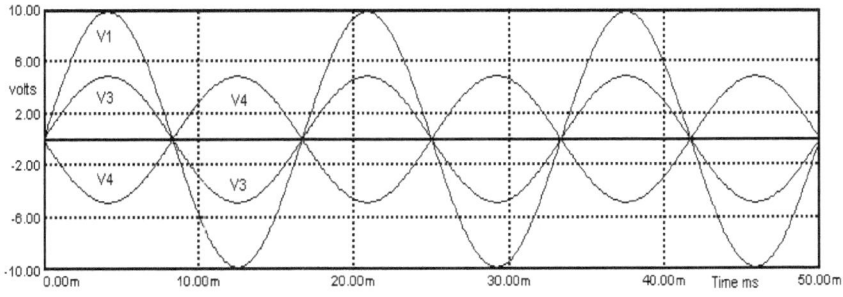

---

[2] For essentially complete design information see
Schade, O.H. "Analysis of Rectifier Operation" Proc. IRE 31,7 July 1943 p341
or Langford-Smith, F. , "Radiotron Designer's Handbook" ISBN B0007D0X9Y

Electric Circuits - Analysis and Design

Spice Program 4121
```
Fig4121.ckt      full wave rectifier
V11  11  0  sin(0  9.8  60  0  0)
Ra   11  1  1u
.Model D1N4148 D(IS=1f RS=16 CJO=2p TT=12n BV=100 IBV=400p)
L1  1  0  1H
L2  3  0  0.25H
L3  0  4  0.25H
K12  L1  L2  0.99
K13  L1  L3  0.99
D1  3  5  D1N4148  OFF
D2  4  5  D1N4148  OFF
C1  5  0  1u
R1  5  0  10K
.TRAN 0.0001 0.05 0 0.0001
.TEMP 27
.PLOT TRAN V(3) V(5) -5,5
.END
```

We ignore the "Spice noise" in the figures.

Figure 41212 V_5 Rectifier action C1 omitted

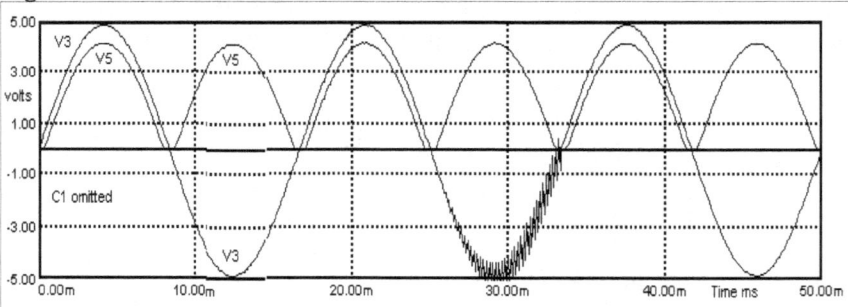

Figure 41213 V_5 Rectifier action C1=1µF

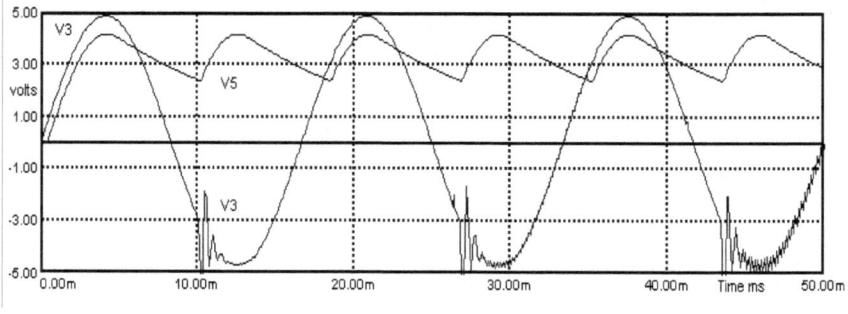

5 Circuit Analysis

The laws we need to solve circuit problems are *vi constraint laws* and the *Kirchhoff connection constraint laws*. In what follows we focus on learning how to analyze circuits in the easiest way we know how. We get ready by discussing several ideas. Then we proceed to explain the node analysis method, which is based on Kirchhoff's current law that the sum of currents at a node equals zero. And, we show how to deal with a voltage source in a branch. This is followed by an explanation of the mesh analysis method, which is based on Kirchhoff's voltage law that the sum of voltages around a mesh equals zero. And, we show how to deal with a current source in a branch. We show how to use the methods via attenuator and low pass filter circuit designs with the same topology.

5.1 Circuit Analysis Process

Voltage-current constraints[1] Ohm's law is the resistor vi constraint. The inductor vi constraint follows from a changing current flowing in a wire creating a changing flux that passes by its own (source) wire inducing a voltage in that same wire. From $Q = CV$ the differential equation relating a capacitor's voltage and current follows immediately.

(1) $\boxed{v(t) = Ri(t)}$ (2) $\boxed{v(t) = L\dfrac{di(t)}{dt}}$ (3) $\boxed{i(t) = C\dfrac{dv(t)}{dt}}$

Kirchhoff connection constraints[2] The mesh method is based on Kirchhoff's Voltage Law (KVL) *The algebraic sum of the voltage differences around every closed circuit contour is zero.* The node method is based on Kirchhoff's Current Law (KCL) *The algebraic sum of all currents entering and leaving one node is zero.*

Writing circuit equations Having decided to use the mesh or node method, equations are written as a sum of voltages or currents. Circuit components are introduced by replacing the equation voltages or currents by the voltage-current constraints. RLC component vi constraints are time domain equations, and so are the circuit equations. The circuits' equations 5 and 12 are integro-differential equations (Figures 501, 502).

---

[1] Chapter 3
[2] Chapter 1

Electric Circuits - Analysis and Design

The laws that solve circuit analysis problems are the *Kirchhoff connection constraint laws (Chapter 1) and the vi constraint laws (Chapter 3).*

Analysis of any circuit starts with Kirchhoff's connection constraint laws. The one mesh circuit of Figure 501 is analyzed by Kirchhoff's voltage law, which states that the sum of voltages around a mesh equals zero.

(4) $v_1(t) = v_{C1}(t) + v_{L1}(t) + v_{R1}(t)$

Figure 501 Series RLC

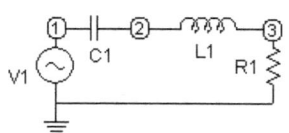

Then the vi constraints for C_1, L_1, and R_1 replace the voltages.

(5) $v_1(t) = \dfrac{1}{C_1} \int i(t)dt + v(0) + L_1 \dfrac{di(t)}{dt} + R_1 i(t)$

At the same time we could choose $v_1(t)$ to have some waveform. However we defer that choice for the moment.

Knowing the Laplace mathematical solution (Chapter 7) we multiply both sides by *exp(–pt)*, and integrate from zero to infinity.

(6) $e^{-pt} v_1(t) = e^{-pt} \dfrac{1}{C_1} \int i(t)dt + e^{-pt} L_1 \dfrac{di}{dt} + e^{-pt} R_1 i(t)$

(7) $\int_0^\infty e^{-pt} v_1(t)dt = \int_0^\infty e^{-pt} \dfrac{1}{C_1} \int i(t)dt + \int_0^\infty e^{-pt} L_1 \dfrac{di}{dt} dt + \int_0^\infty e^{-pt} R_1 i(t)dt$

We assume the integrals converge and that the integral transform transforms i(t) into I(p) and $v_1(t)$ into $V_1(p)$. The integral transform is the Laplace operator $\mathscr{L}$. We define

(8) $I(p) = \mathscr{L}(i(t)) = \int_0^\infty e^{-pt} i(t)dt$ and $V_1(p) = \mathscr{L}(v_1(t)) = \int_0^\infty e^{-pt} v_1(t)dt$

The constants C_1, L_1, and R_1 factor out of the integrals. Evaluating the integrals we get (Section 7.3)

(9) $V_1(p) = \dfrac{1}{pC_1} I(p) + \dfrac{v(0)}{p} + pL_1 I(p) - L_1 i(0) + R_1 I(p)$

(10) $V_1(p) = \dfrac{1}{pC_1} I(p) + pL_1 I(p) + R_1 I(p)$ when v(0) and i(0) equal 0

5 Circuit Analysis

Next we solve for I(p). *This is the solution for the circuit current in the complex frequency domain p.* Observe that this solution is valid for ANY $v_1(t)$ waveform transformed into $V_1(p)$.

(11) $\quad I(p) = \dfrac{1}{\left(\dfrac{1}{pC_1} + pL_1 + R_1\right)} V_1(p)$

Figure 502 Parallel RLC

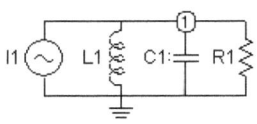

The one node circuit of Figure 502 is readily analyzed by Kirchhoff's current law, which states that the sum of currents at a node equals zero.

(12) $\quad i_1(t) = \dfrac{1}{L_1} \int_0^t v(x)dx + i(0) + C_1 \dfrac{dv(t)}{dt} + \dfrac{1}{R_1} v(t)$

The Laplace Transform of this equation is

(13) $\quad I_1(p) = \dfrac{1}{pL_1} V(p) + pC_1 V(p) + \dfrac{1}{R_1} V(p)$ when $i(0) = 0$, $v(0) = 0$

Next we solve for V(p). *This is the solution for the node voltage in the complex frequency domain p.* Observe that this solution is valid for ANY $i_1(t)$ waveform transformed into $I_1(p)$.

(14) $\quad V(p) = \dfrac{1}{\left(\dfrac{1}{pL_1} + pC_1 + \dfrac{1}{R_1}\right)} I_1(p)$

Important We have shown that the waveform of the input signal that affects the output does not affect the mesh and nodal analysis methods.

In other words – mesh and nodal analysis in the frequency domain does NOT depend on the *waveform* of the input signal(s).

This means the time domain equations can be bypassed, and circuit analysis can start with the transformed circuit equations in the frequency domain.

Analogous to Ohm's law, L and C voltage drops are *pLi* and *i/pC* (equation 10). And, analogous currents are *v/pL* and *pCV* (equation 13)

Electric Circuits - Analysis and Design

The Steady State explained Choose source $v_1(t)$ to have a complex exponential waveform, which is the complex sum of sine and cosine functions.

(15) $v_1(t) = V_M e^{j\omega t}$ $e^{j\omega t} = \sin \omega t + j \cos \omega t$

(16) $V_1(p) = \mathcal{L}[v_1(t)] = \mathcal{L}[V_M e^{j\omega t}] = \int_0^\infty e^{-pt} V_M e^{j\omega t} dt = \dfrac{V_M}{p - j\omega}$

Short the capacitor to delete it from the circuit in order to simplify the mathematics. Then equation 11 becomes

(17) $I(p) = \dfrac{1}{(pL_1 + R_1)} \cdot \dfrac{V_M}{p - j\omega}$

Make a partial fraction expansion (Appendix A1) The trick we use is based on the fact the inverse transform of $1/(p+a)$ is an exponential function (equation 18). This is why we need a partial fraction expansion.

(18) $e^{-at} \Leftrightarrow \dfrac{1}{p+a}$

(19) $I(p) = \dfrac{1}{L} \cdot \dfrac{1}{p + \dfrac{R}{L}} \cdot \dfrac{V_m}{p - j\omega}$

$I(p) = \dfrac{1}{L} \cdot \dfrac{1}{j\omega + \dfrac{R}{L}} \cdot \dfrac{V_m}{p - j\omega} + \dfrac{1}{L} \cdot \dfrac{1}{p + \dfrac{R}{L}} \cdot \dfrac{V_m}{-\dfrac{R}{L} - j\omega}$

$I(p) = \dfrac{V_m}{R + j\omega L} \cdot \dfrac{1}{p - j\omega} - \dfrac{V_m}{R + j\omega L} \cdot \dfrac{1}{p + \dfrac{R}{L}}$

Inverse transform I(p) transforms to i(t) when we substitute an e^{-at} for each of the two $1/(p+a)$ terms. The complete solution consists of a steady-state term, a transient term due to the forcing function.

(20) $i(t) = \mathcal{L}^{-1}[I(p)] = \dfrac{V_m}{R + j\omega L} e^{j\omega t} - \dfrac{V_m}{R + j\omega L} e^{-\frac{R}{L} t}$

After an elapsed time greater than 5 time constants (L/R), the transient e^{-at} term goes to zero, and the current becomes the steady state current. Steady, because i(t) continues unchanged *in magnitude* forever. The basic reason is that the derivative of sin or cos is cos or sin. The waveforms are sinusoids throughout the circuit. I.e. no additional transients are created.

5 Circuit Analysis

Here is why we use Laplace: The (difficult) classical method follows.

Particular integral The e^{pt} waveform of a voltage driving any circuit produces currents with the *same* waveform, because only linear operations produce circuit voltages & currents. For a series RLC circuit the equation is

if $v_1(t) = V_m e^{pt}$, then $i_p(t) = Ie^{pt}$

$$V_m e^{pt} = R_1 i + L\frac{di}{dt} + \frac{1}{C_1}\int_0^t i(x)dx \quad \Rightarrow \quad V_m e^{pt} = \left(R_1 + pL_1 + \frac{1}{pC_1}\right)Ie^{pt}$$

$$I = \frac{V_m}{R_1 + pL_1 + \frac{1}{pC_1}} \quad \Rightarrow \quad i_p(t) = Ie^{pt} = \frac{V_m e^{pt}}{Z(p)} \quad \Rightarrow \quad Z(p) = R_1 + pL_1 + \frac{1}{pC_1}$$

$i_p(t)$ is the *particular* integral solution of the differential equation for the forcing function $v_1(t)$. On the real frequency axis $p=j\omega$, and $i_p(t)$ has constant amplitude V_m at any p. Then we say $i_{p=j\omega}(t)$ is in the *steady state*.

Complementary Integral When the forcing function $v_1(t)=0$ we get

$$0 = R_1 i(t) + L_1 \frac{di(t)}{dt} + \frac{1}{C_1}\int_0^t i(x)dx$$

let $i(t) = Ie^{pt}$

$$0 = \left(R_1 + pL_1 + \frac{1}{pC_1}\right)Ie^{pt} \quad \Rightarrow \quad 0 = p^2 L_1 C_1 + pC_1 R_1 + 1 \quad \text{when } I \neq 0$$

The non-trivial solutions require $I \neq 0$, because if $I=0$ then $i(t)=0$. When $I \neq 0$ the polynomial (in p) must equal 0. The polynomial equals 0 when p takes the values of the polynomial's two roots p_1, p_2. The following (not obvious) equation is the *complementary* integral solution of the equation.

$$i_c(t) = I_1 e^{p_1 t} + I_2 e^{p_2 t}$$

The constants of integration I_1 and I_2 are found by invoking the two *initial conditions*. This is a complicated process using this method so we will not use it. (However, know that the initial conditions are *automatically* produced by the Laplace method.) The complete solution is

$$i(t) = \frac{V_m e^{pt}}{R_1 + pL_1 + \frac{1}{pC_1}} + I_1 e^{p_1 t} + I_2 e^{p_2 t}$$

Electric Circuits - Analysis and Design

5.2 Node Method

The node method *formulates* circuit equations in the form of sums of currents by applying Kirchhoff's current law (KCL) that states that the algebraic sum of all (branch) currents entering and leaving a node is zero (Section 1.3.1). Here is our convention for *branch* currents.

Figure 502

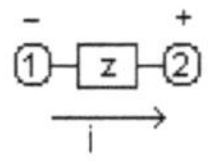

(21a) $\quad v_2 - v_1 = iz$

(21b) $\quad i = y(v_2 - v_1) \Rightarrow y = \dfrac{1}{z}$

Our convention for KCL is that currents entering the node have a positive sign, and currents leaving the node have a negative sign.

5.2.1 Circuit Node Equations

There are five nodes that might produce equations (Figure 503a). However ground voltage $v_0=0$ volts, $v_{40}=v_4-v_0=v_s-0$ so that $v_4=v_s$. Consequently we need to find values for v_1, v_2, v_3. Observe that only current i_1 enters and leaves node 1. Since z_1 and z_2 are in series consider them to be one impedance z_{12}. This action eliminates node 1 for the moment. In other words we defer finding v_1 that equals $v_2-i_1z_2$. In this way we have reduced the problem to analysis of two nodes: 2 and 3.

Node 2 *Our convention is that we assume v_2 is greater than any other voltage when we formulate the node 2 equation.* Thus i_1, i_2, and i_3 flow into node 2 (Figure 503a). Their sum equals zero according to KCL. The sum (equation 22) is the KCL connection constraint.

Figure 503a Node 2 Analysis

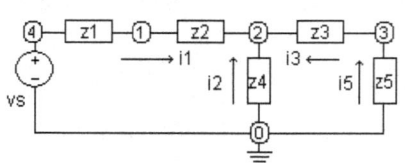

(22) $\quad 0 = i_1 + i_2 + i_3$

Each current equals a $y(v_a-v_b)$ expression that is found by inspection of Figure 503a, while applying equation 21b. So we start with sums of currents and end up with equations for node voltages.

(23a) $\quad i_1 = y_{12}(v_2 - v_4) \qquad i_2 = y_4(v_2 - v_0) \qquad i_3 = y_3(v_2 - v_3)$

(23b) $\quad 0 = y_{12}(v_2 - v_4) + y_4(v_2 - v_0) + y_3(v_2 - v_3)$

5 Circuit Analysis

Straightforward algebraic manipulations put the equation in *standard form* (equation 24b) with two unknowns v_2 and v_3. Observe that each admittance y is connected to node 2. That is why their sum is the coefficient of v_2.

(24a) $\quad v_s = v_4 \quad$ and $\quad v_0 = 0$
(24b) $\quad y_{12}v_s = (y_{12} + y_4 + y_3)v_2 - y_3v_3$

Node 3 *Our convention is that we assume v_3 is greater than any other voltage when we formulate the node 3 equation.* Thus i_5 and i_3 flow into node 3 (Figure 503b).

Figure 503b Node 3 Analysis

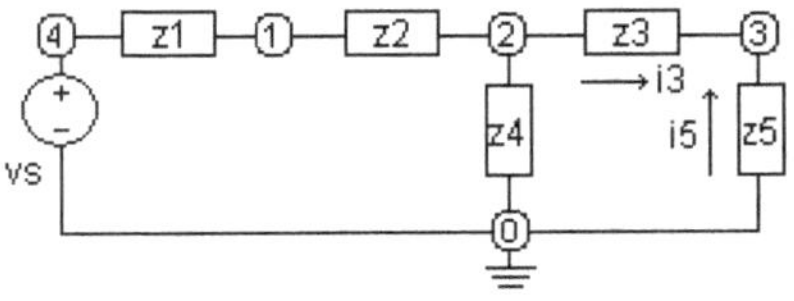

(25) $\quad 0 = i_3 + i_5$

Each current equals a $y(v_a - v_b)$ expression that is found by inspection of Figure 503, while applying equation 21b. Again, note that *we assume v_3 is greater than any other voltage*.

(26a) $\quad i_3 = y_3(v_3 - v_2) \qquad i_5 = y_5(v_3 - v_0)$
(26b) $\quad 0 = y_3(v_3 - v_2) + y_5(v_3 - v_0)$

Algebraic manipulations put the equation in *standard form* (equation 27b) with two unknowns v_2 and v_3. Observe that each admittance y is connected to node 3 that is why their sum is the coefficient of v_3.

(27a) $\quad v_0 = 0$
(27b) $\quad 0 = (y_3 + y_5)v_3 - y_3v_2$

Mathematics The two node equations (24b and 27b) with two unknowns v_2, v_3 are solved by using Cramer's rule.
(24b) $\quad y_{12}v_s = (y_{12} + y_4 + y_3)v_2 \qquad - y_3v_3$
(27b) $\quad \quad 0 = \qquad \quad -y_3v_2 + (y_3 + y_5)v_3$

> There may be an independent or dependent current source in a branch. This is just another current at the two nodes connected to the current source. Add or subtract the current to the KCL node equations according to the current direction. However a voltage source in a branch is another matter (5.2.2).

Electric Circuits - Analysis and Design

5.2.2 Branch with Voltage Source

The voltage source in a branch can be an independent or dependent source. Assume the circuit model has an independent voltage source in one branch (Figure 504). Formulating the sum of the currents at nodes 4 and 1 is straightforward (equations 28a and 28b). This is not so for nodes 2 and 3. What is the current through the voltage source v_s? One procedure for circuits like this is to *assume a value, such as I_x*, for the current flowing through the voltage source v_s. Then we can formulate the KCL equations for nodes 2 and 3 (equations 28c and 28d), as well as a fifth equation 28e relating v_2 to v_3. The I_x trick is not obvious, however it solves the problem. Note: Figure 504 circuit is NOT related to Figure 503 circuit!

Figure 504 Branch with voltage source

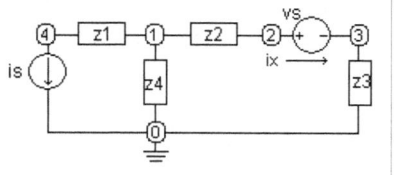

(28a)　node 4　　$i_s = y_1(v_4 - v_1)$
(28b)　node 1　　$0 = (y_1 + y_2 + y_4)v_1 - y_1 v_4 - y_2 v_2$
(28c)　node 2　　$0 = -y_2 v_1 + y_2 v_2 - I_x$
(28d)　node 3　　$0 = y_3 v_3 + I_x$
(28e)　　　　　　$v_2 = v_s + v_3$

Observe that I_x drops out when the v_2 and v_3 node equations are added. Addition leaves three equations. We still need a fourth equation. The fourth equation is equation 28e that relates the voltage source v_s to node voltages v_2 and v_3.

(28a)　　　node 4　　　$i_s = y_1(v_4 - v_1)$
(28b)　　　node 1　　　$0 = (y_1 + y_2 + y_4)v_1 - y_1 v_4 - y_2 v_2$
(28c+d)　 node 2+3　　$0 = -y_2 v_1 + y_2 v_2 + y_3 v_3$
(28e)　　　　　　　　　$v_2 = v_s + v_3$

Substitute 28e into 28c+d to eliminate v_2. Now we have three equations and three unknowns. The rest is mathematics.

(28a)　　　　node 4　　　　$i_s = y_1 v_4 - y_1 v_1$
(28b)　　　　node 1　　　　$y_2 v_s = (y_1 + y_2 + y_4)v_1 - y_1 v_4 - y_2 v_3$
(28c+d,e)　 node 2+3　　　$-y_2 v_s = -y_2 v_1 + (y_2 + y_3)v_3$

5 Circuit Analysis

Problem 501 Reference Figure P501. Use the node voltage method. Formulate the equations in standard form. Solve for i_1, i_2, v_2, and v_3. Then try an equivalent to DC Spice Program Fig209.ckt (Section 8.1).

Problem 502 Reference Figure P502. Use the node voltage method. Formulate the equations. Use Spice to solve for i_1, i_2, i_3, v_2, and v_3.

Figure P501   **Figure P502**

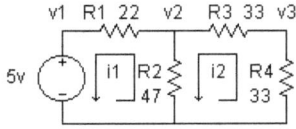
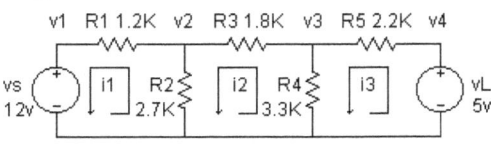

Problem 503 Reference Figure P503. Use the node voltage method. Formulate the equations. Solve for v_1, i_S.

Problem 504 Reference Figure P504. Use the node voltage method. Formulate the equations. Solve for i_S, v_L.

Figure P503   **Figure P504**   **Figure P505**

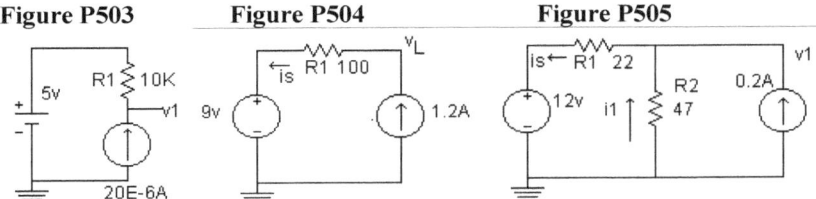

Problem 505 Reference Figure P505. Use the node voltage method. Formulate the equations. Solve for i_S, i_1, v_1.

Figure P506

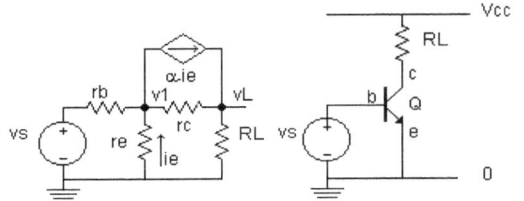

Problem 506 Reference Figure P506. Use the node method. Solve for v_1, v_L, and i_S, i_E.

Problem 507 Refer to Figure P507. Use the node voltage method. Formulate the equations. Solve for i_1, i_2, v_E, and v_C.

Problem 508 Refer to Figure P508. Use the node voltage method. Formulate the equations. Solve for i_1, i_2, and v_2.

Figure P507

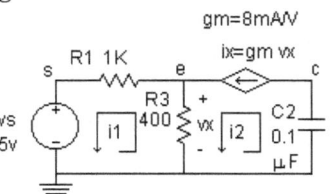

Figure P508

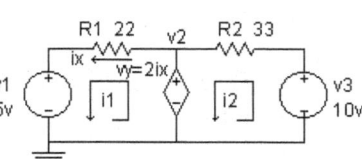

Problem 509 Refer to Figure P509. Use the node voltage method. Formulate the equations. Solve for v_0/v_1.

Figure P509 Op amp inverter

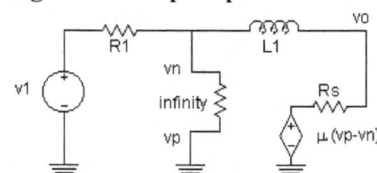

Problem 510 Refer to Figure P500 and P510. Draw the equivalent circuit. Let $\alpha=0.98$, $r_B=400$, $r_E=100$, $r_C=$infinity, $R_X=5K$, $g_M=12mA/v$. Use the node voltage method. Solve for V_B, V_E, I_C, I_B, I_E.

Problem 511 Refer to Figure P400 and P511 Let $\alpha=0.98$, $r_B=400$, $r_E=100$, $r_C=$infinity, $R_X=5K$, $g_M=12mA/v$. Draw the equivalent circuit. Use the node voltage method. Solve for v_C, v_B, i_C, i_B.

Figure P510 Emitter follower

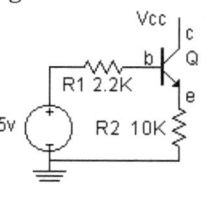

Figure P511 CE amplifier

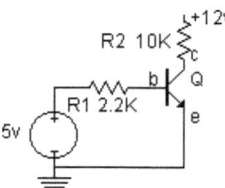

Problem 512 Refer to Figure P512. Use the node voltage method. Formulate the equations. Solve for i_1, i_2, and v_2.

Figure P512

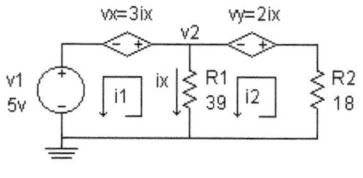

Figure P500

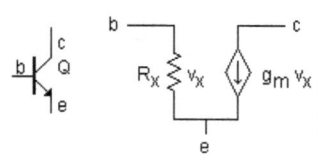

5 Circuit Analysis

5.3 Maxwell's Mesh Current Method

The mesh method formulates circuit equations in the form of sums of voltages by applying Kirchhoff's voltage law (KVL) that states that the algebraic sum of all (branch) voltages around a mesh is zero (Section 1.3.2). Here is our convention for *branch* voltages.

Figure 502

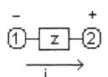

(21a) $\quad v_2 - v_1 = iz$

(21b) $\quad i = y(v_2 - v_1) \Rightarrow y = \dfrac{1}{z}$

5.3.1 Circuit Mesh Equations

Mesh 1 Circling mesh 1 we pass nodes 4, 1, 2, 0, and return to node 4. The voltage differences are v_{4-1}, v_{1-2}, v_{2-0}, and v_{0-4} whose sum equals zero according to KVL (equation 29). The sum is the KVL connection constraint.

Figure 505 Mesh analysis

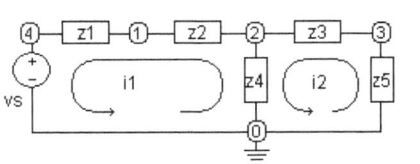

For convenience let $v_{pq} = v_p - v_q$ where p and q are node numbers.

(29) $\quad 0 = v_{41} + v_{12} + v_{20} + v_{04}$

Each voltage difference v_{pq} equals an iz expression that is found by inspection of Figure 505, while applying equation 21a.

(30a) $\quad v_{41} = z_1 i_1 \quad v_{12} = z_2 i_1 \quad v_{20} = z_4(i_1 - i_2) \quad v_{04} = -v_s$

(30b) $\quad 0 = z_1 i_1 + z_2 i_1 + z_4(i_1 - i_2) - v_s$

Straightforward algebraic manipulations put the equation in *standard form* (equation 31) with two unknowns i_1 and i_2. Observe that each impedance z is in mesh 1 that is why their sum is the coefficient of i_1. So we start with sums of voltages and end up with equations for mesh currents.

(31) $\quad v_s = (z_1 + z_2 + z_4)i_1 - z_4 i_2$

Electric Circuits - Analysis and Design

Mesh 2 Circling mesh 2 we pass nodes 2, 3, 0, and return to node 2. The voltage differences are v_{23}, v_{30}, and v_{02} whose sum equals zero according to KVL (equation 32). The sum is the KVL connection constraint.

Figure 505 Mesh analysis

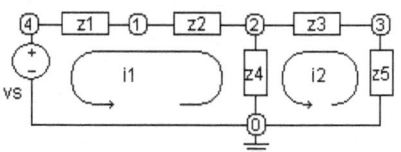

(32) $\quad 0 = v_{23} + v_{30} + v_{02}$

Each voltage difference v_{pq} equals an iz expression that is found by inspection of Figure 505.

(33a) $\quad v_{23} = z_3 i_2 \quad v_{30} = z_5 i_2 \quad v_{02} = z_4(i_2 - i_1)$
(33b) $\quad 0 = z_3 i_2 + z_5 i_2 + z_4(i_2 - i_1)$

Straightforward algebraic manipulations put the equation in *standard form* (equation 34) with two unknowns i_1 and i_2. Observe that each impedance z in mesh 2 is part of the sum that is the coefficient of i_2.

(34) $\quad 0 = -z_4 i_1 + (z_3 + z_4 + z_5) i_2$

Mathematics The two node equations (31 and 34) with two unknowns i_1, i_2 are solved by using Cramer's rule. With experience you will be able to write these down immediately.

(31) $\quad v_s = (z_1 + z_2 + z_4) i_1 - z_4 i_2$
(34) $\quad 0 = -z_4 i_1 + (z_3 + z_4 + z_5) i_2$

There can be an independent or dependent voltage source in a branch. This is just another voltage in the mesh. Add or subtract the voltage to the KVL mesh equations according to the voltage polarity. However a current source in a branch is another matter (5.3.2).

5.3.2 Branch with a Current Source

A branch may include an independent or dependent current source. A circuit model for a common emitter transistor circuit (Figure 506) has a dependent current source in one branch. Formulating the sum of the voltages around mesh 1 is straightforward. This is not so for meshes 2 and 3. What is the voltage drop v_x across the dependent current source? It is not an iR drop. The *trick* we use is to assume a value, such as v_x, for the voltage difference across the current source so that we can proceed directly to formulate the mesh equations. Equation 35d relates the dependent current source to the mesh currents

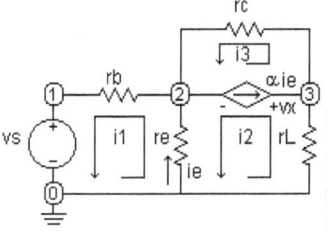

Figure 506 Branch with a dependent current source

(35a) $\quad v_s = (r_e + r_b)i_1 - r_e i_2$
(35b) $\quad 0 = -r_e i_1 + (r_e + r_L)i_2 - v_x$
(35c) $\quad 0 = r_c i_3 + v_x$
(35d) $\quad i_e = i_1 - i_2$

Observe that v_x drops out when the i_2 and i_3 mesh equations 35b and 35c are added. Addition leaves two equations. We still need a third equation. The third equation is equation 35d that relates the dependent current source i_e to mesh currents i_1 and i_2.

(35a) $\qquad v_s = (r_e + r_b)i_1 - r_e i_2$
(35b+c) $\quad 0 = -r_e i_1 + (r_e + r_L)i_2 + r_c i_3$
(35d) $\qquad i_e = i_1 - i_2$

We can eliminate i_3. The KCL equation at node 3 is equation 36a.
(36a) $\quad \alpha i_e = i_3 - i_2$
(36b) $\quad i_3 = i_2 + \alpha i_e = i_2 + \alpha(i_1 - i_2)$
(36c) $\quad i_3 = \alpha i_1 + (1-\alpha)i_2$

Substitute (36c) into (35b+c). Now we have two equations and two unknowns. The rest is mathematics.
(35a) $\qquad\qquad v_s = (r_e + r_b)i_1 - r_e i_2$
(35b+c, 36c) $\quad 0 = (-r_e + r_c \alpha)i_1 + [r_e + r_L + r_c(1-\alpha)]i_2$

Electric Circuits - Analysis and Design

Problem 513 Reference Figure P501. Use the mesh current method. Formulate the equations in standard form. Solve for i_1, i_2, v_2, and v_3. Then try an equivalent to DC Spice Program Fig2091.ckt (Section 8.1).

Problem 514 Reference Figure P502. Use the mesh current method. Formulate the equations. Use Spice to solve for i_1, i_2, i_3, v_2, and v_3.

Figure P501   **Figure P502**

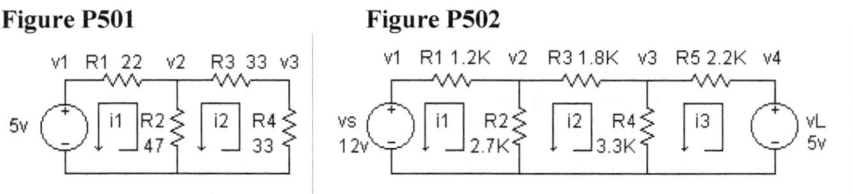

Problem 515 Reference Figure P503. Use the mesh current method. Formulate the equations. Solve for v_1, i_S.

Problem 516 Reference Figure P504. Use the mesh current method. Formulate the equations. Solve for i_S, v_L.

Figure P503   **Figure P504**   **Figure P505**

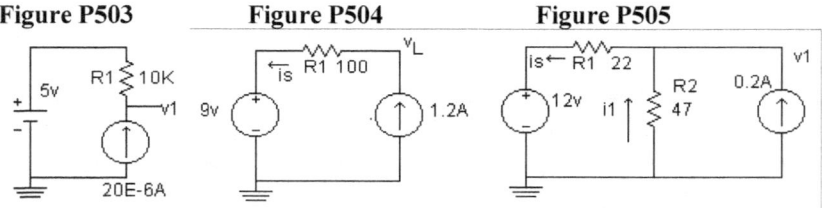

Problem 517 Reference Figure P505. Use the mesh current method. Formulate the equations. Solve for i_S, i_1, v_1.

Figure 506

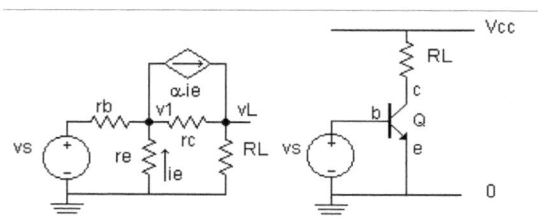

Problem 518 Reference Figure 506. Use mesh method to solve for v_1, v_L, and i_S, i_E.

Problem 519 Refer to Figure P507. Use the mesh current method. Formulate the equations. Solve for i_1, i_2, v_E, and v_C.

Problem 520 Refer to Figure P508. Use the mesh current method. Formulate the equations. Solve for i_1, i_2, and v_2.

5 Circuit Analysis

Figure P507

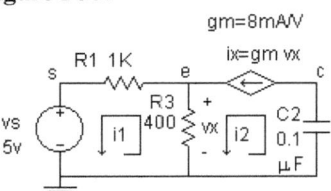

Figure P508

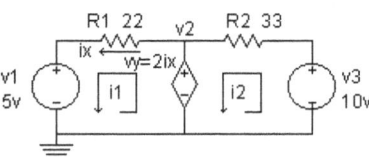

Problem 521 Refer to Figure P509. Use the mesh current method. Formulate the equations. Solve for v_o/v_1.

Figure P509 Op amp inverter

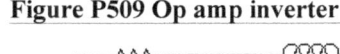
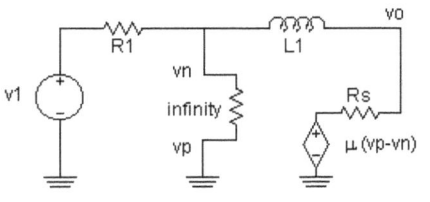

Problem 522 Refer to Figure P500 below and P510. Draw the equivalent circuit. Let $R_X=5K$, $g_M=12mA/v$. Use the mesh current method. Solve for V_B, V_E, I_C, I_B, I_E.

Problem 523 Refer to Figure P500 and P511 Let $R_X=5K$, $g_M=12mA/v$. Draw the equivalent circuit. Use the mesh current method. Solve for v_C, v_B, i_C, i_B.

Figure P510 Emitter follower

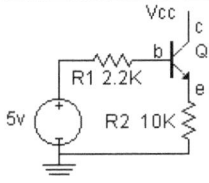

Figure P511 CE amplifier

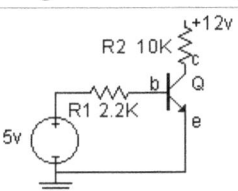

Problem 524 Refer to Figure P512. Use the mesh current method. Formulate the equations. Solve for i_1, i_2, and v_2.

Figure P512

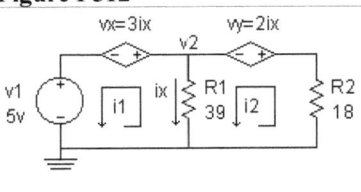

Figure P500

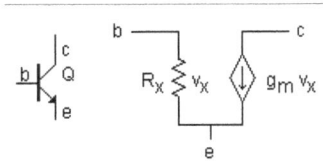

73

Electric Circuits - Analysis and Design

5.4 Attenuator Circuit Equations

The node and mesh analysis methods of Sections 5.2 and 5.3 analyzed a circuit with impedances (admittances) that were not specific. The node voltage and mesh equations we derived are applied here, because the circuits in Figures 507 and 508 have *the same topology* as the circuit in Figure 503. An attenuator has a precise transmission ratio. Since $R_1=R_5=R$ the v_3/v_4 ratio is ½ when the resistor "T" is removed, and node 1 is connected to node 3. Inserting this specific T makes the v_3/v_4 ratio equal 1/20.

Figure 507 Attenuator

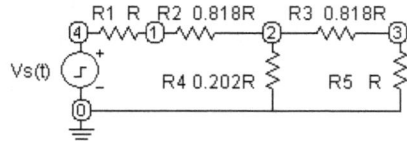

mesh method The vi constraints of attenuator impedances (Figure 507) when substituted in equations 31 and 34 produce equations 37.

(31) *mesh 1* $\quad v_s = (z_1 + z_2 + z_4)i_1 \quad\quad\quad - z_4 i_2$
(34) *mesh 2* $\quad 0 = \quad\quad\quad -z_4 i_1 + (z_3 + z_4 + z_5)i_2$

(37a) *mesh 1* $\quad v_s = (R_1 + R_2 + R_4)i_1 - R_4 i_2$
(37b) *mesh 2* $\quad 0 = -R_4 i_1 + (R_3 + R_4 + R_5)i_2$

node method The vi constraints of attenuator admittances (Figure 507) when substituted in equations 24b and 27b produce equations 38.

(24b) *node 2* $\quad y_{12}v_s = (y_{12} + y_4 + y_3)v_2 \quad\quad -y_3 v_3$
(27b) *node 3* $\quad 0 = \quad\quad\quad -y_3 v_2 + (y_3 + y_5)v_3$

(38a) *node 2* $\quad g_{12}v_s = (g_{12} + g_4 + g_3)v_2 \quad\quad -g_3 v_3$
(38b) *node 3* $\quad 0 = \quad\quad\quad -g_3 v_2 + (g_3 + g_5)v_3$

We defer the solution to Section 5.6. The results are

(39) $\quad \dfrac{v_1(t)}{v_S(t)} = \dfrac{1}{2} \quad\quad \dfrac{v_2(t)}{v_S(t)} = \dfrac{1}{11} \quad\quad \dfrac{v_3(t)}{v_S(t)} = \dfrac{1}{20}$

5.5 Low Pass Filter Circuit Equations

An ideal low pass filter passes sinewaves whose frequency is less than the "cutoff" frequency with zero attenuation, and attenuates to zero sinewaves whose frequency is greater than the "cutoff" frequency. A real filter approximates this performance.

Figure 508 Low Pass Filter

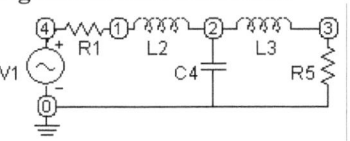

mesh method The vi constraints of filter impedances (Figure 508) when substituted in equations 31 and 34 produce equations 40.

(31) mesh 1 $v_s = (z_1 + z_2 + z_4)i_1 \qquad - z_4 i_2$

(34) mesh 2 $0 = \qquad - z_4 i_1 + (z_3 + z_4 + z_5)i_2$

(40a) mesh 1 $v_s = \left(R_1 + pL_2 + \dfrac{1}{pC_4}\right)i_1 - \dfrac{1}{pC_4} i_2$

(40b) mesh 2 $0 = -\dfrac{1}{pC_4} i_1 + \left(pL_3 + \dfrac{1}{pC_4} + R_5\right)i_2$

node method The vi constraints of filter admittances (Figure 508) when substituted in equations 24b and 27b produce equations 41.

(24b) node 2 $y_{12} v_s = (y_{12} + y_4 + y_3)v_2 \qquad - y_3 v_3$

(27b) node 3 $0 = \qquad - y_3 v_2 + (y_3 + y_5)v_3$

(41a) node 2 $\dfrac{1}{R_1 + pL_2} v_s = \left(\dfrac{1}{R_1 + pL_2} + pC_4 + \dfrac{1}{pL_3}\right)v_2 - \dfrac{1}{pL_3} v_3$

(41b) node 3 $0 = -\dfrac{1}{pL_3} v_2 + \left(\dfrac{1}{pL_3} + \dfrac{1}{R_5}\right)v_3$

> The circuits in Figures 507 and 508 are examples applying a general formulation of equations to any circuit with *the same topology*. The general formulation emphasizes the fact that *Kirchhoff's laws are connection constraints*. We have proceeded this way to demonstrate that KVL and KCL do not need to know what type of components are in the circuit.

Cramer's Rule

The first subscript is the row number. The second subscript is the column number. Determinants are expanded by rows or columns.

Cramer's solutions are expansions by columns where forcing functions replace the column's elements. Note: incorporate minus signs into the a_{ij}'s.

Cramer found responses y_1, y_2 to forcing functions x_1, x_2.

If $x_1 = a_{11}y_1 + a_{12}y_2$ and $x_2 = a_{21}y_1 + a_{22}y_2$
Then $\Delta = a_{11}a_{22} - a_{21}a_{12}$ and

$$y_1 = \frac{\begin{vmatrix} x_1 & a_{12} \\ x_2 & a_{22} \end{vmatrix}}{\begin{vmatrix} a_{11} & a_{12} \\ a_{21} & a_{22} \end{vmatrix}} = \frac{x_1 a_{22} - x_2 a_{12}}{\Delta} \qquad y_2 = \frac{\begin{vmatrix} a_{11} & x_1 \\ a_{21} & x_2 \end{vmatrix}}{\begin{vmatrix} a_{11} & a_{12} \\ a_{21} & a_{22} \end{vmatrix}} = \frac{-x_1 a_{21} + x_2 a_{11}}{\Delta}$$

And, for three responses y_1, y_2, y_3 to forcing functions x_1, x_2, x_3.

If $x_1 = a_{11}y_1 + a_{12}y_2 + a_{13}y_3$
$x_2 = a_{21}y_1 + a_{22}y_2 + a_{23}y_3$
$x_3 = a_{31}y_1 + a_{32}y_2 + a_{33}y_3$

Then $\Delta = a_{11}\Delta_{11} - a_{21}\Delta_{21} + a_{31}\Delta_{31}$ (expansion by column 1)
$\Delta = a_{11}(a_{22}a_{33} - a_{23}a_{32}) - a_{21}(a_{12}a_{33} - a_{13}a_{32}) + a_{31}(a_{12}a_{23} - a_{13}a_{22})$

$y_1 = \dfrac{x_1\Delta_{11} - x_2\Delta_{21} + x_3\Delta_{31}}{\Delta}$ (expansion down column 1, rows 1, 2, 3)

$y_2 = \dfrac{x_1\Delta_{12} - x_2\Delta_{22} + x_3\Delta_{32}}{\Delta}$ (expansion down column 2, rows 1, 2, 3)

$y_3 = \dfrac{x_1\Delta_{13} - x_2\Delta_{23} + x_3\Delta_{33}}{\Delta}$ (expansion down column 3, rows 1, 2, 3)

5 Circuit Analysis

5.6 Solution of Mesh Equations, Attenuator

Up to this point we have formulated equations, however we have not solved them. Use Cramer's rule to solve the mesh current equations. The two mesh general solution proceeds as follows.

(42) $\quad$ mesh 1 $\quad v_S = r_{11}i_1 - r_{12}i_2$
$\qquad$ mesh 2 $\quad 0 = -r_{21}i_1 + r_{22}i_2$

(43) $\quad \Delta_R = r_{11}r_{22} - r_{12}r_{21}$

$$(44a) \quad i_1 = \frac{\begin{vmatrix} v_S & r_{12} \\ 0 & r_{22} \end{vmatrix}}{\Delta_R} \qquad (34b) \quad i_2 = \frac{\begin{vmatrix} r_{11} & v_S \\ r_{21} & 0 \end{vmatrix}}{\Delta_R}$$

Attenuator solution

Figure 507 Attenuator

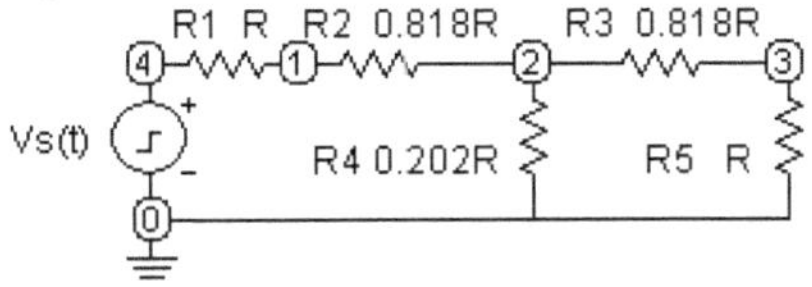

(37a) $\quad$ mesh 1 $\quad v_S = (R_1 + R_2 + R_4)i_1 - R_4 i_2$
(37b) $\quad$ mesh 2 $\quad 0 = -R_4 i_1 + (R_3 + R_4 + R_5)i_2$

(45a) $\quad v_S = (R + 0.818R + 0.202R)i_1 - 0.202 R i_2 \quad = 2.02 R i_1 - 0.202 R i_2$
(45b) $\quad 0 = -0.202 R i_1 + (0.818R + 0.202R + R)i_2 \quad = -0.202 R i_1 + 2.02 R i_2$

(46) $\quad \Delta_R = r_{11}r_{22} - r_{12}r_{21} = 2.02 \times 2.02 - 0.202 \times 0.202 = 4.04 R^2$

$$(47a) \quad i_1 = \frac{\begin{vmatrix} v_S & r_{12} \\ 0 & r_{22} \end{vmatrix}}{\Delta_R} = \frac{\begin{vmatrix} v_S & -0.202R \\ 0 & 2.02R \end{vmatrix}}{4.04 R^2} = \frac{2.02R}{4.04 R^2} v_S = \frac{v_S}{2R}$$

$$(47b) \quad i_2 = \frac{\begin{vmatrix} r_{11} & v_S \\ r_{21} & 0 \end{vmatrix}}{\Delta_R} = \frac{\begin{vmatrix} 2.02R & v_S \\ -0.202R & 0 \end{vmatrix}}{4.04} = \frac{0.202R}{4.04 R^2} v_S = \frac{v_S}{20R}$$

Electric Circuits - Analysis and Design

5.7 Solution of Node Equations. Low Pass Filter

Also use Cramer's rule to solve the node equations. The two node general solution proceeds as follows.

(48) $\quad$ node 2 $\quad i_S = y_{22}v_2 - y_{23}v_3$
$\qquad$ node 3 $\quad 0 = -y_{32}v_2 + y_{33}v_3$

(49) $\quad \Delta_Y = y_{22}y_{33} - y_{23}y_{32}$

$$(50a) \quad v_2 = \frac{\begin{vmatrix} i_S & -y_{23} \\ 0 & y_{33} \end{vmatrix}}{\Delta_Y} \qquad (50b) \quad v_3 = \frac{\begin{vmatrix} y_{22} & i_S \\ -y_{32} & 0 \end{vmatrix}}{\Delta_Y}$$

Low Pass Filter Solution

Figure 508 Low Pass Filter

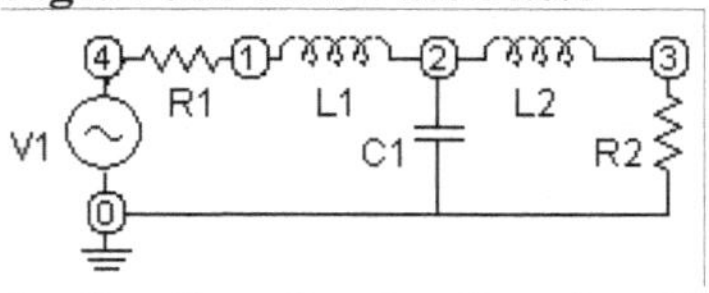

let $R = R_1 = R_2$, $L = L_1 = L_2$, $C = C_1$

(41a) node 2 $\quad i_1 = \dfrac{1}{R+pL}v_1 = \left(\dfrac{1}{R+pL} + pC + \dfrac{1}{pL}\right)v_2 - \dfrac{1}{pL}v_3$

(41b) node 3 $\qquad\qquad 0 = -\dfrac{1}{pL}v_2 + \left(\dfrac{1}{pL} + \dfrac{1}{R}\right)v_3$

(51) $\Delta_Y = \left(\dfrac{1}{R+pL} + pC + \dfrac{1}{pL}\right)\left(\dfrac{1}{pL} + \dfrac{1}{R}\right) - \left(\dfrac{1}{pL}\right)^2 = \dfrac{1}{pLR}\left(2 + pCR + p^2LC\right)$

(52) $v_2 = \dfrac{\begin{vmatrix} i_S & -y_{23} \\ 0 & y_{33} \end{vmatrix}}{\Delta_Y} = \dfrac{\dfrac{1}{R+pL}v_S \cdot \left(\dfrac{1}{pL} + \dfrac{1}{R}\right)}{\dfrac{1}{pLR}\left(2 + pCR + p^2LC\right)} = \dfrac{v_S}{\left(2 + pCR + p^2LC\right)}$

(53) $v_3 = \dfrac{\begin{vmatrix} y_{22} & i_S \\ -y_{32} & 0 \end{vmatrix}}{\Delta_Y} = \dfrac{\dfrac{1}{R+pL}v_S \cdot \dfrac{1}{pL}}{\dfrac{1}{pLR}\left(2 + pCR + p^2LC\right)} = \dfrac{Rv_S}{(R+pL)\left(2 + pCR + p^2LC\right)}$

5 Circuit Analysis

if $z_C = z_L$ at ω_0, then $\omega_0 L = \dfrac{1}{\omega_0 C}$ or $LC = \dfrac{1}{\omega_0^2}$

$CR = \dfrac{\omega_0}{\omega_0} CR = \omega_0 CR \dfrac{1}{\omega_0} = \dfrac{R}{\omega_0 L} \dfrac{1}{\omega_0} = \dfrac{1}{Q_S} \dfrac{1}{\omega_0}$ and $\dfrac{L}{R} = \dfrac{\omega_0 L}{R} \dfrac{1}{\omega_0} = Q_S \dfrac{1}{\omega_0}$

(54) $T(p) = \dfrac{1}{\left(\dfrac{p^2}{\omega_0^2} + \dfrac{1}{Q_S}\dfrac{p}{\omega_0} + 2\right)\left(1 + Q_S \dfrac{p}{\omega_0}\right)}$

Select values for a specific design.

If $L = 470\mu H$ and $C = 1000 pF$, then

(55) $Z_o = \omega_0 L = \dfrac{L}{\sqrt{LC}} = \sqrt{\dfrac{L}{C}} = \sqrt{\dfrac{470u}{1000p}} = 685\Omega$

(56) $f_0 = \dfrac{\omega_0}{2\pi} = \dfrac{1}{2\pi}\dfrac{1}{\sqrt{LC}} = \dfrac{1}{2\pi}\dfrac{1}{\sqrt{470\times10^{-6}\times1000\times10^{-12}}} = 232 KHz$

(57) $Q = \dfrac{Z_0}{R} \rightarrow Q = \dfrac{685}{330} = 2.1,\ Q = \dfrac{685}{680} = 1,\ Q = \dfrac{685}{1500} = 0.46$

Spice program 5081 calculates the frequency response (equation 54). Voltage v drives three low pass filters in parallel to show how resistor values affect the transfer function.

```
Fig5081.ckt  low pass filter
V 1 0 AC 1 0 PULSE(0 4 0 0 0 1000u 2000u)
R1 1 2 1500    ;Q=0.46
L1 2 3 470u
L2 3 4 470u
C1 3 0 1000p
R2 4 0 1500
R11  1 12 680   ;Q=1
L11 12 13 470u
L12 13 14 470u
C11 13  0 1000p
R12 14  0 680
R21  1 22 330   ;Q=2.1
L21 22 23 470u
L22 23 24 470u
C21 23  0 1000p
R22 24  0 330
.AC DEC 200 1000 1e+007
.PLOT AC VDB(4) VDB(14) VDB(24)  -50,0
.TEMP 27
.end
```

Electric Circuits - Analysis and Design

Figure 50811 Low Pass Filter Frequency Response

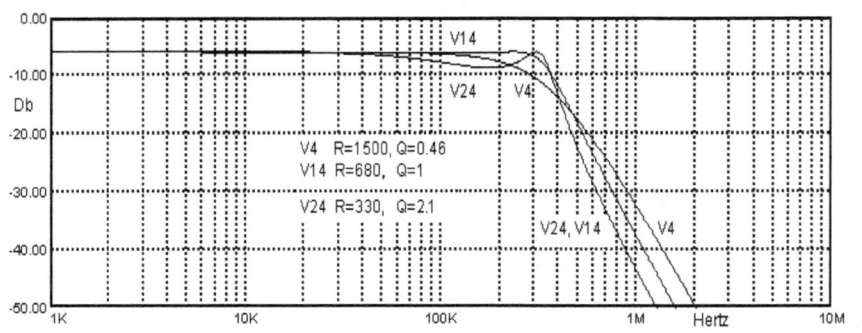

Spice program 508 also calculates transient response when we add TRAN and PLOT TRAN lines. Note how the plots depend on R.

```
Fig1111.ckt   low pass filter
V 1 0 AC 1 0 PULSE(0 4 0 0 0 1000u 2000u)   ;figure 11112
*V 1 0 AC 1 0 PULSE(0 4 0 0 0 10u 20u)   ;figure 11113
.TRAN 1e-007 2e-005 0
.PLOT TRAN V(4) V(14) V(24) 0,2.5
```

Figure 50812 Low Pass Filter Transient Response to Step Function

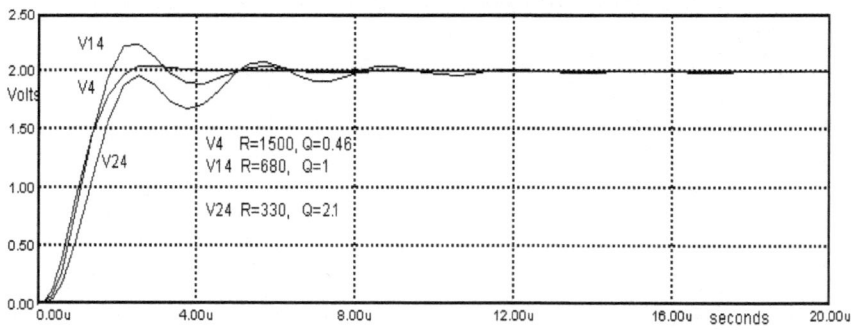

Figure 50813 Low Pass Filter Transient Response to 10μs Pulse

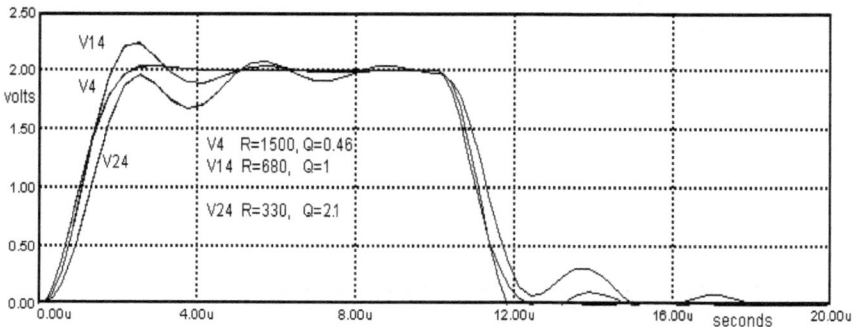

5 Circuit Analysis

5.8 Practice Analyzing Circuits

We use the node and mesh methods to gain experience analyzing circuits. We recommend that you derive the equations prior to reading the text.

The circuit in Figure e211 has two unknown mesh currents, and one unknown node. We use the node method. (Try writing the mesh equations that are a mess.) The currents in and out of node 1 add to zero (equation 58). We solve for the only unknown node 1 voltage v_1 (equation 59b). Then we find i_1 and i_s (equations 60) to verify the currents add to zero (0.2+0.11−0.31=0).

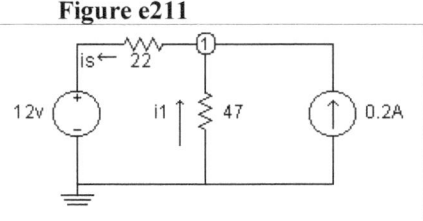

Figure e211

node 1 is the one unknown node

(58) $\quad 0 = \left(\dfrac{1}{22\Omega}\right)(v_1 - 12v) + \left(\dfrac{1}{47\Omega}\right)(v_1 - 0v) + 0.2 A$

(59a) $\quad \left(\dfrac{1}{22\Omega}\right)(12v) - 0.2 A = \left[\left(\dfrac{1}{22\Omega}\right) + \left(\dfrac{1}{47\Omega}\right)\right](v_1)$

(59b) $\quad 0.545 - 0.2 = \dfrac{1}{15} v_1 \quad i \Rightarrow \quad v_1 = 5.175 volts$

(60a) $\quad i_1 = \dfrac{v_1}{47\Omega} = \dfrac{5.175 V}{47\Omega} = 0.11 A \quad$ (60b) $\quad -i_s = \dfrac{v_1 - 12}{22\Omega} = 0.31$

check $\quad i_s = i_1 + 0.2 \quad \rightarrow \quad 0.31 \equiv 0.11 + 0.2$

Example 212 Solve for v_1 in Figure e212. Use node or mesh method.

Example 213 Solve for v_1 & i_s in Figure e213. Use node or mesh method.

Figure e212   Figure e213

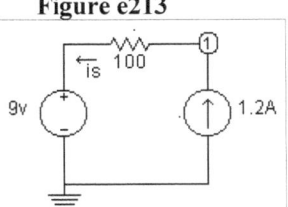

81

Electric Circuits - Analysis and Design

Example 501 This *diode connected transistor* has many applications. Three meshes and one node indicates using the node method. In this circuit (Figure e501) the node 1 voltage $v_1=v_{be}=v_\pi$. Reminder: our convention is that current entering a node has a positive sign. Three currents enter the node, and their sum flows to the source.

(61) $\quad i = \dfrac{v_{be}-0}{r_\pi} + g_m v_{be} + \dfrac{v_{be}-0}{r_o} = v_{be}\left(\dfrac{1}{r_\pi} + g_m + \dfrac{1}{r_o}\right)$

Figure e501

Example 502 Four meshes and two unknown nodes 2, 3 indicates using the node method. Note the complexity of the determinant Δ_y. Let p=0 to see that the gain $v_3/v_4 = g_m R_2$. Let p>>1 to see that R_2 drops out so that the gain is $v_3/v_4 = 1/pR_1[C_1+C_3+(C_1C_3)/C_2]$.

Figure e502

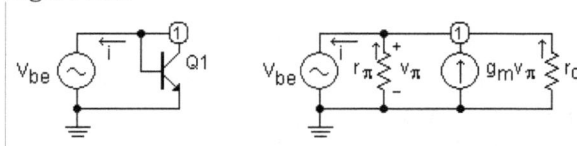

Node 2
(62a) $\quad i_{r1} = g_1(v_2-v_4) \qquad i_{c1} = pC_1(v_2-0) \qquad i_{c2} = pC_2(v_2-v_3)$
(62b) $\quad 0 = g_1(v_2-v_4) + pC_1(v_2-0) + pC_2(v_2-v_3)$
(62c) $\quad g_1 v_4 = (g_1 + pC_1 + pC_2)v_2 - pC_2 v_3$
Node 3
(63a) $\quad i_{r2} = g_2(v_3-0) \qquad i_{c3} = pC_3(v_3-0) \qquad i_{c2} = pC_2(v_3-v_2)$
(63b) $\quad 0 = g_m v_2 + g_2(v_3-0) + pC_3(v_3-0) + pC_2(v_3-v_2)$
(63c) $\quad 0 = -(pC_2 - g_m)v_2 + (g_2 + pC_2 + pC_3)v_3$

Solve for v_3 (equations 48c and 49c)
(64) $\quad \Delta_y = (g_1+pC_1+pC_2)(g_2+pC_2+pC_3) - pC_2(pC_2-g_m)$
$\qquad = (g_1+pC_1+pC_2)(g_2+pC_3) + (g_m+g_1+pC_1)pC_2$
$\qquad = g_1 g_2 + p(g_2[C_1+C_2] + g_1[C_2+C_3] + g_m C_2) + p^2(C_1C_2+C_2C_3+C_3C_1)$

(65) $\quad v_3 = \dfrac{\begin{vmatrix} g_1+pC_1+pC_2 & g_1 v_4 \\ -(pC_2-g_m) & 0 \end{vmatrix}}{\Delta_y} = \dfrac{(pC_2-g_m)g_1 v_4}{\Delta_y}$

5 Circuit Analysis

(66) $T(0) = \dfrac{v_3}{v_4} = \dfrac{(-g_m)g_1}{g_1 g_2} = -g_m R_2$

(67) $T(p \gg 1) = \dfrac{v_3}{v_4} = \dfrac{(pC_2)g_1}{p^2(C_1 C_2 + C_2 C_3 + C_3 C_1)} = \dfrac{1}{pR_1(C_1 + C_3 + C_3 C_1 / C_2)}$

Example 503 Ladder circuit Ladder circuit analysis starts with a current i flowing through the resistor at the end of the ladder. The binary ladder (Figure e503) is not only a useful circuit; its simplicity allows us to focus on a special analytical method. With i flowing in the load resistor R the voltage $v_0 = iR$, and $v_1 = 2Ri$.

> Here we would like to introduce an idea: the *resistance to the right* of the v_1 node equals 2R.

Figure e503 Binary ladder

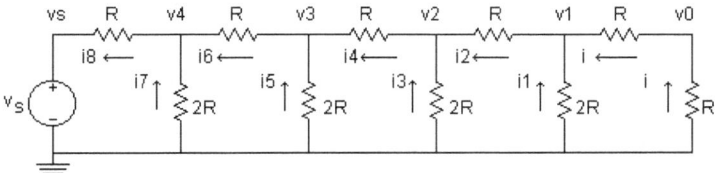

Moving to the left of the v_1 node, the resistance to the right now includes the 2R from node v_1 to the reference node, or ground. Since 2R in parallel with 2R equals R the resistance to the right that includes node v_1 equals R. Clearly $i_1 = v_1/2R = 2Ri/2R = i$: two currents equal to i flow to the left from node v_1 that is why $i_2 = i_1 + i = 2i$. Now you can calculate the other node voltages.

Example 504 Reference Figure e503. Show that the current ratios are 1/2/4/8/16/32 where $v_0/v_S = 1/32$, and $i_6 = 8i$.

Reference Figure e504. Solve for i_S and v_3 as functions of v_S.

Figure e504

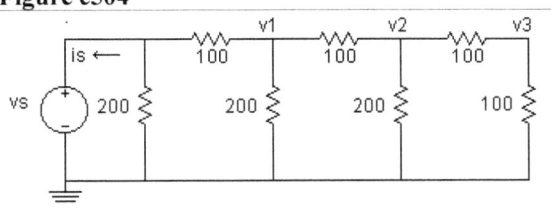

83

Electric Circuits - Analysis and Design

Example 505 Delta/wye Delta-to-wye and wye-to-delta equations (Figure e505) convert circuits such as bridges (Figure e506) to series-parallel form.

$$(68) \quad \begin{aligned} R_{1y} &= \frac{R_{2d} R_{3d}}{R_{1d} + R_{2d} + R_{3d}} & R_{1d} &= \frac{R_{1y} R_{2y} + R_{2y} R_{3y} + R_{3y} R_{1y}}{R_{1y}} \\ R_{2y} &= \frac{R_{3d} R_{1d}}{R_{1d} + R_{2d} + R_{3d}} & R_{2d} &= \frac{R_{1y} R_{2y} + R_{2y} R_{3y} + R_{3y} R_{1y}}{R_{2y}} \\ R_{3y} &= \frac{R_{1d} R_{2d}}{R_{1d} + R_{2d} + R_{3d}} & R_{3d} &= \frac{R_{1y} R_{2y} + R_{2y} R_{3y} + R_{3y} R_{1y}}{R_{3y}} \end{aligned}$$

Figure e505 Delta and wye circuits

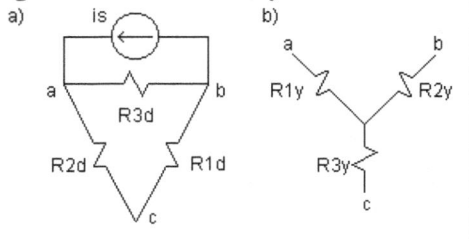

Example 506 Bridge Circuit

Figure e506 Bridge Circuit

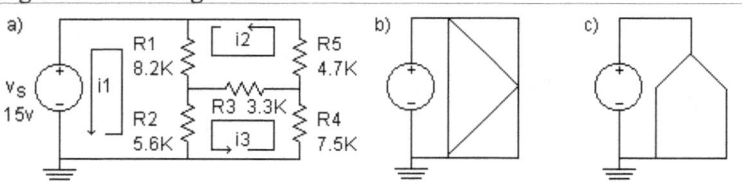

Use the wye-delta equations to convert the bridge circuit to series-parallel form as in Figures e506b and e506c.

Convert the left Y to delta (Figure e506b) If $R_{1Y} = R_1$, $R_{2Y} = R_2$, $R_{3Y} = R_3$, then show that the Y converts to Delta $R_{1D} = 11.15K$, $R_{2D} = 16.33K$, $R_{3D} = 27.72K$. Check your work by converting Delta back to Y.

Convert the upper delta to Y (Figure e506c) If $R_{1D} = R_1$, $R_{2D} = R_5$, $R_{3D} = R_3$, then show that the upper delta converts to Y $R_{1Y} = 0.96K$, $R_{2Y} = 1.67K$, $R_{3Y} = 2.38K$. Check your work by converting Y back to Delta.

5 Circuit Analysis

Complex Numbers

The words complex and imaginary are potentially misleading, because complex numbers are not complicated and imaginary operators are not part of someone's imagination. Both words are labels: they are technical terms used to designate a class of numbers. A complex number z is represented by an ordered pair of real numbers x and y written as (x,y).

Multiplication by −1 and √−1 A number can be represented as a distance on a number line. We define steps to the right as positive so that distance AB=+4. Multiply +4 by −1 to get −4 that is the distance AC. Multiply AC by −1 to get back to AB. Clearly multiplication by −1, in effect, *rotates* AB and AC by 180°.

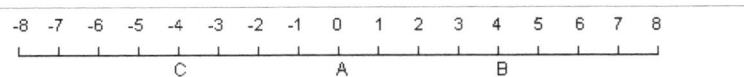

If +4 is multiplied by √−1 the result is 4√−1. Multiply 4√−1 by √−1 to get −4. Therefore two multiplication's by √−1 rotates AB by 180°. And so multiplication by √−1 effects a 90° rotation of AB.

The world has agreed that numbers such as 4√−1 are *imaginary* numbers. To save writing, √−1 is replaced by *i* in the mathematical literature. However the EE world uses i to designate current. That is why the EE world uses j for √−1.

A *complex number* is a real number plus an imaginary number such a 3+j4.

In this way any point in a plane can be represented by a complex number z=x+jy (Figure CN1).

Figure CN1 Complex numbers in Cartesian and polar coordinates

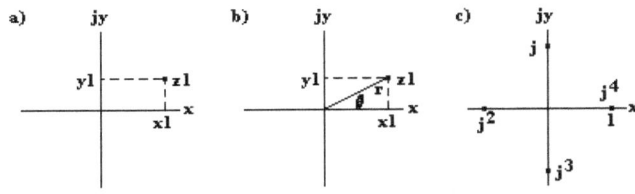

85

Electric Circuits - Analysis and Design

Sinusoidal alternating current and voltage

An alternating voltage at a node is positive with respect to ground, and at a different time the voltage at the node is negative with respect to ground.

An alternating current flows one direction in a wire, and at a different time it flows in the opposite direction.

In other words the sign of the voltage or current is alternately positive or negative. Specifically our present interest is in *sinusoidal* waveforms (Figure 605). For example.

$v(t) = V_m \cos(\omega t + \theta)$ *and, or* $v(t) = V_m \sin(\omega t + \theta)$
$i(t) = I_m \cos(\omega t + \theta)$ *and, or* $i(t) = I_m \sin(\omega t + \theta)$

where $\omega = 2\pi f$ *and period* $T = \dfrac{1}{f}$

v(t) or i(t) is the instantaneous voltage or current at time t.

V_m or I_m is the maximum or peak value or amplitude of the waveform (positive or negative).

T is the time duration of one complete waveform cycle or period, and frequency f is the number of cycles repeated in one second (f = 1/T cycles per second).

θ is the phase shift relative to time t = 0.

Figure 605 Two cycles of a sinewave

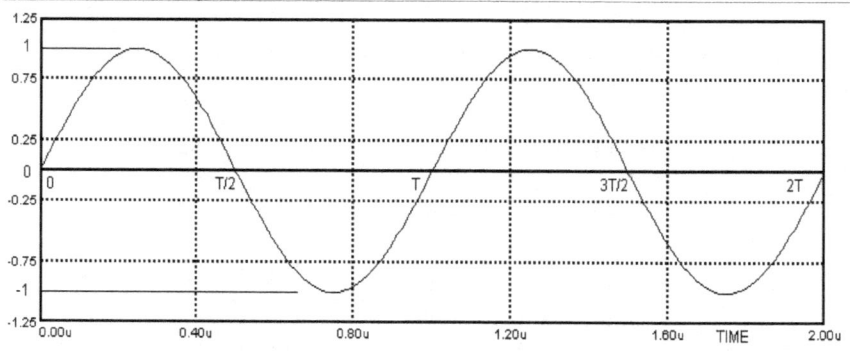

6 Frequency Response

The L and C impedances are frequency dependent. That means circuit behavior is a function of frequency. The world says a circuit has a *frequency response*. We are in the frequency domain and the circuit is in the steady state (page 62 and Chapter 8). The frequency responses of four circuits are analyzed so you will know how to find frequency responses of any circuit. The four circuits are resistor R, RC low pass filter, RC phase equalizer, and RLC series resonant circuit.

Many circuit design goals are specified as a frequency response transfer function T(p) from an input node to an output node. A specified transient response is another type of design objective (Chapter 7).

The R, L, C impedances are R, pL, and 1/pC. Complex frequency p is required for transient responses, whereas real frequency ω is required for frequency responses. Nevertheless we write p instead of jω to make fewer errors, and to simplify the writings.

(1a) $z_R = R$ (1b) $z_L = j\omega L \text{ or } pL$ (1c) $z_C = \dfrac{1}{j\omega C} \text{ or } \dfrac{1}{pC}$

Spice assists when creating designs and solving problems. In this modern world there is no need to hand calculate numerical solutions. We show how to write Spice programs that calculate numerical solutions (Chapter 12). Spice programs that create performance plots. Furthermore the Spice programs in this chapter are another form of the Bode method (Chapter 10).

Spice programs have a multitude of features you can spend your life learning about. We only use the "spice-text" feature of these Spice programs that are in fact complex user friendly wrap arounds to the basic Berkeley Spice.

Identifying numbers A program is given the name Fig6011.ckt when it is the first program written about the circuit in Figure 601. Plots of results from program Fig6011.ckt are labeled Fig60111, Fig60112, etc. A second program for Figure 601 is given the name Fig6012.ckt. Plots are labeled Fig60121, Fig60122, and so forth. In this way we know how circuit schematics, Spice programs, and plots are related.

Note: Hspice programs require a *.option post* line. Refer to Hspice Help.

Electric Circuits - Analysis and Design

6.1 Resistor Circuit

DC is circuit behavior at zero frequency. A resistors-only circuit has a frequency response that is independent of frequency.

For example, derivation of the node equations 3a and 3b is straightforward. If we need a solution in the form of general expressions, then we apply Cramer's rule and solve equations 3 for voltages and currents.

We prefer to use Spice to calculate specific solutions. That is why we have written Spice program `Fig6011.ckt`. Accurate numbers are available in the Spice *numeric output* produced by the dot print statement.

Figure 601a　　　　　　　　　**Figure 601b For Spice**

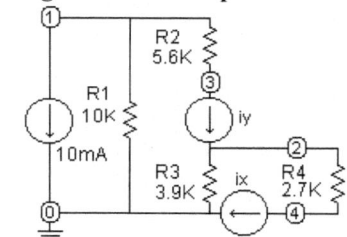

Source i_s exits node 1 (Figure 601a), while i_1 and i_2 enter node 1 (equation 2b). Since i_2 enters node 1, it is exiting node 2. That is why it is negative in equation 2c.

(2a) $i_1 = g_1(v_1 - 0)$ 　　 $i_2 = g_2(v_1 - v_2)$ 　　 $i_3 = g_3(v_2 - 0)$ 　　 $i_4 = g_4(v_2 - 0)$
(2b) $0 = -i_s + i_1 + i_2$
(2c) $0 = -i_2 + i_3 + i_4$

Replace currents with vi constraints and collect terms (Figure 601a).
(3a) $i_s = (g_1 + g_2)v_1 - g_2 v_2$
(3b) $0 = -g_2 v_1 + (g_2 + g_3 + g_4)v_2$

Problem 601 Reference equations 3. Use Cramer to solve for v_1 and v_2.

---

6 Frequency Response

Figure 60111 Voltage-current vi constraint v=iR

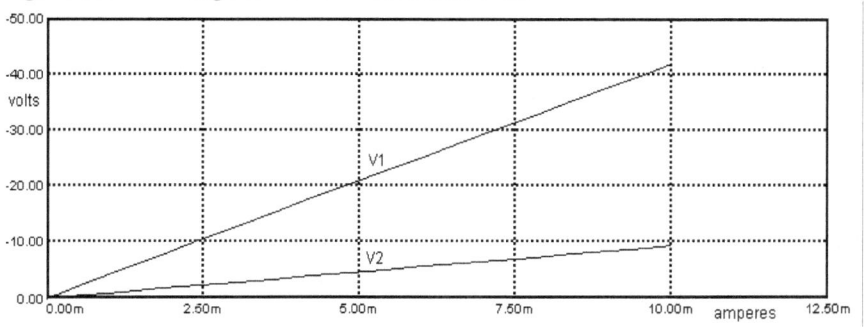

The Spice program for the Resistor circuit.

```
Fig6011.ckt    ;title which is mandatory first line
     ;spaces are ignored
Is 1 0 DC 0    ;current source
R1 1 0 10K     ;the circuit
R2 1 3 5.6K
Vy 3 2 DC 0    ;iy ammeter
R3 2 0 3.9K
R4 2 4 2.7K
Vx 4 0 DC 0    ;ix ammeter

.DC Is 0 10m 1m    ;range of the current source
.PLOT DC V(1) V(2) 0,-50 ;what to plot
.PRINT DC I(Vy) I(Vx) ;create numeric file
.PRINT DC V(1) V(2)
.TEMP 27
.end      ;mandatory last line
```

Numeric output
```
I(IS) V(1)   V(2)   I(Vy)   I(Vx)
 (V)   (V)   (mA)   (mA)
0.010 -41.845 -9.278  -5.815  -3.436
```

Check
Show that Ix=I(Vx)=V2/R4
Show that Iy=I(Vy)=(V1-V2)/R2

89

6.2 RC Low Pass Filter

If a steady state sinusoidal input to a circuit is $V_S \sin \omega t$, then a change in frequency ω changes the values of all L and C impedances $j\omega L$ and $1/j\omega C$, as well as the currents and voltages in the circuit. Clearly the magnitude and phase of $T(p) = T(j\omega)$ (equation 8) varies with frequency.

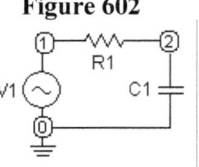

Figure 602

Formulate the node 2 equation.
(4a) $i_1 = g_1(v_2 - v_1)$ $i_2 = pC_1(v_2 - 0)$
(4b) $0 = i_1 + i_2$

Replace currents (4b) with vi constraints (4a) and collect terms.
(5) $0 = -g_1 v_1 + (g_1 + pC_1)v_2$

Solve for v_2.

(6) $v_2 = \dfrac{g_1 v_1}{g_1 + pC_1} = \dfrac{g_1 v_1}{g_1 + pC_1} \times \dfrac{R_1}{R_1} = \dfrac{v_1}{1 + pC_1 R_1}$

An alternative solution uses the transmission T(p) from node 1 to node 2.

(7) $T(p) = \dfrac{v_2}{v_1} = \dfrac{z_{C1}}{z_{C1} + z_{R1}} = \dfrac{1}{1 + \dfrac{z_{R1}}{z_{C1}}} = \dfrac{1}{1 + pC_1 R_1}$

(8) $T(j\omega) = \dfrac{1}{1 + j\omega C_1 R_1} = \dfrac{1}{1 + j2\pi f C_1 R_1}$

The transmission T(p) is a typical ratio of impedances (7) formed from z_{R1} and z_{C1}. A frequency response is a plot of the magnitude or phase of T(p) as ω is varied (Figures 60211 and 60212). Frequency ω is varied by changing the frequency ω of steady state sinusoidal input $V_S \sin \omega t$, while holding V_S constant. This is what a Spice .PLOT AC statement does in conjunction with the .AC and V_1 statements (*Spice Program 6021*).

Magnitude To understand how and why $T(j\omega)$ varies with frequency we note some properties of z_R and z_C. If $z_R = R_1 = 1000$ ohms, then the magnitude is 1000 ohms at any frequency. On the other hand z_C decreases as $1/\omega$ as frequency increases. The arithmetic simplifies if the 2π of $2\pi f$ cancels (9). This is why we select $C_1 = (1/20\pi)$ µf $= 10^{-6}/20\pi$.

6 Frequency Response

(9) $x_C = |z_C| = \dfrac{1}{\omega C_1} = \dfrac{20\pi \times 10^6}{2\pi f} = \dfrac{10 \times 10^6}{f}$

The capacitive reactance x_C at $f=10^2$ Hz is 100,000 ohms, and at 10^6 Hz x_C is 10 ohms (equations 10 and 11).

(10) $x_{C\_10^2} = \dfrac{10 \times 10^6}{10^2} = 10^5 \, ohms$ \qquad (11) $x_{C\_10^6} = \dfrac{10 \times 10^6}{10^6} = 10 \, ohms$

At some frequency $\omega_0 = 2\pi f_0$, $x_C = 1000\,ohms = R_1$. In other words the *magnitudes* of impedances z_R and z_C are defined to be equal at f_0. In this example x_C line crosses the z_R line at $f_0 = 10\,KHz$ (Figure 60221).

(11) $R_1 = \dfrac{1}{\omega_0 C_1}$

Phase A sinewave is shifted in phase as well as attenuated (Figure 60231). Phase θ increases from 0 to -90 degrees (Figure 60212) as ω ranges from 0 to infinity.

Figure 602b

(12) If $T(j\omega) = \dfrac{1}{1 + j\omega C_1 R_1}$ then $\tan \theta = \dfrac{\omega C_1 R_1}{1}$

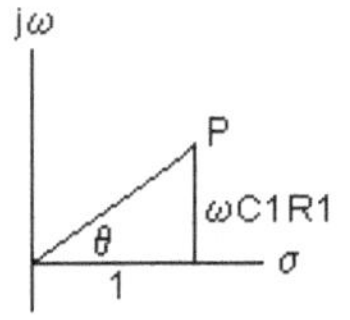

Figure 60231 V_2 is 50KHz sinewave attenuated by 14dB, phase shifted by 78.6°

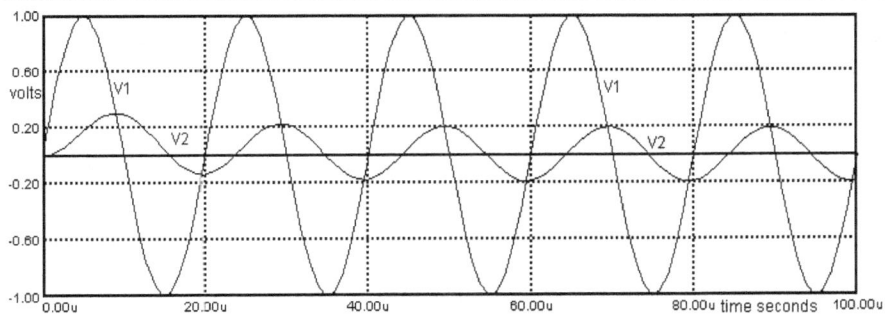

Problem 602 Reference Figure P603. Calculate equations analogous to equations 7 and 8.

Problem 603 Reference Figure P603. Write a Spice program and plot figures for transmission magnitude and phase analogous to the RC Spice program Fig6021.Ckt.

Figure P603

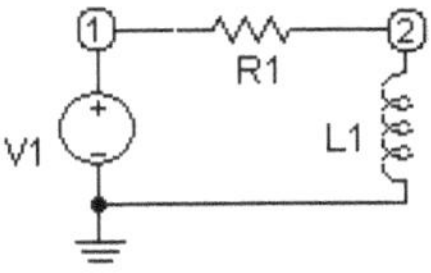

The Spice program for the RC circuit.
```
Fig6021.ckt   factor 1/(p+a)
V1 1 0 AC 1 0   ; volts
R1 1 2 1000
C1 2 0 .0159155u
*.PLOT AC VDB(2) -40,10
.AC DEC 200 10 1e+007
.PLOT AC VP(2) -100,0
.PRINT AC VDB(2) VP(2)  ;for numeric data
.TEMP 27
.end
```

Figure 602

Write the program (Spice is not case sensitive)
`Fig6021.ckt          factor 1/(p+a)`
 First line The first line of every program is assumed by Spice to be a title statement. The title statement can include any words.

`V1 1 0 AC 1 0   ; volts`
 Voltage source Signal generator V_1 is connected to nodes 1 and 0. The v_1 signal is a AC sinewave with 1 volt magnitude and 0 phase. A semicolon starts a comment ; `volts`.

`R1 1 2 1000`
`C1 2 0 .0159155u`
 Circuit components Resistor R_1 is connected to nodes 1 and 2. Capacitor C_1 is connected to nodes 2 and 0.

`.AC DEC 200 10 1e+007`
 AC control statement Dot AC DEC defines a frequency range from 10 Hz to 10^7 Hz with points to be calculated every 200Hz.

`.PLOT AC VDB(2) -40,10`
`*.PLOT AC VP(2) -100,0`
`.PRINT AC VDB(2) VP(2)  ;for numeric data`
 Plot the data dot plot AC executes the dot AC statement calculating the v_2 from 10Hz to 10^7Hz (V_1 is a one volt sinewave for all frequencies). The asterisk * that defines a line as a comment, inactivates that line. Therefore we can plot magnitude OR phase.

`.TEMP 27`
 Temperature Dot temp (.temp) defines temperature as 27 degrees C.
`.end`
 Last line The last line of every program is .end (dot end).

6 Frequency Response

Spice Program 6021 Calculates Frequency Response
```
Fig6021.ckt factor 1/(p+a)
V1 1 0 AC 1 0   ; volts
R1 1 2 1000
C1 2 0 .0159155u
*.PLOT AC VDB(2) -40,10
.AC DEC 200 10 1e+007
.PLOT AC VP(2) -100,0
.PRINT AC VDB(2) VP(2) ;for numeric data
.TEMP 27
.end
```

Figure 602

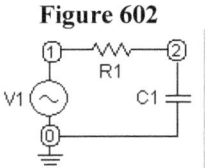

V_1 is a voltage source connected from node 1 to node 0. AC 1 0 means the source is sinusoidal with 1volt peak amplitude and 0 degrees phase. The dot AC (.AC) line defines a frequency range of 10Hz to 10MHz. Spice starts by selecting 10Hz, and calculating the v_2/v_1 ratio (i.e. $|T(j2\pi 10)|$). Then Spice increments the frequency by 200 Hz to 210Hz, and calculates $|T(j2\pi 210)|$. The process is repeated every 200 Hz up to 10MHz. The dot plot lines produce Figures 60211 and 60212.

Figure 60211 Magnitude of $T(p) = V_2/V_1$

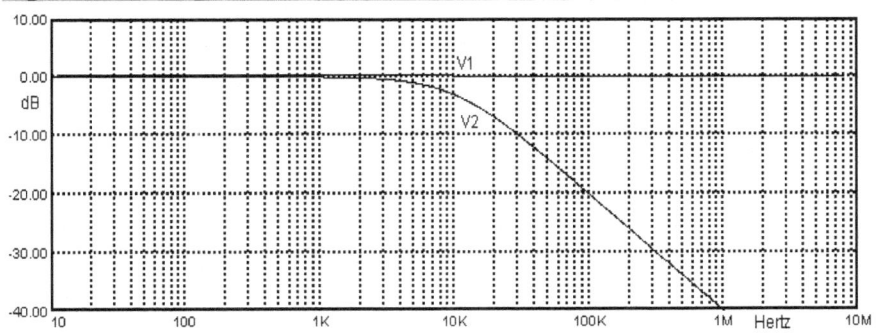

Figure 60212 Phase of $T(p) = V_2/V_1$

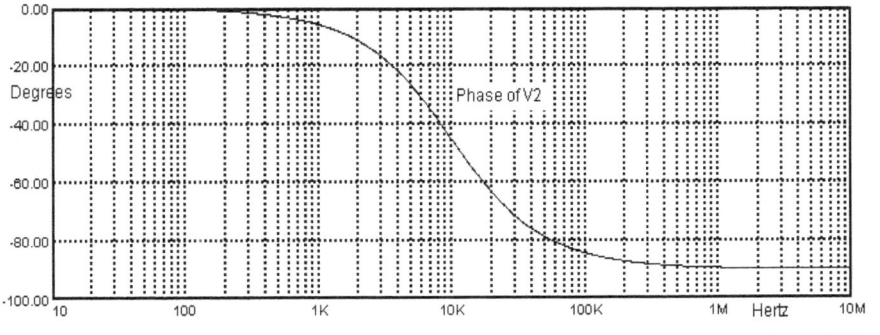

Electric Circuits - Analysis and Design

Spice Program 6022 Calculates Impedance Magnitudes

```
Fig6022.ckt
V1 1 0 AC 1000
R1 1 0 1K
V2 2 0 AC 1000
C1 2 0 0.0159155u
.AC DEC 10 100 1e+007
.TEMP 27
.PLOT AC IDB(V1) IDB(V2) 20,-30
.end
```

Tricks of the trade The current IDB(V2) *increases* with frequency, because $1/\omega C_1$ decreases with frequency. We want the dB scale to be replaced by an ohms scale. Clearly maximum current represents minimum ohms. So we manually write ohms in lieu of dB. Let 20dB=10Ω, 10dB=100Ω, 0dB=1KΩ, −10dB=10KΩ, −20dB=100KΩ, and −30dB =1000KΩ=1MΩ. And so the Spice plot spans 20dB to −30dB (10Ω to 10^6Ω).

Figure 60221 Magnitude of R_1 and x_{C1}

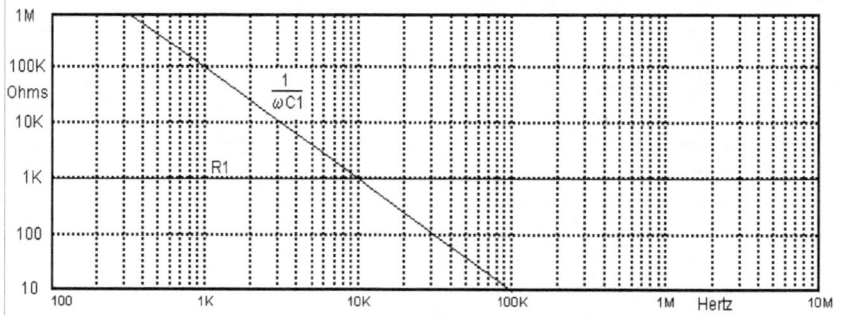

Observe that |T(p)| is −3dB at f_0 (Figure 60211) or $1/\sqrt{2}$ (equation 13).

$$(13) \quad T(j\omega_0) = \frac{v_2}{v_1} = \frac{\frac{1}{pC_1}}{R_1 + \frac{1}{pC_1}} = \frac{1}{1 + j\omega_0 C_1 R_1} = \frac{1}{1+j1} = \frac{1}{\sqrt{2}} e^{-j\frac{\pi}{4}}$$

The magnitude of $x_{C1} \gg R_1$ when $f < f_0$, and the magnitude of $x_{C1} \ll R_1$ when $f > f_0$. Consequently the *asymptotes* of T(p) are *0dB/octave when $f < f_0$*, and *−6dB/octave when $f > f_0$* , because x_{C1} halves when frequency f doubles. (Figure 60221).

6 Frequency Response

6.3 RC Phase Equalizer

Figure 603

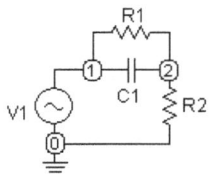

Know that feedback circuits can produce unintended oscillations when the phase shift around the feedback loop is 360 degrees and the gain is greater than 1. The 360 degrees phase shift is usually produced by the amplifier (180°) and 3 or more poles (Section 7.9). Each pole can contribute up to 90 degrees at infinity for a potential total of 270 degrees. Therefore, if at some high frequency 180° is reached, and the gain function is greater than 0dB the circuit will oscillate. One solution to this problem is the phase equalizer (Figure 603) that is designed to retard the phase advance (Figure 60312) as the gain function goes through 0dB.

Figure 60312 Transfer Function of Phase Equalizer, Phase Degrees

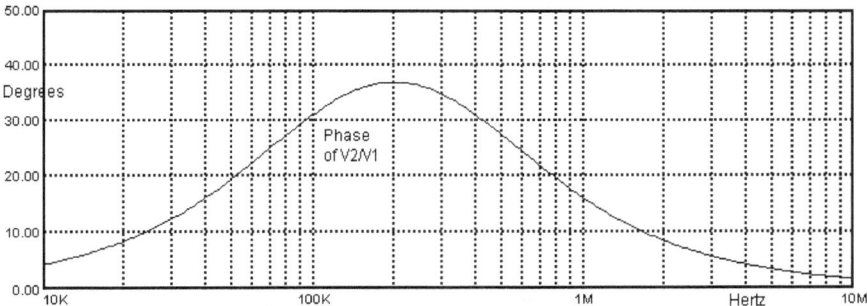

A phase shift is positive when the corresponding gain function is increasing. A zero separated from a pole by 2 octaves provides a 12dB gain increase (see Equation 14 and Figure 604).

$$(14) \quad T_1(p) = \frac{p + \omega_0}{p + 4\omega_0}$$

Figure 604 Sum of terms of $T_1(p)$ Bode Plot (Chapter 10)

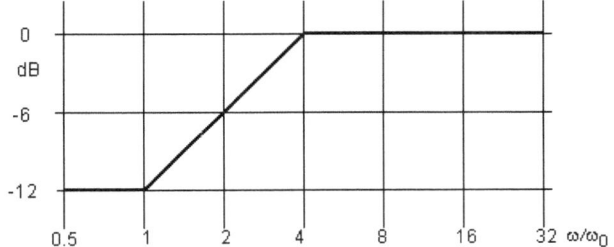

95

Electric Circuits - Analysis and Design

We decompose $T_1(p)$ into its parts by manipulating its expression

(15) $\quad T_1(p) = \dfrac{p+\omega_0}{p+4\omega_0} = \dfrac{p+\omega_0}{\omega_0} \times \dfrac{1}{4} \dfrac{4\omega_0}{p+4\omega_0} = \dfrac{1}{4} \times \left(1+\dfrac{p}{\omega_0}\right) \times \dfrac{1}{1+\dfrac{p}{4\omega_0}}$

$T_1(p)$ is the dB sum of three terms: a −12dB constant, a zero at ω_0, and a pole at $4\omega_0$.

Figure 605 Constant, zero, and pole of $T_1(p)$ Bode Plots

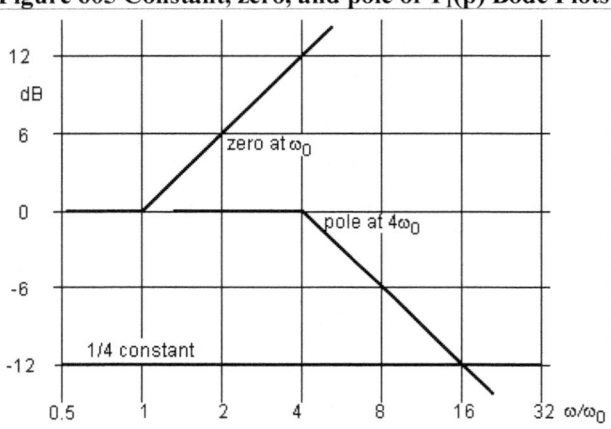

Synthesis We can convert the abstract form of $T_1(p)$ to reveal circuit components (Figure 603b). Each term has to be R or g, and pC or 1/pC.

(16a) $\quad T_1(p) = \dfrac{p+\omega_0}{p+4\omega_0} = \dfrac{1}{1+\dfrac{3\omega_0}{p+\omega_0}} = \dfrac{1}{1+\dfrac{3}{1+\dfrac{p}{\omega_0}}}$

Figure 603b

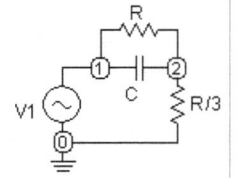

(16b) $\quad T_1(p) = \dfrac{1}{1+\dfrac{3}{1+pRC}} \quad$ where $\omega_0 = \dfrac{1}{RC}$

(16c) $\quad T_1(p) = \dfrac{1}{1+\dfrac{3}{1+pRC}} = \dfrac{1}{1+\dfrac{3g}{g+pC}} = \dfrac{\dfrac{R}{3}}{\dfrac{R}{3}+\dfrac{1}{\dfrac{1}{R}+pC}}$

96

6 Frequency Response

Analysis Analysis confirms the design.

Formulate the node 2 equation (Figure 603).
(17a) $i_1 = (g_1 + pC_1)(v_2 - v_1)$ $i_2 = g_2(v_2 - 0)$
(17b) $0 = i_1 + i_2$

Figure 603

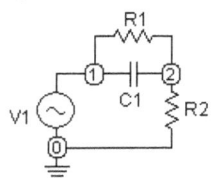

Replace currents with vi constraints and collect terms.
(18) $0 = -(g_1 + pC_1)v_1 + (g_2 + g_1 + pC_1)v_2$

Solve for v_2.

(19) $v_2 = \dfrac{(g_1 + pC_1)v_1}{g_2 + g_1 + pC_1} = \dfrac{g_1}{g_2 + g_1} \dfrac{1 + pC_1 R_1}{1 + pC_1(R_1 \| R_2)} v_1 \Rightarrow R_1 \| R_2 = \dfrac{R_1 R_2}{R_1 + R_2}$

Let $R_1 = 3R_2$ and $\omega_0 = 1/R_1 C_1$. Then $g_1 + g_2 = 4/R_1$, and the transmission T(p) from node 1 to node 2 is as follows.

(20) $T(p) = \dfrac{v_2}{v_1} = \dfrac{1}{4} \dfrac{1 + \dfrac{p}{\omega_0}}{1 + \dfrac{p}{4\omega_0}} = \dfrac{p + \omega_0}{p + 4\omega_0}$

The frequency response is a plot of the magnitude or phase of T(p) as ω is varied (Figures 60311 and 60312). Frequency ω is varied by changing the frequency ω of steady state sinusoidal input $V_1 \sin \omega t$, while holding V_1 constant. This is what a Spice .PLOT AC statement does in conjunction with the .AC and V1 statements.

Phase The zero increases phase from 0° towards 90°. Two octaves later the pole decreases phase from 0° towards −90°. the algebraic sum of the two phase plots is shown in Figure 60312.

Problem 604 Reference Figure P605. Calculate equations analogous to equation 20.

Figure P605

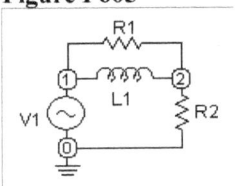

Problem 605 Reference Figure P605. Write a Spice program and plot figures for transmission magnitude and phase analogous to the RC phase equalizer Spice program Fig6031.Ckt.

Electric Circuits - Analysis and Design

Spice program for the RC Phase Equalizer

```
Fig6031.ckt
V1 1 0 AC 1 0
R1 1 2 3000
C1 1 2 530.5p
R2 2 0 1000
.AC DEC 100 10000 1e+007
*.PLOT AC VP(20) 0,50
.PLOT AC VDB(20) -15,10
.TEMP 27
.end
```

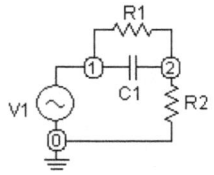

Figure 603

Write the program (Spice is not case sensitive)

`Fig6031.ckt`
 First line The first line of every program is assumed by Spice to be a title statement. The title statement can include any words.
`V1 1 0 AC 1 0`
 Voltage source Signal generator v_1 is connected to nodes 1 and 0. The v_1 signal is a AC sinewave with 1 volt magnitude and 0 phase.

`R1 1 2 3000`
`C1 1 2 530.5p`
`R2 2 0 1000`
 Circuit components Resistor R_1 is connected to nodes 1 and 2. Capacitor C_1 is connected to nodes 1 and 2, and resistor R_2 is connected to nodes 2 and 0.

`.AC DEC 100 10000 1e+007`
 AC control statement Dot AC DEC defines a frequency range from 10^4Hz to 10^7 Hz with points to be calculated every 100Hz.

`.PLOT AC VDB(2) -15,10`
`*.PLOT AC VP(20) 0,50`
 Plot the data dot plot AC executes the dot AC statement calculating the v_2 from 10^4Hz to 10^7Hz every 100Hz (v_1 is a one volt sinewave for all frequencies). The asterisk * that defines a line as a comment in effect inactivates the line. Therefore we can plot magnitude *or* phase.

`.TEMP 27`
 Temperature Dot temp defines temperature as 27 degrees C.
`.end`
 Last line The last line of every program is .end (dot end).

6 Frequency Response

Spice program 6031 Calculates Transfer Function

```
Fig6031.ckt

V1 1 0 AC 1  :pole and zero
R1 1 2 3000
C1 1 2 530.5p
R2 2 0 1000

*.PLOT AC VP(20) 0,50
.AC DEC 100 10000 1e+007
.TEMP 27
.PLOT AC VDB(20) -15,10
.end
```

Figure 60311 Transfer Function of Phase Equalizer, Magnitude dB

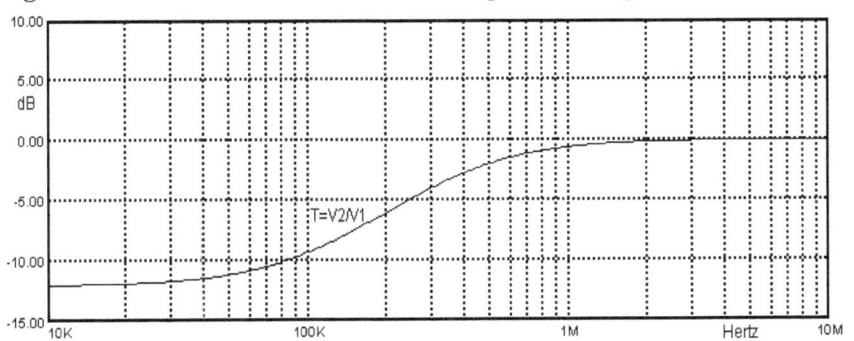

Figure 60312 Transfer Function of Phase Equalizer, Phase Degrees

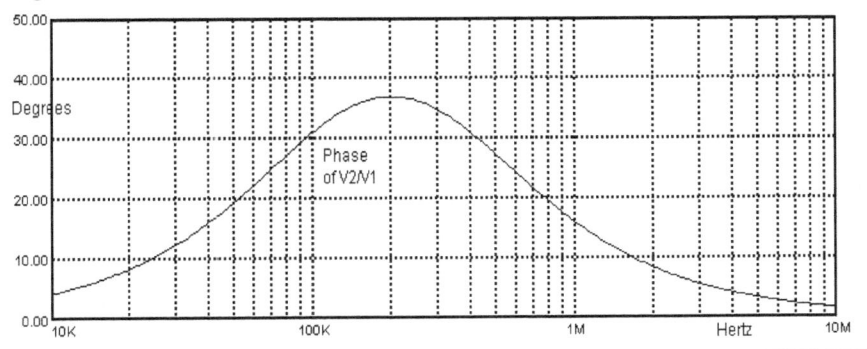

6.4 RLC Series Resonant Circuit

Resonant circuit properties are useful, because they vary dramatically with frequency. The properties are said to be selective. Selectivity is closely related to the quality factor $Q_{Series} = \omega L/R$. An important resonant circuit is the series RLC circuit (Figure 606).

Figure 606 Series RLC

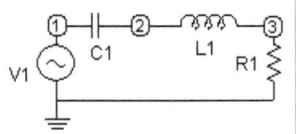

(21) $Z(p) = R + pL + \dfrac{1}{pC}$

The magnitude of impedance Z is a minimum at *resonant frequency* f_0.

(22) if $p = j\omega$ then $Z(j\omega) = R + j\omega L + \dfrac{1}{j\omega C} = R + j\left[\omega L - \dfrac{1}{\omega C}\right] = R + jX(\omega)$

(23) if $X(\omega_0) = 0$ then $\omega_0 L - \dfrac{1}{\omega_0 C} = 0 \Rightarrow \omega_0^2 = \dfrac{1}{LC}$ and $Z(j\omega_0) = R$

Selectivity is defined as the −3dB bandwidth. At the −3dB frequencies ω_1 and ω_2 the impedance magnitude of $Z(j\omega)$ equals $R\sqrt{2}$.

(24a) $+R = \omega_2 L - \dfrac{1}{\omega_2 C}$ $\qquad -R = \omega_1 L - \dfrac{1}{\omega_1 C}$

(24b) $LC\omega_2^2 - RC\omega_2 - 1 = 0$ $\qquad LC\omega_1^2 + RC\omega_1 - 1 = 0$

(24c) $\left(\dfrac{\omega_2}{\omega_0}\right)^2 - \dfrac{1}{Q_S}\dfrac{\omega_2}{\omega_0} - 1 = 0$ $\qquad \left(\dfrac{\omega_1}{\omega_0}\right)^2 + \dfrac{1}{Q_S}\dfrac{\omega_1}{\omega_0} - 1 = 0$

$\omega_0^2 = \dfrac{1}{LC} \qquad Q_S = \dfrac{\omega_0 L}{R} = \dfrac{1}{R\omega_0 C} = \dfrac{1}{R}\sqrt{\dfrac{L}{C}} = \dfrac{z_0}{R}$

The solutions to the quadratic equations 24c are

(25a) $\dfrac{\omega_2}{\omega_0} = \left[1 + \dfrac{1}{4Q_S^2}\right]^{\frac{1}{2}} + \dfrac{1}{2Q_S}$ $\qquad$ (25b) $\dfrac{\omega_1}{\omega_0} = \left[1 + \dfrac{1}{4Q_S^2}\right]^{\frac{1}{2}} - \dfrac{1}{2Q_S}$

(26) $\omega_2 - \omega_1 = \dfrac{\omega_0}{Q_S}$

The three frequencies ω_0, ω_1, and ω_2 are properties of the impedance Z. If $Q_S = 20$, then $\omega_2 = 1.025\omega_0$, $\omega_1 = 0.975\omega_0$, and $\omega_2 - \omega_1 = 0.050\omega_0 = \omega_0/20$

6 Frequency Response

Emphasis Q is the ratio of $\omega_0 L/R = z_0/R$ where z_0 is the *characteristic impedance*. You will use various forms of Q every day when you work with resonant circuits.

Selectivity comparison Without digressing into the Spice program that produced Figure 60711, compare plots of the RLC $-6 \times Q$ dB/octave slope (equation 27a) to the RC circuit -6dB/octave slope (equation 27b).

$$(27a) \quad T_{RLC}(p) = \frac{R}{R + pL + \frac{1}{pC}} \approx \frac{R}{pL} = \frac{\omega_0}{p} \frac{R}{\omega_0 L} = \frac{\omega_0}{Qp}$$

$$(27b) \quad T_{RC}(p) = \frac{\frac{1}{pC}}{R + \frac{1}{pC}} \approx \frac{1}{pCR} = \frac{\omega_0}{p}$$

Figure 60711 comparison of RC and RLC Selectivity

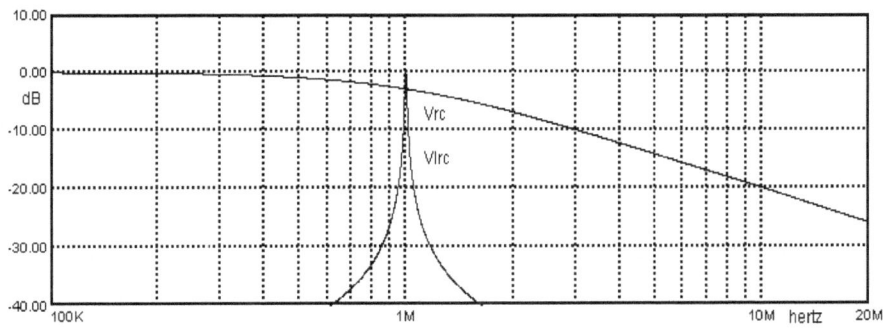

More Tricks of the trade - Impedance The RLC series impedance plot over a frequency range (Figure 60621) for Q values 0.1 (R=1000Ω), 1 (R=100Ω), and 10 (R=10Ω) is produced by Spice by calculating current that is inversely proportional to impedance. Then plotting *negative* dB produces the reciprocal that is the impedance.

Problem 606 Show that the impedance magnitude equals R√2 and R=±X(ω_0) at the "3dB" frequencies ω_1 and ω_2.

Problem 607 Start from equations 24a and derive equations 24c. Show that equations 25 are solutions to equations 24c.

Spice program for the Series RLC Circuit

```
Fig6061.ckt
V1 1 0 AC 1 0 ; volts
C1 1 2 1000p
L1 2 3 20u
R1 3 0 20
.AC DEC 200 100000 1e+008
.PLOT AC VDB(1) VDB(2) VDB(3) -30,20
.TEMP 27
.end
```

Write the program   (Spice is not case sensitive)

Fig6061.ckt

> *First line* The first line of every program is assumed by Spice to be a title statement. The title statement can include any words.

V1 1 0 AC 1 0

> *Voltage source*   Signal generator v_1 is connected to nodes 1 and 0. The v_1 signal is a AC sinewave with 1 volt magnitude and 0 phase.

C1 1 2 1000p
L1 2 3 20u
R1 3 0 20

> *Circuit components* Capacitor C_1 is connected to nodes 1 and 2. Inductor L_1 is connected to nodes 2 and 3, while resistor R_1 is connected to nodes 3 and 0.

.AC DEC 200 100000 1e+008

> *AC control statement* Dot AC DEC defines a frequency range from 10^5 Hz to 10^8 Hz with points to be calculated every 200Hz.

.PLOT AC VDB(1) VDB(2) VDB(3) -30,20

> *Plot the data* dot plot AC executes the dot AC statement calculating the v_1, v_2, v_3 from 10^5Hz to 10^8Hz (v_1 is defined as a one volt sinewave for all frequencies).

.TEMP 27

> *Temperature* Dot temp (.temp) defines temperature as 27 degrees C.

.end

> *Last line* The last line of every program is .end (dot end).

6 Frequency Response

Spice program 6061 Calculates Transfer Function

```
Fig6061.ckt     series RLC

V1 1 0  AC 1  0  ; volts

R1 3 0 20
L1 2 3 20u
C1 1 2 1000p

*.PLOT AC VDB(1) VDB(2) VDB(3) -30,20
.AC DEC 200 100000 1e+008
.TEMP 27
.PLOT AC VP(1) VP(2) VP(3) -90,180
.end
```

Figure 60611 Series RLC Transfer Functions v_2/v_1, v_3/v_1, Magnitude

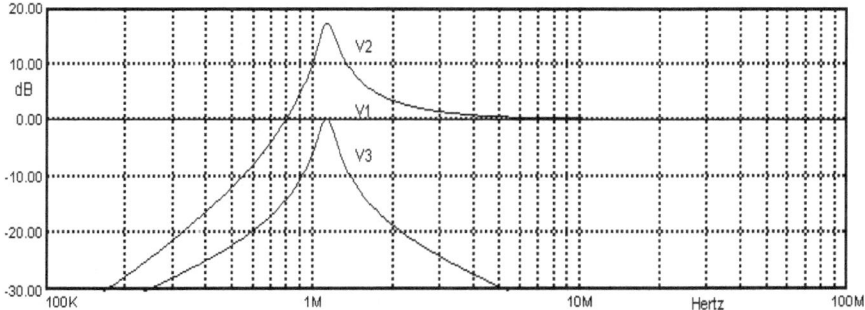

Figure 60612 Series RLC Transfer Functions v_2/v_1, v_3/v_1, Phase

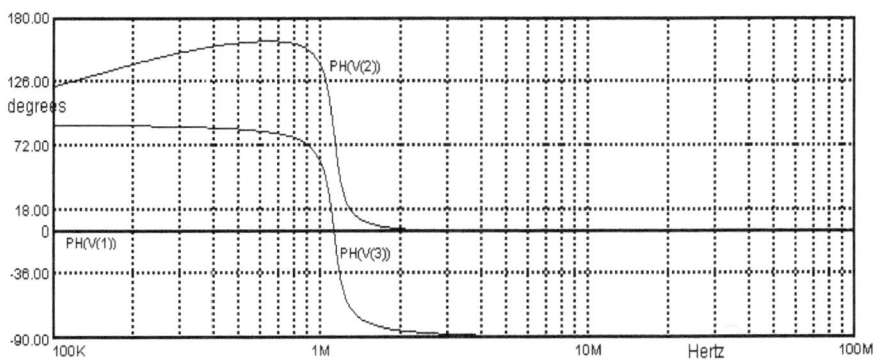

Electric Circuits - Analysis and Design

Spice program 6062 Calculates Currents (i=v/z)

```
Fig6062.ckt series RLC
V1  1  0 AC 1 0   ; volts
V11 11 0 AC 1 0   ; volts
V21 21 0 AC 1 0   ; volts
* names/nodes/values
R1  3  0  10
L1  2  3  15.9u
C1  1  2  1590p
R11 13 0  100
L11 12 13 15.9u
C11 11 12 1590p
R21 23 0  1000
L21 22 23 15.9u
C21 21 22 1590p
.AC DEC 200 10000 1e+008
.TEMP 27
.PLOT AC IDB(V1) IDB(V11) IDB(V21) -20,-70
.end
```

Figure 60621 RCL impedance |Z| for various R values (Q implied)

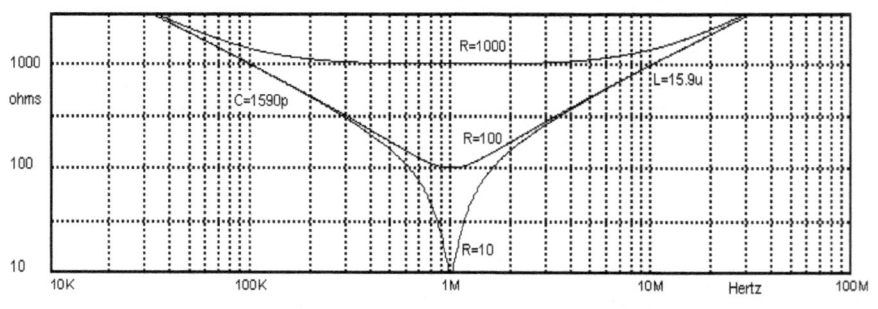

Problem 608 Reference Figure P609. Calculate T_{BP} for $Q_p=1$, 10, and 50

$$T_{BP}(\lambda) = \frac{1}{Q_p} \frac{\lambda}{1+\frac{1}{Q_p}\lambda+\lambda^2} \quad Q_p = \frac{R}{\omega_0 L}, \ \omega_0^2 = \frac{1}{LC}, \ \lambda = \frac{p}{\omega_0}, \ T(p) = \frac{v_2}{v_1}$$

Problem 609 Reference Figure P605. Write a Spice program and plot figures for transmission magnitude and phase analogous to the RLC Spice program Fig6061.Ckt.

Figure P609 Band Pass

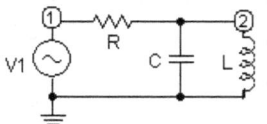

7 Transient Response

Transient response of electric circuits requires solution of ordinary and partial differential equations

The Laplace Transform is a terrific tool for solving ordinary and partial differential equations, because it produces the frequency response equations, including initial conditions, as well as the transient response equations. The Laplace transform transforms the ordinary differential equations of electric circuits into elementary algebraic equations. The algebraic equations are manipulated to solve for the variables of interest. Then the manipulated equations are inverse transformed back to the time domain *as a solution* of the original problem represented by the differential equations.

The time domain equations of a resistor-only circuit are algebraic. Resistors do not store energy, and so there are no transient states that dissipate stored energy in a resistor-only circuit. There are no initial conditions, because there are no constants of integration.

However, when energy storing capacitors and, inductors are included in a circuit everything changes.

Circuit equation (1) of an RLC circuit (Figure 701) is an integro-differential equation. Terms in the time domain equation are no longer exclusively algebraic; some terms may include differentials and others may include integrals.

Figure 701 Series RLC circuit

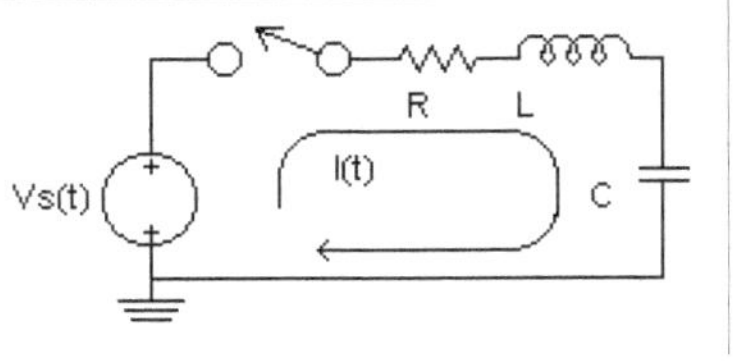

(1) $$v_S(t) = Ri(t) + L\frac{di}{dt} + \frac{1}{C}\int_0^t i(x)dx$$

Two initial conditions at time t=0 are possible, because there are two constants of integration. Inductor L stores energy when a current flows through it. Capacitor C stores energy when a voltage is across it. The $v_S(t)$ waveform is not restricted. You are faced with the apparently formidable problem of finding solutions for simultaneous sets of such equations, which the Laplace transform provides an amazingly easy way to solve.

7.1 The Laplace Transform

Let $f(t)$ be a real or complex valued function of time t for t>0, and $p=\sigma+j\omega$ be a complex variable used as a parameter. Then the Laplace transform of $f(t)$ is defined as

(2) $\quad F(p) = \mathcal{L}[f(t)] = \int_0^\infty f(t) e^{-pt} dt$

The symbol $\mathcal{L}$ is the Laplace transformation *operator* that invokes the Laplace transform. Equations in the *time domain variable t* are transformed into equations in the *complex frequency domain variable p*.

Inverse transform If we use the Laplace transform to solve a problem in the p domain, then we need an inverse transform to return to the time domain.

(3) $\quad f(t) = \mathcal{L}^{-1}[F(p)] = \dfrac{1}{2\pi i} \int_{\sigma-j\infty}^{\sigma+j\infty} F(p) e^{tp} dp$

A return from the *complex frequency domain p* to the time domain t is achieved by performing the inverse operation. The operation is known as the Inverse Laplace Transform (equation 3).

This integral for calculating the inverse transform is a "contour integration in the p plane." This process is part of the mathematical theory of functions of complex variables that is presented in mathematics texts.[1]

We use a well known trick that allows us to avoid calculating this formidable integral for almost all problems we might ever encounter.

The trick is straightforward. Expand the algebraic solution into a sum of partial fractions (Appendix A1). Then use the inverse of the transform of the exponential.

(4) $\quad \mathcal{L}[e^{-at}] = \dfrac{1}{p+a} \quad \Rightarrow \quad e^{-at} \Leftrightarrow \dfrac{1}{p+a}$

---

[1] Joel L. Shiff, *The Laplace Transform*, ISBN 0387 986 987

7.2 Transforms Simplify Functions

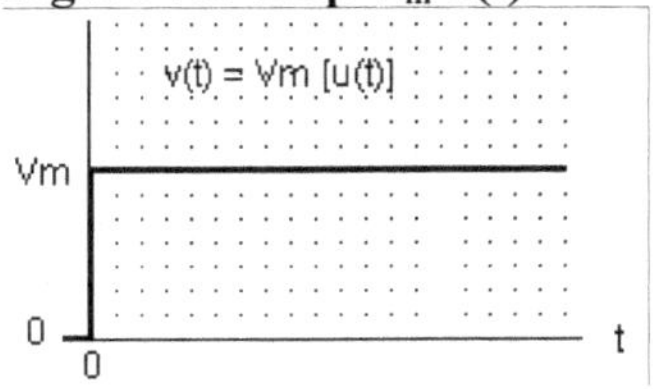

Many complicated functions of real variable t directly transform into elementary functions of a complex variable p.

Figure 703 Step V_m u(t)

Transform of the step function u(t) The unit step function (Figure 703,) transforms to the algebraic expression $1/p$.

$$u(t) = 1 \quad \text{for all } t$$

$$F(p) = \mathscr{L}[u(t)] = \int_0^\infty u(t)e^{-pt}\,dt = \int_0^\infty 1 \times e^{-pt}\,dt = \left.\frac{e^{-pt}}{-p}\right|_0^\infty = \frac{0}{-p} - \frac{e^{-0}}{-p} = \frac{1}{p}$$

where $e^{-\infty} = 0$ means $\lim_{t \to \infty} e^{-pt} = 0$

(5) $u(t) \Leftrightarrow \dfrac{1}{p}$

You can also read the u(t) transform pair as "one over p is the Laplace transform of one", or "one is the Inverse Laplace transform of one over p".

Transform of exp(–at) Most transcendental functions transform into algebraic expressions. Consider the exponential function.

$$F(p) = \mathscr{L}[e^{-at}u(t)] = \int_0^\infty e^{-at}u(t)e^{-pt}\,dt = \int_0^\infty e^{-(p+a)t}\,dt = \left.\frac{e^{-(p+a)t}}{-(p+a)}\right|_0^\infty$$

$$= \frac{0}{-(p+a)} - \frac{e^{-0}}{-(p+a)} = \frac{1}{p+a}$$

(6) $e^{-at}u(t) \Leftrightarrow \dfrac{1}{p+a}$

The transform of u(t) follows from this result by letting a = 0. This is a useful method for generating other transform pairs.

Problem 701 From the definition of the Laplace transform calculate
a) $F(p) = \mathscr{L}[f(t)]$ for $f(t) = 3t$
b) $F(p) = \mathscr{L}[f(t)]$ for $f(t) = 7te^{-3t}$
c) $F(p) = \mathscr{L}[f(t)]$ for $f(t) = \cosh \omega t$
d) $F(p) = \mathscr{L}[f(t)]$ for $f(t) = t + e^{-at}$

Electric Circuits - Analysis and Design

Transforms of damped sin ωt and cos ωt The sin and cos functions in exponential format reveal that the transform of an exponential can be used as a shortcut.

Damped sine function

$$F(p) = \mathcal{L}[e^{-\sigma t} \sin \omega t \times u(t)]$$

$$F(p) = \int_0^\infty \frac{e^{-(\sigma-i\omega)t} - e^{-(\sigma+i\omega)t}}{2i} u(t) e^{-pt} dt = \int_0^\infty \frac{e^{-(p+\sigma-i\omega)t} - e^{-(p+\sigma+i\omega)t}}{2i} dt$$

$$F(p) = \frac{1}{2i}\left(\frac{1}{p+\sigma-i\omega} - \frac{1}{p+\sigma+i\omega}\right) = \frac{\omega}{(p+\sigma)^2 + \omega^2}$$

(7) $\quad e^{-\sigma t} \sin \omega t \times u(t) \Leftrightarrow \dfrac{\omega}{(p+\sigma)^2 + \omega^2}$

Damped cosine function

$$F(p) = \mathcal{L}[e^{-\sigma t} \cos \omega t \times u(t)]$$

$$F(p) = \int_0^\infty \frac{e^{-(\sigma-i\omega)t} + e^{-(\sigma+i\omega)t}}{2} u(t) e^{-pt} dt = \int_0^\infty \frac{e^{-(p+\sigma-i\omega)t} + e^{-(p+\sigma+i\omega)t}}{2} dt$$

$$F(p) = \frac{1}{2}\left(\frac{1}{p+\sigma-i\omega} + \frac{1}{p+\sigma+i\omega}\right) = \frac{p+\sigma}{(p+\sigma)^2 + \omega^2}$$

(8) $\quad e^{-\sigma t} \cos \omega t \times u(t) \Leftrightarrow \dfrac{p+\sigma}{(p+\sigma)^2 + \omega^2}$

Transform of the ramp function We have no shortcuts for the ramp function integration. Here we integrate by parts.

$$\text{ramp}: \quad F(p) = \mathcal{L}[\frac{t}{T} u(t)] = \int_0^\infty \frac{t}{T} u(t) e^{-pt} dt = \frac{1}{T}\int_0^\infty t \times e^{-pt} dt = \frac{1}{T}\int_0^\infty u\, dv$$

let $\quad u = t, \; dv = e^{-pt} dt \quad \Rightarrow \quad du = dt, \; v = -\dfrac{1}{p} e^{-pt}$

$$F(p) = \frac{1}{T}\int_0^\infty u\, dv = \frac{1}{T} uv \Big|_0^\infty - \frac{1}{T}\int_0^\infty v\, du$$

$$F(p) = -\frac{1}{T} t \frac{1}{p} e^{-pt}\Big|_0^\infty + \frac{1}{pT}\int_0^\infty e^{-pt} dt = 0 + \frac{1}{pT} \frac{-1}{p} e^{-pt}\Big|_0^\infty = \frac{1}{T}\frac{1}{p^2}$$

(9) $\quad t\, u(t) \Leftrightarrow \dfrac{1}{p^2}$

7.3 Transforms Simplify Operations

The vi constraints for R, L, and C components include multiplication by a constant, differentiation, and integration. We need the Laplace transforms of these operations.[2] We use integration by parts to find the transform of a derivative and an integral. However, first we show that multiplication by a constant carries over from the real domain to the complex domain. This is relevant, because constants such as R, L, and C are coefficients of terms.

Transform of multiplication by a constant

if $f(t) \Leftrightarrow F(p)$, then

$$\mathscr{L}[K \cdot f(t)] = \int_0^\infty K \cdot f(t) e^{-pt} dt = K \cdot \int_0^\infty f(t) e^{-pt} dt = K \cdot F(p)$$

(10) $\quad Kf(t) \Leftrightarrow KF(p)$

Transform of a first derivative

if $f(t) \Leftrightarrow F(p)$ and $\dfrac{df(t)}{dt} u(t) \Leftrightarrow F_1(p)$, then

$$F_1(p) = \mathscr{L}\left[\dfrac{df(t)}{dt} u(t)\right] = \int_0^\infty \dfrac{df(t)}{dt} u(t) e^{-pt} dt = \int_0^\infty \dfrac{df(t)}{dt} e^{-pt} dt$$

if $u = e^{-pt}$, $dv = \dfrac{df(t)}{dt} dt$ then $du = -pe^{-pt} dt \quad v = f(t)$

integrating by parts: $F_1(p) = \int_0^\infty u\,dv = uv\Big|_0^\infty - \int_0^\infty v\,du$

$$F_1(p) = e^{-pt} f(t)\Big|_0^\infty - \int_0^\infty f(t)(-p) e^{-pt} dt$$

$$= [e^{-p\infty} f(\infty) - e^{-p0} f(0)] + p \int_0^\infty f(t) e^{-pt} dt$$

$$= [0 - f(0)] + pF(p) = pF(p) - f(0)$$

(11) $\quad \dfrac{df(t)}{dt} u(t) \Leftrightarrow pF(p) - f(0)$

---

[2] Compare to Appendix A4 Easy Method for Evaluating Laplace Transforms

Transform of an integral

if $f(t) \Leftrightarrow F(p)$ and $\int_0^t f(x)dx \Leftrightarrow F_1(p)$, then

$$F_1(p) = \mathcal{L}[\int_0^t f(x)dx \; u(t)] = \int_0^\infty u(t) \int_0^t f(x)dx \; e^{-pt} dt = \int_0^\infty \int_0^t f(x)dx \; e^{-pt} dt$$

if $u = \int_0^t f(x)dx$ and $dv = e^{-pt}dt$ then $du = f(t)dt$ and $v = -\frac{1}{p}e^{-pt}$

$$F_1(p) = \int_0^\infty u\,dv = vu\Big|_0^\infty - \int_0^\infty v\,du$$

$$F_1(p) = \left[-\frac{1}{p}e^{-pt}\int_0^t f(x)dx\right]_0^\infty - \int_0^\infty \left(-\frac{1}{p}\right)f(t)e^{-pt}dt$$

$$= \left(-\frac{1}{p}e^{-p\infty}\int_0^\infty f(x)dx\right) - \left(-\frac{1}{p}e^{-p\times 0}\int_0^0 f(x)dx\right) + \frac{1}{p}\int_0^\infty f(t)e^{-pt}dt$$

$$= \left(-\frac{1}{p}\times 0 \int_0^\infty f(x)dx\right) - \left(-\frac{1}{p}\times 1 \times 0\right) + \frac{1}{p}F(p) = 0 - 0 + \frac{F(p)}{p}$$

(12) $\quad u(t)\int_0^t f(x)dx \Leftrightarrow \frac{F(p)}{p}$

Transform of a time delay c (translation in time)

If $F(p) = \int_0^\infty f(t)u(t)e^{-pt}dt$, then let $F_c(p) = \int_0^\infty f(t-c)u(t-c)e^{-pt}dt$

let $\quad \tau = t - c$

$$F_c(p) = \int_0^\infty f(\tau)u(\tau)e^{-p(\tau+c)}d\tau = e^{-pc}\int_0^\infty f(\tau)u(\tau)e^{-p\tau}d\tau = e^{-pc}F(p)$$

If $f(t)u(t) \Leftrightarrow F(p)$ then

(13) $\quad f(t-c)u(t-c) \Leftrightarrow e^{-pc}F(p)$

Problem 702 Use the Laplace transform to solve for i(t) when initial conditions equal zero.

$$L\frac{di}{dt} + Ri + \frac{1}{C}\int_0^t i(t)dt = V_m \sin \omega t$$

7.4 RL Circuit Transient Response[3]

The Laplace transform operator is represented by the script letter $\mathcal{L}$. If the Laplace transform of f(t) is F(p) then F(p) is calculated from the integral transform (equation 2).

Figure 702 Series RL

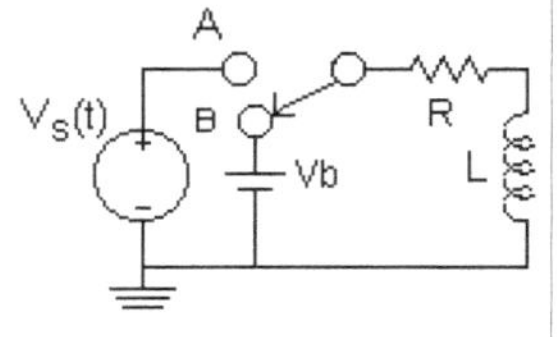

Define the Laplace Transform of any function f(t), from the time domain t to the complex frequency domain p, as a function F(p).

(2) $\quad F(p) = \mathcal{L}[f(t)] = \int_{0}^{\infty} f(t)e^{-pt} dt$

The Laplace transform operator is implemented by multiplying each term in equation 14 (Figure 702) by exp(–pt), and integrating from 0 to ∞.

(14) $\quad v_S(t) = Ri(t) + L\dfrac{di}{dt} \qquad i(0) = \dfrac{V_b}{R}$

(15) $\quad e^{-pt} v_S(t) = e^{-pt} Ri(t) + e^{-pt} L \dfrac{di}{dt}$

(16) $\quad \int_{0}^{\infty} e^{-pt} v_S(t) dt = \int_{0}^{\infty} e^{-pt} Ri(t) dt + \int_{0}^{\infty} e^{-pt} L \dfrac{di}{dt} dt$

We say the Laplace operator transforms $v_s(t)$ into $V_s(p)$, and i(t) into I(p).

(17) $\quad define \ V_s(p) = \int_{0}^{\infty} e^{-pt} v_S(t) dt \qquad I(p) = \int_{0}^{\infty} e^{-pt} i(t) dt$

Then the equation 14 transforms into equation 18 by applying the theorem for differentials (Section 7.3).

(18) $\quad V_S(p) = RI(p) + pLI(p) - Li(0)$
(19) $\quad V_S(p) = (R + pL)I(p) - Li(0)$

Now we can solve for I(p).

(20) $\quad I(p) = \dfrac{1}{(R + pL)}(V_S(p) + Li(0))$

Where i(0) is the current in L at time t=0 with the switch in position B.

---

[3] Recommendation: before you start - read A3 Oliver Heaviside's Method

Electric Circuits - Analysis and Design

Select a source We cannot proceed until we specify the $v_S(t)$ waveform, and transform it into $V_S(p)$. We arbitrarily select a sinewave as the forcing function. The source $v_S(t)$ is applied to the circuit at time t=0, because u(t) in effect moves the switch arm from position B to A at time t=0. Note how the exponential simplifies the math. Do not use trig functions sin and cos! (see superposition in 7.8)

if $v_S(t) = V_M e^{j\omega t} u(t)$, then $V_S(p) = \dfrac{V_M}{p - j\omega}$ and

(21) $I(p) = \dfrac{V_S(p)}{R + pL} + \dfrac{Li(0)}{R + pL} = \dfrac{1}{L} \cdot \dfrac{1}{p + \dfrac{R}{L}} \cdot \dfrac{V_M}{p - j\omega} + \dfrac{i(0)}{p + \dfrac{R}{L}}$

Make a partial fraction expansion (Appendix A1) We need a partial fraction expansion, because the trick we use is based on the fact the inverse transform of $1/(p+a)$ is an exponential function (equation 4).

(4) $e^{-at} \Leftrightarrow \dfrac{1}{p + a}$

(22) $I(p) = \dfrac{1}{L} \cdot \dfrac{1}{p + \dfrac{R}{L}} \cdot \dfrac{V_m}{p - j\omega} + \dfrac{i(0)}{p + \dfrac{R}{L}}$

$I(p) = \dfrac{1}{L} \cdot \dfrac{1}{j\omega + \dfrac{R}{L}} \cdot \dfrac{V_m}{p - j\omega} + \dfrac{1}{L} \cdot \dfrac{1}{p + \dfrac{R}{L}} \cdot \dfrac{V_m}{-\dfrac{R}{L} - j\omega} + \dfrac{i(0)}{p + \dfrac{R}{L}}$

$I(p) = \dfrac{V_m}{R + j\omega L} \cdot \dfrac{1}{p - j\omega} - \dfrac{V_m}{R + j\omega L} \cdot \dfrac{1}{p + \dfrac{R}{L}} + \dfrac{V_b}{R} \cdot \dfrac{1}{p + \dfrac{R}{L}}$

Inverse transform A return from the *complex frequency domain* p to the time domain t is achieved by using the trick. The trick allows us to avoid calculating the inverse integral (equation 3). I(p) transforms to i(t) when we substitute an e^{-at} for each of the three $1/(p+a)$ terms. The complete solution consists of a steady-state term, a transient term due to the forcing function, and a transient term due to the natural response to the initial condition $i(0)=V_b/R$.

(23) $i(t) = \dfrac{V_m}{R + j\omega L} e^{j\omega t} - \dfrac{V_m}{R + j\omega L} e^{-\frac{R}{L}t} + \dfrac{V_b}{R} e^{-\frac{R}{L}t}$

$\boxed{\text{Time constant } \tau = L/R = 1/\omega_0.}$

112

7 Transient Response

Laplace Transforms of vi Constraints

Time Domain:

component	R	L	C
voltage $v(t) =$	$Ri(t)$	$L\dfrac{di(t)}{dt}$	$\dfrac{1}{C}\displaystyle\int_0^t i(x)dx + v(0)$
current $i(t) =$	$\dfrac{1}{R}v(t)$	$\dfrac{1}{L}\displaystyle\int_0^t v(x)dx + i(0)$	$C\dfrac{dv(t)}{dt}$

Complex Frequency Domain: The responses i(t) or v(t) are not known. How do we know if the equations are Laplace transformable? *We do not know, however we proceed on the assumption they are Laplace transformable.*

Assume: $v(t) \Leftrightarrow V(p)$ $i(t) \Leftrightarrow I(p)$

component	R	L	C
voltage $V(p) =$	$RI(p)$	$L[pI(p) - i(0)]$	$\dfrac{I(p)}{pC} + \dfrac{v(0)}{p}$
current $I(p) =$	$\dfrac{V(p)}{R}$	$\dfrac{V(p)}{pL} + \dfrac{i(0)}{p}$	$C[pV(p) - v(0)]$

Here they are again written as voltage and current transform pairs: f(t) ⇔ F(p).

Voltage Transforms

$v(t) \Leftrightarrow V(p)$

$Ri(t) \Leftrightarrow RI(p)$

$L\dfrac{di(t)}{dt} \Leftrightarrow L[pI(p) - i(0)]$

$\dfrac{1}{C}\displaystyle\int_0^t i(x)dx + v(0) \Leftrightarrow \dfrac{I(p)}{pC} + \dfrac{v(0)}{p}$

Current Transforms

$i(t) \Leftrightarrow I(p)$

$\dfrac{v(t)}{R} \Leftrightarrow \dfrac{V(p)}{R}$

$\dfrac{1}{L}\displaystyle\int_0^t v(x)dx + i(0) \Leftrightarrow \dfrac{V(p)}{pL} + \dfrac{i(0)}{p}$

$C\dfrac{dv(t)}{dt} \Leftrightarrow C[pV(p) - v(0)]$

Electric Circuits - Analysis and Design

The Exponential Function The famous mathematician Leonhard Euler (1707-1783) defined the exponential function when he proved that sum of a power series solution to a first order differential equation is his number e raised to the $-\alpha x$ power ($e^{-\alpha x}$). He also proved the important property that the derivative of $e^{-\alpha x}$ is $-\alpha e^{-\alpha x}$. Consider a parallel RC circuit.

$$0 = Gv(t) + C\frac{dv(t)}{dt} \quad \Rightarrow \quad -\frac{1}{RC}dt = \frac{dv}{v}$$

Integrate both sides using x as a dummy variable.

$$-\frac{1}{RC}\int_0^t dx = \int_0^t \frac{dv(x)}{v(x)} \rightarrow -\frac{1}{RC}(t-0) = \ln v(t) - \ln v(0)$$

$$-\frac{1}{RC}t = \ln\frac{v(t)}{v(0)} \quad \Rightarrow \quad v(t) = v(0)e^{-\frac{t}{RC}}$$

Plots of e^{-x} and $1 - e^{-x}$

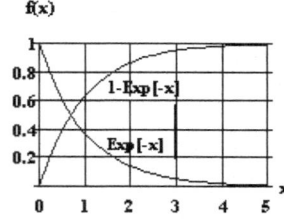
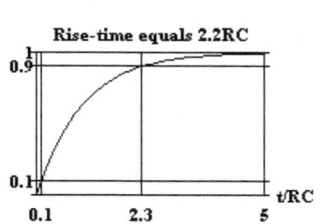

Step function u(t) (Figure 703) The step function is the implied forcing function when a switch is closed or opened.

Definition of the unit step function

$$u(t) = \begin{cases} 0 & (t < 0) \\ \text{undefined}(t = 0) \\ 1 & (t > 0) \end{cases}$$

Time Constant The step function transient response of circuits that reduce to one R and one L, or one R and one C takes the form of the exponential function $\exp(-x)$ and $1 - \exp(-x)$ where

$x = \dfrac{t}{\tau}$ and the time constant $\tau = \dfrac{L}{R}$ or $\tau = RC$

7.5 RC Circuit Transient Response

We examine the series RC circuit response to *step function u(t)*.

The step function $v_1(t) = V_m u(t)$ (Figure 703) is in effect an "infinite" dv/dt at time t=0. Since $i_C = C dv/dt$ there must be an infinite current through the capacitor (Figure 704). In fact, the capacitor current is limited to V_m/R. This leaves V_m volts across R, and zero voltage across C that is another way to see that C's voltage starts from zero. *In other words, the capacitor C initially responds to a step function at t=0 as if C is a short circuit.*

Figure 703 Step V_m u(t)

As the current builds the di/dt decreases to zero exponentially as time elapses so that in the final analysis all of the voltage V_m is across C (Figure 70411).

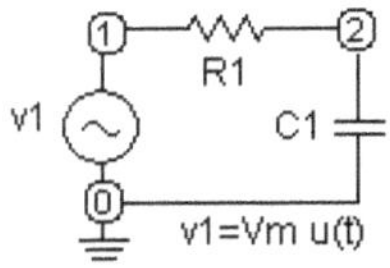

The Laplace transform analysis shows the response to the step function u(t) waveform is a current with an e^{-x} waveform. The current starts at V_m/R_1 and decreases to 0, because C is charged to V_m volts after the transient expires. We go directly to the transformed mesh equation where R and 1/pC are the impedances.

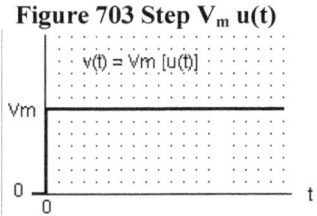

Figure 704

v1=Vm u(t)

(24) $V_1(p) = RI(p) + \dfrac{1}{pC} I(p) = \left(R + \dfrac{1}{pC} \right) I(p) \quad \text{let } v(0^+) = 0$

(25) $I(p) = \dfrac{1}{R + \dfrac{1}{pC}} V_1(p) = \dfrac{1}{1 + \dfrac{1}{pRC}} \dfrac{V_m}{pR} = \dfrac{1}{p + \dfrac{1}{RC}} \dfrac{V_m}{R}$

(26) $i(t) = \dfrac{V_m}{R} e^{-at} u(t) \quad \text{where } a = \dfrac{1}{RC}$

In this RC circuit the pole (Section 7.9) is at −1/RC, and the exponential is exp(−t/RC).

The frequency ω, time t relationship is

> Time constant τ = RC = 1/ω₀.

Electric Circuits - Analysis and Design

As a check we calculate R and C voltages, which add to V_m.

(27) $v_C(t) = \dfrac{1}{C}\int_0^t i(x)dx = \dfrac{1}{C}\cdot\dfrac{V_m}{R}\int_0^t e^{-ax}dx = \dfrac{V_m}{RC}\dfrac{-1}{a}e^{-ax}\Big|_0^t = V_m\left(1-e^{-at}\right)$

(28) $v_R(t) = iR = \dfrac{V_m}{R}\operatorname{Re}^{-at} = V_m e^{-at}$

(29) $v_R(t) + v_C(t) = V_m e^{-at} + V_m\left(1 - e^{-at}\right) = V_m$

The capacitor voltage has the $1-e^{-x}$ waveform, and the resistor voltage has the e^{-x} waveform. Their sum is 1. The circuit time constant is $\tau = RC$.

Spice program Calculates the RC Transient Response

```
Fig7041.ckt series RC circuit step
*PULSE( Vbase Vmax Tdelay Trise Tfall Twidth Tperiod )
V1 1 0 Pulse(0 0.9 0.2n 0n 0n 50n 100n)
R1 1 2 1000
C1 2 0 100f
.TRAN 1e-012 1e-009 0
.PLOT TRAN V(1) V(2) 0,1
.TEMP 27
.end
```

Figure 70411 Series RC circuit transient response to a step function

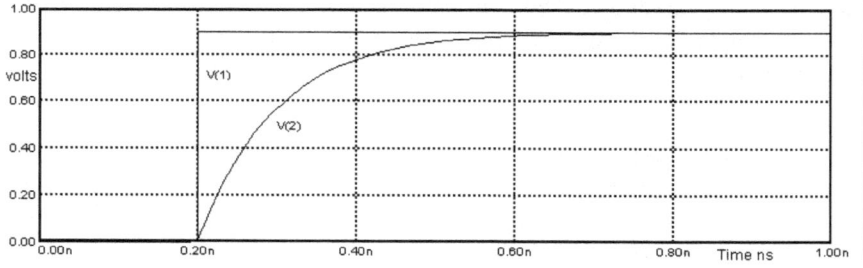

Figure P703

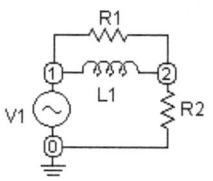

Figure P704

Problem 703 Reference Figure P703. Write Spice program FigP703.ckt. Plot the v_2 transient response.

Problem 704 Reference Figure P704. Write Spice program FigP704.ckt. Plot the v_2 transient response.

116

7 Transient Response

The Spice program for the RC circuit.

```
Fig7041.ckt series RC circuit step
*PULSE( Vbase Vmax Tdelay Trise Tfall Twidth Tperiod )
V1 1 0 Pulse(0 0.9 0.2n 0n 0n 50n 100n)
R1 1 2 1000
C1 2 0 100f
.TRAN 1e-012 1e-009 0
.PLOT TRAN V(1) V(2) 0,1
.TEMP 27
.end
```

Figure 704

Write the program (Spice is not case sensitive)

`Fig7041.ckt series RC circuit step`
 First line The first line of every program is assumed by Spice to be a title statement. The title statement can include any words.

The asterisk * defines a line as a comment.
```
* PULSE(Vbase Vmax Tdelay Trise Tfall Twidth Tperiod)
V1 1 0 Pulse(0    0.9  0.2n   0n    0n    50n   100n)
```
 Voltage source Pulse generator v_1 is connected to nodes 1 and 0. The V1 pulse is 50ns wide so that a 1ns plot (Figure 70411) shows only the initial step.

```
R1 1 2 1000
C1 2 0 100f
```
 Circuit components Resistor R_1 is connected to nodes 1 and 2. Capacitor C_1 is connected to nodes 2 and 0 ($\tau = RC = 0.1$ns).

`.TRAN 1e-012 1e-009 0`
 TRAN control statement Dot TRAN defines a time range from 0 to 1ns (`1e-009`) that is incremented every 1ps (`1e-012`).

`.PLOT TRAN V(1) V(2) 0,1`
 Plot the data dot plot TRAN executes the dot TRAN statement calculating the v_2 from 0ns to 1ns every 1ps (v_1 is a 0.9 volt step).

`.TEMP 27`
 Temperature Dot temp (.temp) defines temperature as 27 degrees C.

`.end`
 Last line The last line of every program is .end (dot end).

117

7.6 RLC Circuit Transient Response

RLC circuit responses vary according to the circuit damping or Q (there may be more than one Q in complex circuits). The responses take the forms $e^{-\alpha t}+e^{-\beta t}$, $e^{-\alpha t}$, or $e^{-\alpha t}\sin \beta t$. The cases are presented by analyzing a series RLC circuit with one each R, L, and C in series (Figure 705). Assume the initial stored energy is zero.

Figure 705 RLC Circuit

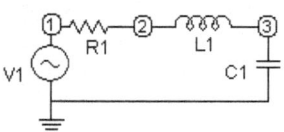

(30) $\quad v_1(t) = v_L + v_R + v_C = L\dfrac{di(t)}{dt} + Ri(t) + \dfrac{1}{C}\int_0^t i(x)dx$

(31) $\quad V_1(p) = \left[pL + R + \dfrac{1}{pC}\right]I(p) \quad [v(0)=0, i(0)=0]$

(32) $\quad I(p) = \dfrac{1}{pL + R + \dfrac{1}{pC}} V_1(p) \quad \Rightarrow \quad Z(p) = pL + R + \dfrac{1}{pC}$

(33) $\quad \text{if } v_1(t) = V_m u(t) \text{ then } V_1(p) = \dfrac{V_m}{p}$

(34) $\quad I(p) = \dfrac{1}{pL + R + \dfrac{1}{pC}} \cdot \dfrac{V_m}{p} = \dfrac{1}{p^2 + p\dfrac{R}{L} + \dfrac{1}{LC}} \cdot \dfrac{V_m}{L} = \dfrac{V_m}{L}\dfrac{1}{(p+p_1)(p+p_2)}$

(35) where $p_1, p_2 = -\alpha \mp \beta = -\dfrac{R}{2L}\left[1\pm\sqrt{1-\dfrac{4L}{R^2C}}\right] = -\dfrac{\omega_0}{2Q_s}\left[1\pm\sqrt{1-4Q_s^2}\right]$

There are three cases. We select the under damped case where Q>½.

(36) $\quad I(p) = \dfrac{V_m}{L}\dfrac{1}{(p+p_1)(p+p_2)} = \dfrac{V_m}{L}\dfrac{1}{p_1-p_2}\left(\dfrac{1}{p+p_2} - \dfrac{1}{p+p_1}\right)$

$= \dfrac{V_m}{L}\dfrac{1}{2j\gamma}\left(\dfrac{1}{p+\alpha-j\gamma} - \dfrac{1}{p+\alpha+j\gamma}\right) \quad \text{where } \beta = j\gamma$

(37) $\quad i(t) = \dfrac{V_m}{L}\dfrac{1}{2j\gamma}\left(e^{-(\alpha-j\gamma)t} - e^{-(\alpha+j\gamma)t}\right) = \dfrac{V_m}{L}\dfrac{1}{2j\gamma}e^{-\alpha t}\left(e^{j\gamma t} - e^{-j\gamma t}\right)$

$= \dfrac{V_m}{\gamma L}e^{-\alpha t}\sin \gamma t$

7 Transient Response

Spice program Calculates the RLC Transient Response

```
Fig7051.ckt series RLC

* The trick to viewing responses for various Q's is to enter
* one circuit for each Q value.

V 1 0 PULSE(0 2.5 0 0 0 1000u 2000u)

R1 1 2 10    ; Q=10 under damped
L1 2 3 15.9u
C1 3 0 1590p

R11 1 12 100   ; Q=1 close to critical
L11 12 13 15.9u
C11 13 0 1590p

R21 1 22 1000 ; Q=0.1 over damped
L21 22 23 15.9u
C21 23 0 1590p

.TRAN 1e-009 1e-005 0 1e-008
.TEMP 27
.PLOT TRAN V(3) V(13) V(23) 0,5
.end
```

Figure 70511 Series RLC Transient Response

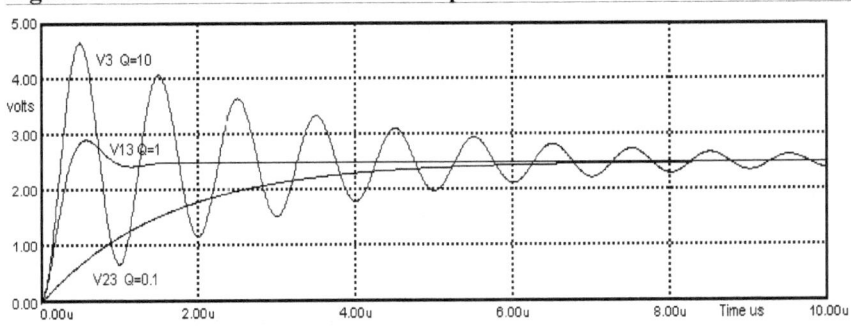

Figure P705 Band pass

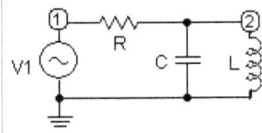

Problem 705 Reference Figure P705. Write Spice program Figp7051.ckt. Plot the transient response when Q=5.

119

Spice program for the Series RLC Circuit

```
Fig7051.ckt    series RLC
V 1 0 PULSE(0 2.5 0 0 0 1000u 2000u)

R1 1 2 10    ; Q=10 under
L1 2 3 15.9u
C1 3 0 1590p

.TRAN 1e-009 1e-005 0 1e-008
.TEMP 27
.PLOT TRAN V(3) 0,5
.end
```

Write the program   (Spice is not case sensitive)

```
Fig7051.ckt    series RLC
```
    First line The first line of every program is assumed by Spice to be a title statement. The title statement can include any words.

```
V 1 0 PULSE(0 2.5 0 0 0 1000u 2000u)
```
    Voltage source Pulse generator v_1 is connected to nodes 1 and 0.

```
R1 1 2 10    ; Q=10 under
L1 2 3 15.9u
C1 3 0 1590p
```
    Circuit components Resistor R_1 is connected to nodes 1 and 2. Inductor L_1 is connected to nodes 2 and 3. Capacitor C_1 is connected to nodes 3 and 0.

```
.TRAN 1e-009 1e-005 0 1e-008
```
    TRAN control statement Dot TRAN defines a time range from 0 to 10μs that is calculated every 1ns. Start time is 0s.

```
.PLOT TRAN V(3) 0,5
```
    Plot the data dot plot TRAN executes the dot TRAN statement calculating v_3 from 0ns to 10μs (`1e-005`) every 1ns (`1e-009`).

```
.TEMP 27
```
    Temperature Dot temp (.temp) defines temperature as 27 degrees C.

```
.end
```
    Last line The last line of every program is .end (dot end).

7.7 Transient State and Steady-State

The Laplace transform solution to an RL circuit driven by a sinewave has three terms.

$$(23) \quad i(t) = \underbrace{\frac{V_m}{R + j\omega L} e^{j\omega t}}_{steady-state} - \underbrace{\frac{V_m}{R + j\omega L} e^{-\frac{R}{L}t}}_{transient} + \underbrace{i(0) e^{-\frac{R}{L}t}}_{transient}$$

Some other circuit solution will have n terms. The number does not matter. What matters is that in any circuit some terms decrease to negligible values after a number of time constants such as 5, while the magnitude of other terms is constant as time elapses. The first type are *transient state* terms and the second type are *steady state* terms. Many circuit solutions *do not have* steady state terms, while resistor only circuit solutions *do not have* transient terms. The types of terms are influenced by the selected source as well as the components in the circuit. The RL circuit solution happens to have one steady-state term and two transient terms.

Problem 706 Reference Figure P706. Assume initial conditions equal zero. Formulate the p-domain mesh equations. Solve for $v_4(p)$.

Problem 707 Reference Figure P707. Let $v_s(t)$= PULSE(0 4 0 0 0 1000u 2000u). Let L = 100µH, C = 0.01µF, R = 100Ω. Write a Spice program and run it.

Problem 708 Reference Figure P708. Assume initial conditions equal zero. Let Is(t)=I_Mu(t). Calculate Z(p) when R=2√(L/C). Solve for $v_2(p)$.

Problem 709 Reference Figure P709. Assume initial conditions equal zero. Let Is(t)=I_Mu(t). Calculate Z(p). Solve for $v_2(p)$.

Figure P706   **Figure P708** 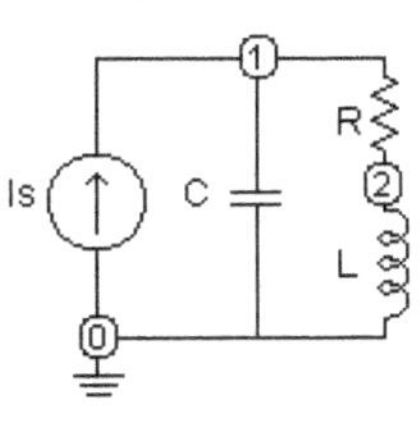   **Figure P709**

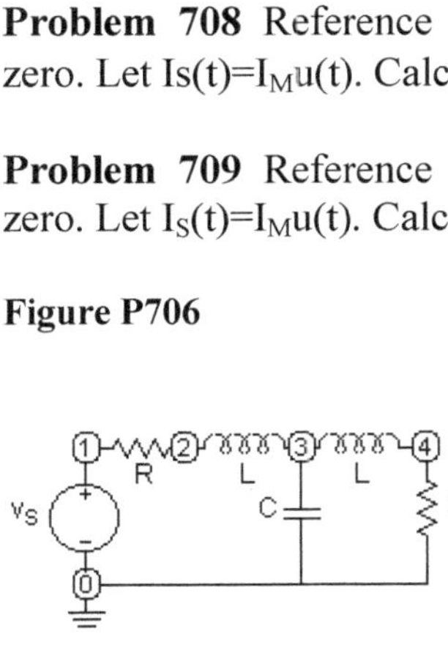

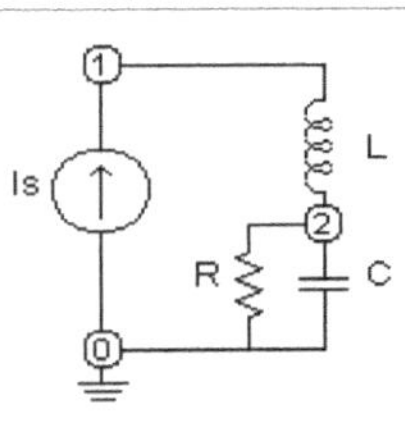

Electric Circuits - Analysis and Design

Problem 710 Reference Figure P710. If i(0)=0, then V(p)=I$_S$(p)/Y according to the Laplace transforms. If i$_S$(t)=I$_M$exp(−at) find its transform I$_S$(p) and solve for v$_1$(p) expanded into partial fraction format.

Problem 711 Reference Figure P711. Let v$_S$(t)=V$_M$ exp(−jωt) u(t) and i(0)=0. Find i(t).

Problem 712 Reference Figure P712. Let v$_1$(t)= V$_M$sin ωt u(t), v(0)=0, i(0)=0. Find v$_3$(p) for real equal roots.

Figure P710   Figure P711   Figure P712

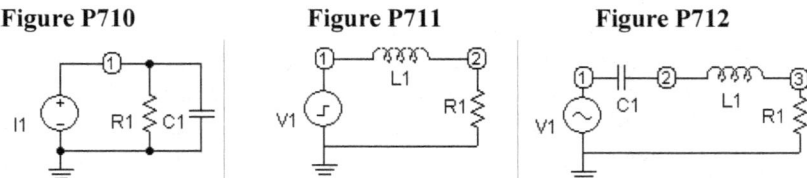

Problem 713 Refer to Figure P713. Write expressions v$_S$(t) for the signal source waveforms as sums of terms. Hint u(t-a) = 1 for t > a.

Figure P713

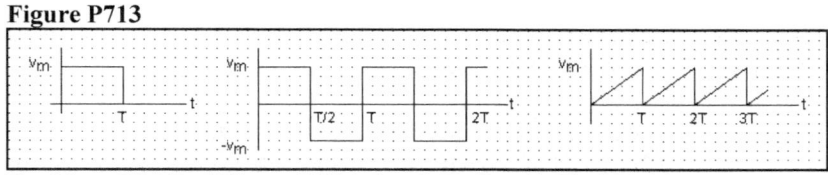

Problem 714 Ref Figure P714. Let i$_S$(t)=I$_M$u(t). Find V$_2$(p).
Problem 715 Ref Figure P715. Select source. Find V$_2$(p).

Figure P714   Figure P715

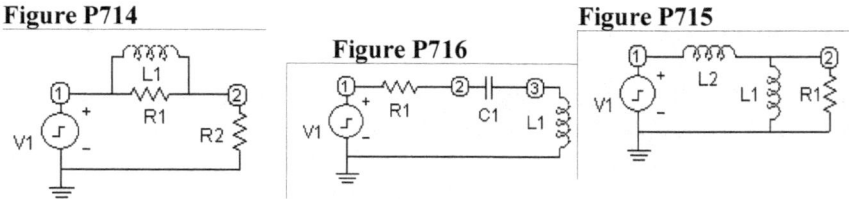

Problem 716
Ref Figure P716. Let v$_S$(t)=V$_M$u(t). Find V$_3$(p).

Problem 717 Reference Figure P717. Let v$_S$(t)=V$_M$ u(t). Find V$_2$(p).

Figure P717

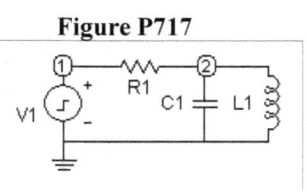

7 Transient Response

7.8 Real and Complex Frequencies

Real Frequency Sinusoidal waveforms repeat every 2π radians of the sinusoidal angle θ. The waveforms are said to be periodic with period 2π radians (Figure 706). If the 2π radian period elapses in T seconds, then nT periods occur in one second (nT=1). If $T=10^{-3}$ sec, then $n=10^3$, or 1000 periods occur per second. If $T=10^{-6}$sec, then $n=10^6$ Hz where a Hertz is one period per second. In practice n is referred to as *frequency f*. The frequency f=1/T. The sinusoidal angle θ is a fraction, or multiple, t/T of the period 2π.

(38) $\quad \sin \theta = \sin 2\pi \dfrac{t}{T} = \sin 2\pi f t \quad where \quad f = \dfrac{1}{T}$

Figure 706 Sinusoidal Waveforms

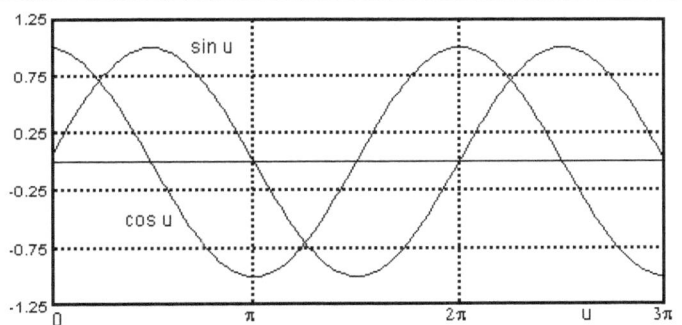

The waveform with period T does not have to be sinusoidal. It can be any single valued function of time t.

Suppose a battery's voltage is a constant 12v. Then the period T is infinite (as long in time as you please), because it does not end. I.e. T→∞, so that f→0.

On the other hand how short a period T can be is determined by the implementing technology. Clocks in modern cpu chips have periods of say 400 picoseconds (400×10^{-12} sec). The corresponding frequency is 2.5 Giga Hertz (2.5×10^9).

Euler's relation The famous mathematician Leonhard Euler (1707-1783) discovered the following very famous and incredibly useful relation, while developing the theory of infinite series.

123

Electric Circuits - Analysis and Design

(39) $\quad e^{i\theta} = \cos\theta + i\sin\theta \qquad$ where Euler's $e = 2.71828182\cdots$

if $i = j$ and $\theta = \omega t$, then

(40) $\quad e^{j\omega t} = \cos\omega t + j\sin\omega t$

Reminder: In mathematics the letter i equals the square root of -1. In electric circuits the letter j equals the square root of -1, because i traditionally represents current in electric circuits.

Superposition Use of a real frequency exponential source of sinusoids, such as $V_m e^{j\omega t}$, in a linear system is justified by the principal of superposition. For example

(41a) $\quad V_m \cos\omega t = Ri_1 + L\dfrac{di_1}{dt} + \dfrac{1}{C}\int_0^t i_1(x)dx$

(41b) $\quad jV_m \sin\omega t = R(ji_2) + L\dfrac{d(ji_2)}{dt} + \dfrac{1}{C}\int_0^t ji_2(x)dx$

(41c) $\quad V_m e^{j\omega t} = V_m \cos\omega t + jV_m \sin\omega t$

(41d) $\quad V_m e^{j\omega t} == R(i_1 + ji_2) + L\dfrac{d(i_1 + ji_2)}{dt} + \dfrac{1}{C}\int_0^t (i_1 + ji_2)dx$

(41e) $\quad V_m e^{j\omega t} = Ri + L\dfrac{di}{dt} + \dfrac{1}{C}\int_0^t i(x)dx \qquad$ where $i = i_1 + ji_2$

Complex frequency Not all sources are sinusoids. More complicated sources are accommodated by expanding the idea of real frequency f, or ω, to complex frequency p. If the complex frequency variable p=σ+jω, then the imaginary axis of the p plane is the real frequency axis, because ω=2πf. That is the physical interpretation justifying the use of p=σ+jω.

(42) let $V_m e^{pt} = (V_{m1} + jV_{m2})e^{(\sigma + j\omega)t}$

$\qquad = (V_{m1}\cos\omega t - V_{m2}\sin\omega t)e^{\sigma t} + j(V_{m1}\sin\omega t + V_{m2}\cos\omega t)e^{\sigma t}$

Each term is sum of sinusoids multiplied by $e^{\sigma t}$. If σ is negative the terms are damped sinusoids. If σ is zero the terms have constant amplitudes. If σ is positive the terms are growing sinusoids. If p=σ+jω=0+j0=0 the terms reduce to step function u(t) (Figure 703). You can create almost any source waveform (Appendix A8).

7 Transient Response

7.9 The Complex Frequency Plane

We showed in the prior discussion that $p=\sigma+j\omega$ has physical significance. The values of the complex frequency $p=\sigma+j\omega$ can be represented geometrically by points in the p plane (Figure 707). The points have rectangular coordinates $(\sigma, j\omega)$. The real values of p are represented by the horizontal axis that has coordinate σ. The imaginary values of p are represented by the vertical axis that has coordinate $j\omega$ (Appendix A2).

Poles and Zeros The impedance $z=R+pL$ of the series RL circuit is a polynomial in p that has a *zero* at $-R/L$ on the negative real axis. This zero is a *pole* of the current that ultimately appears in the terms of the circuit's time response (equations 22, 23, pages 112, 121). The series RLC circuit is a more complex example emphasizing the fact p is a complex frequency variable. The impedance z of the series RLC circuit is a rational function of p, which is the ratio of two polynomials (equation 43).

> Any function arising from a linear lumped parameter electric circuit is a rational function of p. (Think about Cramer's rule.)

$$(43) \quad z = z_{RLC}(p) = pL + R + \frac{1}{pC} = \frac{1}{pL}\left(p^2 + p\frac{R}{L} + \frac{1}{LC}\right) = \frac{1}{L}\frac{(p-p_1)(p-p_2)}{p}$$

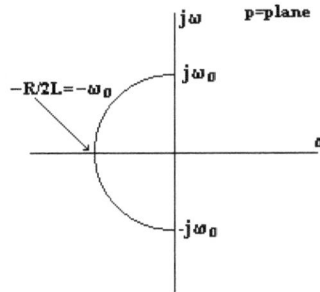

Figure 707 The p plane

A polynomial of degree n has n zeros that means it can be written as a product of factors (equation 43). The polynomial equals zero when p takes the value of one of the n zeros. If the polynomial is in the numerator, then the zeros are *zeros* of rational function z. If the polynomial is in the denominator, then the zeros are *poles* of rational function z. Any rational function such as z tends to infinity as p approaches a pole of z.

In an electric circuit the poles of functions determine the circuit's contribution to the transient response, as well as the steady-state response (e.g. equation 23). The poles and zeros are determined by the values of the circuit's RLC components and any active devices. For example, for a series RLC circuit the admittance y is

Electric Circuits - Analysis and Design

(44) $y = \dfrac{1}{z} = \dfrac{1}{pL+R+\dfrac{1}{pC}} = \dfrac{1}{L}\dfrac{p}{p^2+p\dfrac{R}{L}+\dfrac{1}{LC}} = \dfrac{1}{L}\dfrac{p}{(p-p_1)(p-p_2)}$

(45) $p_1 = -\dfrac{R}{2L} + \sqrt{\left(\dfrac{R}{2L}\right)^2 - \dfrac{1}{LC}}$ and $p_2 = -\dfrac{R}{2L} - \sqrt{\left(\dfrac{R}{2L}\right)^2 - \dfrac{1}{LC}}$

Poles p_1 and p_2 can be real or complex numbers that reinforces the complex frequency idea..

The poles p_1 and p_2 move as component values change. As R/2L increases from 0 the poles move off of the imaginary (real frequency) axis to follow the locus of a circle of radius ω_0 (Figure 707). There are three cases.

1. Under damping   Zero resistance is the limit of under damping

(46) If $R = 0$ and $p = \pm j\omega_0$, then $p_1 = j\sqrt{\dfrac{1}{LC}} = j\omega_0$ and $p_2 = -j\omega_0$

(47) $p_1, p_2 = -\dfrac{R}{2L} \pm j\sqrt{\dfrac{1}{LC} - \left(\dfrac{R}{2L}\right)^2}$ when $\left(\dfrac{R}{2L}\right)^2 < \dfrac{1}{LC}$

and $|p_1|^2 = |p_2|^2 = \left(\dfrac{R}{2L}\right)^2 + \left(\sqrt{\dfrac{1}{LC} - \left(\dfrac{R}{2L}\right)^2}\right)^2 = \dfrac{1}{LC} = \omega_0^2$

As R increases the poles move off of the $j\omega$ axis onto a circle of radius ω_0. (Figure 707).

2. Critical damping.   When R/2L equals ω_0 the poles meet on the negative real axis as they become equal.

(48) If $\left(\dfrac{R}{2L}\right)^2 = \dfrac{1}{LC} = \omega_0^2$, then $p_1 = p_2 = -\dfrac{R}{2L} = -\omega_0$

3. Over damping   Further increase in R/2L makes the poles unequal as one pole moves towards zero and the other pole moves towards infinity.

(49) $p_1, p_2 = -\dfrac{R}{2L} \pm \sqrt{\left(\dfrac{R}{2L}\right)^2 - \dfrac{1}{LC}}$ when $\left(\dfrac{R}{2L}\right)^2 > \dfrac{1}{LC}$

and $\lim\limits_{R\to\infty} p_1 = 0$ and $\lim\limits_{R\to\infty} p_2 = -\infty$

8 The Steady-State and the ω Phasor Method

We explain the phasor idea so that you know about it. Then you can set it aside, because the idea is a special case of the Laplace transform. We chose not to burden you with special cases.

The Laplace transform solution to an RL circuit driven by a sinewave in the form $e^{j\omega t}$ has three terms (equation 23 page 112).

$$(23) \quad i(t) = \underbrace{\frac{V_m}{R+j\omega L}e^{j\omega t}}_{steady-state} - \underbrace{\frac{V_m}{R+j\omega L}e^{-\frac{R}{L}t}}_{transient} + \underbrace{i(0)e^{-\frac{R}{L}t}}_{transient}$$

Connecting a *sinusoidal* source such as $V_m \cos \omega t$ to a circuit at time t=0 generates transients, because switching on the source is a transient event. However $V_m e^{j\omega t}$ remains after all transients have diminished to zero. Why? When σ=0 in p=σ+jω the exponential $e^{pt} = e^{j\omega t}$ has constant amplitude.

> A circuit is in the *steady-state* after all transients have decreased to zero.

Time domain equations integrate and differentiate sinusoids, as well as add sinusoids, to produce sinusoids with the same frequency[1]. Consequently in the steady state all voltages and currents in the circuit are sinusoids with constant amplitude and the same frequency. If frequency changes, then at many internal nodes amplitudes change.

Steady-state analysis uses a transformed circuit in the p=jω domain whose elements have impedances R, jωL, and 1/jωC. Furthermore, the systematic mesh current and node voltage analysis methods derived from KCL (1.3.1 page 11) and KVL (1.3.2) are applicable, because they are valid for any value of the complex frequency variable p such as jω.

Traditional AC analysis uses classical methods with sin x and cos x forcing functions. We show by example that the classical methods using sin x and cos x functions for solving electric circuit equations are not efficient.

---

[1] The time derivative of cos ωt is −ω×sin ωt and the time derivative of sin ωt is ω×cos ωt.

Electric Circuits - Analysis and Design

8.1 Traditional AC Analysis

Our purpose here is to demonstrate via a specific example the need for useful methods to replace inefficient traditional AC analysis. Up to the years immediately following World War 2 only a few undergraduates were exposed to the Laplace transform method of electric circuit analysis. In effect EE undergraduates learned that impedance was a generalization of the resistance concept, and that the impedance of R, L, and C was R, $j\omega L$, $1/j\omega C$ respectively. The presentation leading to these conclusions proceeded something like the following.

We start with the series RL circuit driven by a sinusoidal voltage source. Kirchhoff's voltage law and the vi constraints for L and R permit formulation of the circuit equation.

(1a) $\quad L\dfrac{di}{dt} + Ri = V_m \cos(\omega t + \theta_v) \quad \Rightarrow \quad$ (1b) $\quad \dfrac{di}{dt} + \dfrac{R}{L}i = \dfrac{V_m}{L}\cos(\omega t + \theta_v)$

The rest is applied mathematics. Ordinary differential equation theory multiplies the equation by a factor, μ in this case, which is defined to make the left side of the equation a perfect differential.

To convert equation 1b into a perfect differential we proceed as follows.

$$\mu\dfrac{di}{dt} + \mu\dfrac{R}{L}i \quad \Rightarrow \quad \dfrac{d}{dt}(\mu i) = \mu\dfrac{di}{dt} + \dfrac{d\mu}{dt}i$$

(2) let $\quad \mu\dfrac{R}{L} = \dfrac{d\mu}{dt} \quad \Rightarrow \quad \dfrac{d\mu}{\mu} = \dfrac{R}{L}dt \quad \Rightarrow \quad \ln\mu = \dfrac{R}{L}t \quad \Rightarrow \quad \mu = e^{\frac{R}{L}t}$

The multiplier $\mu = \exp(Rt/L)$ creates the desired perfect differential that facilitates the solution of the differential equation.

$$\dfrac{d}{dt}(\mu i) = \mu \dfrac{V_m}{L}\cos(\omega t + \theta_v)$$

$$d(\mu i) = \mu \dfrac{V_m}{L}\cos(\omega t + \theta_v)dt$$

$$\mu i(t) = \int_0^t \mu \dfrac{V_m}{L}\cos(\omega t + \theta_v)dt$$

(3) $\quad i(t) = e^{-\frac{R}{L}t}\dfrac{V_m}{L}\int_0^t e^{\frac{R}{L}t}\cos(\omega t + \theta_v)dt$

8 The ω Phasor Method

A straightforward way to integrate this integral is to convert the cos function to its exponential form, because the integral of exp(ax) is exp(ax)/a.

$$\int e^{\frac{R}{L}t} \cos(\omega t + \theta_v) dt = \int e^{\frac{R}{L}t} \frac{1}{2}[e^{j(\omega t + \theta_v)} + e^{-j(\omega t + \theta_v)}] dt$$

$$= \frac{1}{2} \int [e^{\frac{R}{L}t + j(\omega t + \theta_v)} + e^{\frac{R}{L}t - j(\omega t + \theta_v)}] dt$$

$$= \frac{1}{2} \int [e^{j\theta_v} e^{(\frac{R}{L} + j\omega)t} + e^{-j\theta_v} e^{(\frac{R}{L} - j\omega)t}] dt$$

$$= \frac{1}{2} \frac{e^{j\theta_v} e^{(\frac{R}{L} + j\omega)t}}{\frac{R}{L} + j\omega} + \frac{1}{2} \frac{e^{-j\theta_v} e^{(\frac{R}{L} - j\omega)t}}{\frac{R}{L} - j\omega}$$

Now multiply by V_m/L, and solve for I.

$$\frac{V_m}{L} \int = \frac{V_m}{2} \frac{e^{\frac{R}{L}t} e^{j\omega t} e^{j\theta_v}}{R + j\omega L} + \frac{V_m}{2} \frac{e^{\frac{R}{L}t} e^{-j\omega t} e^{-j\theta_v}}{R - j\omega L}$$

If $Z = R \pm j\omega L = |Z|e^{\pm j\theta_z}$, then $|Z|^2 = (R + j\omega L)(R - j\omega L) = R^2 + (\omega L)^2$

$$\frac{V_m}{L} \int = \frac{V_m}{2} e^{\frac{R}{L}t} \left[\frac{e^{j\omega t + j\theta_v}}{|Z|e^{+j\theta_z}} + \frac{e^{-j\omega t - j\theta_v}}{|Z|e^{-j\theta_z}}\right]$$

$$\frac{V_m}{L} \int = \frac{V_m}{|Z|} e^{\frac{R}{L}t} \times \frac{1}{2} \left[\frac{e^{j\omega t + j\theta_v}}{e^{+j\theta_z}} + \frac{e^{-j\omega t - j\theta_v}}{e^{-j\theta_z}}\right]$$

$$i(t) = e^{-\frac{R}{L}t} \frac{V_m}{L} \int = e^{-\frac{R}{L}t} \frac{V_m}{|Z|} e^{\frac{R}{L}t} \times \left[\frac{e^{j(\omega t + \theta_v - \theta_z)} + e^{-j(\omega t + \theta_v - \theta_z)}}{2}\right]$$

(4a) $i(t) = \dfrac{V_m}{\sqrt{R^2 + (\omega L)^2}} \cos(\omega t + \theta_v - \theta_z)$

(4b) $i(t) = \dfrac{V_m}{R + j\omega L} \cos(\omega t + \theta_v)$

We conclude that the impedance of R is R and the impedance of L is jωL. Solution of a series RC circuit leads to the conclusion that the impedance of C is 1/jωC. The word impedance replaces the word resistance to eliminate any awkwardness that arises by associating resistance with L and C. (Impedance impedes current flow.)

> At some point in time someone must have said "there must be an easier way." **There is**.

129

Electric Circuits - Analysis and Design

Sinusoidal alternating current and voltage

An alternating voltage at a node is positive with respect to ground, and at a different time voltage at the node is negative with respect to ground.

An alternating current flows one direction in a wire, and at a different time it flows in the opposite direction.

In other words the sign of the voltage or current is alternately positive or negative. Specifically our present interest is in *sinusoidal* waveforms (Figure 801). For example.

$v(t) = V_m \cos(\omega t + \theta)$ *and, or* $v(t) = V_m \sin(\omega t + \theta)$
$i(t) = I_m \cos(\omega t + \theta)$ *and, or* $i(t) = I_m \sin(\omega t + \theta)$
where $\omega = 2\pi f$ *and period* $T = \dfrac{1}{f}$

v(t) or i(t) is the instantaneous voltage or current at time t.

V_M or I_M is the maximum or peak value or amplitude of the waveform (positive or negative),

T is the time duration of one complete waveform cycle or period, and frequency f is the number of cycles repeated in one second (f = 1/T cycles per second).

θ is the phase shift relative to time t = 0. θ = 0 in Figure 801.

Figure 801 Two cycles of a sinewave

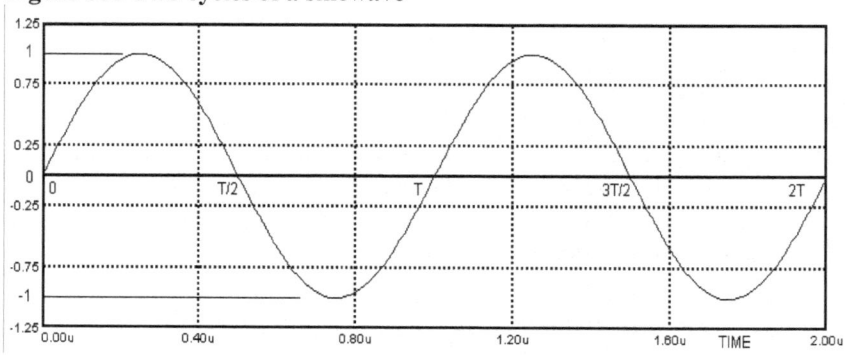

8 The ω Phasor Method

8.2 ω Phasor Transforms

Traditional RL circuit analysis is very complicated (8.1). Clearly the process is rapidly mired in more complexity when the circuit has many components, nodes, and meshes. This emphasizes the fact that sinusoidal analysis based upon sin ωt and cos ωt forcing functions and classical methods is not the way to solve these problems.

The sinusoidal time function f(t) is transformed into the phasor F by the following process. We refer to F as the ω-phasor corresponding to f(t). (See 8.4 for an example.) The phasor transform $P[f(t)]$ is designed to remove trigonometric functions from AC analysis, because we now know how cumbersome they are.

If $f(t) = F_M \cos(\omega t + \theta)$, then $f(t) = \text{Re}[F_M e^{j(\omega t + \theta)}] = \text{Re}[(F_M e^{j\theta})e^{j\omega t}]$

(5) $P[f(t)] = F_M e^{j\theta} = F$

> Note that the ω-phasor is the coefficient of $e^{j\omega t}$ - not $e^{j\theta}$.

> The ω phasor method is limited to steady-state analysis.

Equation 5 defines F as an ω phasor. P is our phasor transform operator, which is analogous to the Laplace transform operator. Here is the phasor transform of the derivative and integral.

$$P\left[\frac{df(t)}{dt}\right] = P\left[\frac{d\,\text{Re}[Fe^{j\omega t}]}{dt}\right] = P\left[\text{Re}\frac{d[Fe^{j\omega t}]}{dt}\right] = P[\text{Re}[j\omega F e^{j\omega t}]]$$

$$= j\omega \cdot P[\text{Re}[Fe^{j\omega t}]] = j\omega \cdot P[f(t)] = j\omega F$$

(6) $P\left[\dfrac{df(t)}{dt}\right] = j\omega F$

(7) $P\left[\int f(x)dx\right] = \dfrac{1}{j\omega} F$ by a similar process

The phasor transform operator P transforms time t to p=jω, and hides the source of ω, which is the frequency of a sinusoidal source.

> An ω-phasor F derived from a sinusoid is a complex expression that is the coefficient of $e^{j\omega t}$ in $\text{Re}[F\,e^{j\omega t}]$.

8.3 RLC Steady-State Impedances

The impedance Z is defined as the ratio of V/I, which are two phasors. Z is not a phasor because there is no z(t) that corresponds to Z. Stating this in another way, the inverse transform of Z(p) is never executed. Inverse transforms of Z(p)I(p) into some v(t) and V(p)/Z(p) into some i(t) are executed. For similar reasons admittance Y is not a phasor. Derivations of impedance expressions for R, L, and C components follow. The phasor is the coefficient of $e^{j\omega t}$ in the Re[...] expression.

Define phasor I by the phasor transform $I = P[i(t)]$

(8) $\quad I = P[i(t)] = I_m e^{j\theta} \quad$ where $i(t) = I_m \cos(\omega t + \theta) = \text{Re}[I_m e^{j\theta} e^{j\omega t}]$

Sinusoidal current through a resistor R

(9a) $\quad v(t) = Ri(t) = RI_m \cos(\omega t + \theta) = \text{Re}\left[RI_m e^{j(\omega t + \theta)}\right] = \text{Re}\left[RI_m e^{j\theta} e^{j\omega t}\right]$

(9b) $\quad V = P[v(t)] = RI_m e^{j\theta}$

(9c) $\quad Z = \dfrac{V}{I} = \dfrac{RI_m e^{j\theta}}{I_m e^{j\theta}} = R$

Sinusoidal current through an inductance L

(10a) $\quad v(t) = L\dfrac{di(t)}{dt} = \omega L I_m \sin(\omega t + \theta) = \omega L I_m \cos(\omega t + \theta + 90°)$

$\qquad = \text{Re}\left[\omega L I_m e^{j(\omega t + \theta + 90°)}\right] = \text{Re}\left[\omega L I_m e^{j90°} e^{j\theta} e^{j\omega t}\right]$

(10b) $\quad V = P[v(t)] = \omega L I_m e^{j\theta} e^{j90°} = \omega L I_m e^{j\theta}(\cos 90° + j\sin 90°) = j\omega L I_m e^{j\theta}$

(10c) $\quad Z = \dfrac{V}{I} = \dfrac{j\omega L I_m e^{j\theta}}{I_m e^{j\theta}} = j\omega L$

Sinusoidal voltage across a capacitor C

(11a) $\quad i(t) = C\dfrac{dv(t)}{dt} = \omega C V_m \sin(\omega t + \theta) = \omega C V_m \cos(\omega t + \theta + 90°)$

$\qquad = \text{Re}\left[\omega C V_m e^{j(\omega t + \theta + 90°)}\right] = \text{Re}\left[\omega C V_m e^{j90°} e^{j\theta} e^{j\omega t}\right]$

(11b) $\quad I = P[i(t)] = \omega C V_m e^{j\theta} e^{j90°} = \omega C V_m e^{j\theta}(\cos 90° + j\sin 90°) = j\omega C V_m e^{j\theta}$

(11c) $\quad Z = \dfrac{V}{I} = \dfrac{V_m e^{j\theta}}{j\omega C V_m e^{j\theta}} = \dfrac{1}{j\omega C}$

> R, L, C "steady-state" impedances are R, $j\omega L$, and $1/j\omega C$.

8.4 Phasor AC analysis

When you want to analyze a circuit's sinusoidal steady state behavior you start with a phasor transform of the circuit equations. Then you proceed as follows to analyze the steady state of the RLC circuit with switch closed (Figure 802).

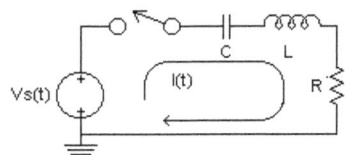

Figure 802 Series RLC Circuit

We use equations 5, 6, 7 to find the phasor transforms of i(t), and the derivative and integral of i(t). Coefficients R, L, and C are just constant multipliers. *The Laplace Transform where $p = j\omega$ achieves the same result.*

If $P[i(t)] = I$ the ω phasor i(t), then

$$v_s(t) = Ri(t) + L\frac{di}{dt} + \frac{1}{C}\int_0^t i(x)dx$$

(12a) $P[v_s(t)] = P[\,Ri(t) + L\frac{di}{dt} + \frac{1}{C}\int_0^t i(x)dx\,]$

(12b) $V_m = RI + j\omega LI + \frac{1}{j\omega C}I = \left[R + j\omega L + \frac{1}{j\omega C}\right]I$

(13) $I = \dfrac{V_m}{R + j\omega L + \dfrac{1}{j\omega C}}$

> ***Practice*** *As a practical matter you can analyze a circuit by skipping the phasor transform process to set up the node or mesh equations directly (equation 12b, and examples 7 and 8).*

Problem 801 *Derive* $P\left[\int_0^t f(x)dx\right] = \dfrac{1}{j\omega}F$

Problem 802 Find the phasor transform F of f(t).
1) $v = V_m \cos(\omega t + d)$ 2) $i = I_m \cos(\omega t - a)$ 3) $v = 1.8\cos(2\pi \cdot 10^6 t + \pi/4)$

Problem 803 Reference Figure 802. Show that the current
$I = I_m e^{j\theta} = I = 0.00329 V_m e^{j70.7°}$ when
$V_s = V_m$ volts, $\omega = 2\pi 10^6$, $R = 100\Omega$, $C = 500\,pF$, $L = 5\mu H$

Electric Circuits - Analysis and Design

Emphasis Use Spice to avoid number crunching. For example the Spice *numeric output file* for Fig8031.ckt lists f = 1.000MHz, $VReal_R$ =110mV, $VImag_R$ = 310mV, $VMag_R$ = 330mV, $VPhase_R$ = 70.8 degrees. Data such as this is calculated every 200Hz over the range 10KHz to 10MHz. Spice has a data plotting capability over the dot AC frequency range (Figure 80311).

Spice program 8031

```
Fig8031.ckt RLC circuit

*source voltage
Vs  1 0   AC 1
R   3 0 100    ;ohms
L   2 3 5u    ;micro H
C   1 2 500p   ;picofarads

.AC DEC 200 10000 1e+007
.TEMP 27
.PLOT AC V(3) 0,1.5
.PLOT AC I(VSI) 0M,15M
.PRINT AC I(VS) VR(3) VI(3) VM(3) VP(3)
.end
```

Figure 803 RLC Circuit

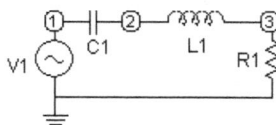

Figure 80311 RLC Circuit Resistor Voltage and Current Magnitudes

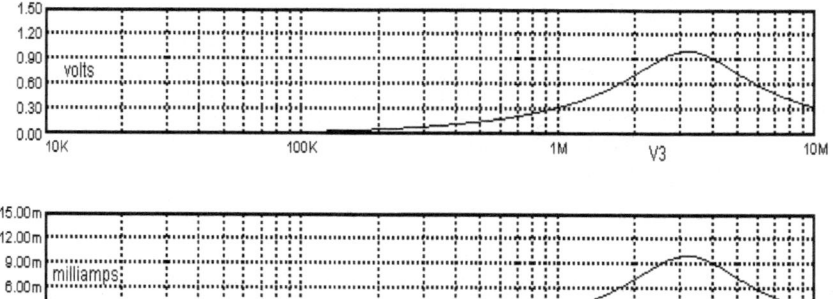

8 The ω Phasor Method

Examples of jω AC Analysis Just number crunching.

> We prefer to use Spice.

1. Given frequency f, current i and source voltage v find the component values.

if $v = 200\cos(1500t)$ and $i = 10\cos(1500t - 50°)$, then

$$z = \frac{v}{i} = \frac{200e^{j0°}}{10e^{-j50°}} = 20e^{j50°}$$

$$z = 20\cos 50° + j20\sin 50° = 12.86 + j15.32 \text{ ohms} = R + jX_L$$

$R = 12.86 \text{ ohms} \qquad L = \frac{X_L}{\omega} = \frac{15.32}{1500} = 10.21 mH$

2. In a series RC circuit current leads voltage by 60°. If R=100 ohms, C=0.1μf, then find Z and frequency.

$\tan 60° = \frac{X_C}{R} \quad \Rightarrow \quad X_C = R\tan 60° = 100 \times 0.1732 = 173.2 \text{ ohms}$

$X_C = \frac{1}{\omega C} \quad \Rightarrow \quad f = \frac{1}{2\pi C X_C} = \frac{1}{2\pi \times 0.1 \cdot 10^{-6} \times 173.2} = 9.19 \cdot 10^3 \text{ Hertz}$

$Z = R - jX_C = 100 - j173.2$

3. In a series RL circuit find R and L given current i and source voltage v.

if $v = 25\cos(1000t + 33.13°)$ and $i = 5\cos(1000t - 20°)$ then

$$z = \frac{v}{i} = \frac{25e^{j33.13°}}{5e^{-j20°}} = 5e^{j53.13°}$$

$\tan 53.13° = \frac{4}{3} = \frac{\omega L}{R} = \frac{1000L}{R} \quad \Rightarrow \quad R = \frac{3}{4}1000L$

$|z| = \sqrt{R^2 + X_L^2} = \sqrt{R^2 + (1000L)^2}$

$5 = \sqrt{\left(\frac{3}{4}1000L\right)^2 + (1000L)^2} = \frac{5}{4}1000L \quad \Rightarrow \quad L = 4 \cdot 10^{-3} H$

$R = \frac{3}{4}1000L = 3 \text{ ohms}$

Electric Circuits - Analysis and Design

4. In a series RC circuit R=2200 ohms, C=180pF, $z=2640e^{j\theta}$. Find θ and ω.

$$|z| = \sqrt{R^2 + X_C^2} \quad \Rightarrow \quad X_C = \sqrt{|z|^2 - R^2} = \sqrt{2640^2 - 2200^2} = 1460 \ ohms$$

$$\theta = \arctan\frac{X_C}{R} = \arctan\frac{1460}{2200} = 33.56°$$

$$\omega = \frac{1}{CX_C} = \frac{10^{12}}{180 \times 1460} = 3.8 \cdot 10^6 \quad and \quad f = \frac{\omega}{2\pi} = 0.6 \cdot 10^6$$

5. In a series RC circuit R=1000 ohms, C=160pF. Current leads voltage by 45°. Find z, ω, and f.

$$\frac{X_C}{R} = \tan 45° = 1 \quad \Rightarrow \quad X_C = R$$

$$\omega = \frac{1}{CX_C} = \frac{1}{CR} = \frac{10^{12}}{160 \times 1000} = 6.25 \cdot 10^6 \quad and \quad f = \frac{\omega}{2\pi} = 0.995 \cdot 10^6$$

$$z = R - jX_C = R - jR = \sqrt{2} \, Re^{j45°} = 1414 \, e^{j45°}$$

6. A battery v_1=3.3V and a cos source v_2=4cos ωt drive a series RL circuit. R=330 ohms, L=180µH, $\omega=10^7$. Find the currents.

$$i_1 = \frac{v_1}{R} = \frac{3.3v}{330} = 10mA \quad (dc \ current)$$

$$i_2 = \frac{v_2}{z} = \frac{4e^{j0}}{R + j\omega L} = \frac{4e^{j0}}{330 + j10^7 \times 180 \cdot 10^{-6}}$$

$$= \frac{4e^{j0}}{330 + j1800} = \frac{4e^{j0}}{1830e^{j79.6°}}$$

$$= 2.186mA \, e^{-j79.6°}$$

8 The ω Phasor Method

7. Single phase sinusoidal AC circuit analysis is performed in the real frequency ω or f domain, where ω = 2πf. The R, L, C impedances are Z_R = R, Z_L = jωL, and Z_C = 1/jωC. The analysis process requires accurate manipulation of complex numbers (Appendix A2). The KVL equation for the one mesh RLC circuit (Figure 803) is a sum of IZ_n terms representing the R, L, C voltages. Observe how we go directly to the jω mesh equation.

(19) $\quad V = RI + j\omega LI + \dfrac{1}{j\omega C} I = \left(R + j\omega L + \dfrac{1}{j\omega C} \right) I$

$$I = \dfrac{V}{R + j\omega L + \dfrac{1}{j\omega C}} = \dfrac{V}{Z}$$

Figure 803 RLC Circuit

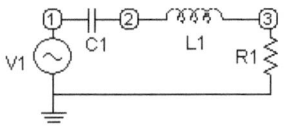

Find the frequency f_0 producing the maximum current $I = e^{j\theta}$.
$V = 1\,volt,\ R = 100\Omega,\ C = 500\,pF,\ L = 5\mu H$

$$I_{max} = \dfrac{V}{Z} = \dfrac{V}{R + j\omega_0 L + \dfrac{1}{j\omega_0 C}} = \dfrac{V}{R} \quad \text{when} \quad \omega_0 L - \dfrac{1}{\omega_0 C} = 0$$

(20) $\quad \omega_0^2 = \dfrac{1}{LC} \Rightarrow f_0 = \dfrac{1}{2\pi} \dfrac{1}{\sqrt{LC}} = \dfrac{1}{2\pi} \dfrac{1}{\sqrt{5 \cdot 10^{-8}}} = \dfrac{10}{\pi} 10^6 = 3.183\,MHz$

Find the current $I = e^{j\theta}$ and mesh impedance $Z = R + j\omega L + \dfrac{1}{j\omega C}$

$Z = 100 + j2\pi \cdot \dfrac{10}{\pi} \cdot 10^6 \cdot 5 \cdot 10^{-6} + \dfrac{1}{j2\pi \cdot (10/\pi) \cdot 10^6 \cdot 500 \cdot 10^{-12}}$

$Z = 100 + j100 - j100 = 100\ ohms$

(21) $\quad I = \dfrac{V}{Z} = \dfrac{1\,volt}{100} = 10\ milliamperes$

Find what is referred to as the characteristic impedance z_0.

(22) $\quad z_0 = \sqrt{\dfrac{L}{C}} = \sqrt{\dfrac{5 \cdot 10^{-6}}{500 \cdot 10^{-12}}} = 100\ ohms$

Electric Circuits - Analysis and Design

8. The KCL equation for node 1 of the two node circuit (Figure 804) is a sum of $Y_N V$ terms. Observe how the numbers are manipulated.

$$I = j\omega CV + \frac{V}{R+j\omega L} = \left(j\omega C + \frac{1}{R+j\omega L}\right)V$$

(23) $\quad V = \dfrac{I}{j\omega C + \dfrac{1}{R+j\omega L}} = \dfrac{I}{Y}$

Figure 804 Series-Parallel RLC Circuit

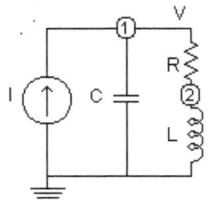

Find the voltage V at the resonant frequency w_0.

$$I = 20mA,\ R = 20\Omega,\ C = 500\,pF,\ L = 5\mu H$$

$$\omega_0 = \frac{1}{\sqrt{LC}} = \frac{1}{\sqrt{5 \cdot 10^{-6} \times 500 \cdot 10^{-12}}} = \frac{10^8}{5} = 20 \cdot 10^6$$

$$Y = j\omega_0 C + \frac{1}{R+j\omega_0 L} = j20 \cdot 10^6 \times 500 \cdot 10^{-12} + \frac{1}{20 + j20 \cdot 10^6 \times 5 \cdot 10^{-6}}$$

$$Y = j10^{-2} + \frac{1}{20+j100} \frac{20-j100}{20-j100} = j\frac{1}{100} + \frac{20-j100}{10400}$$

$$Y = j\frac{1}{100} - j\frac{1}{104} + \frac{20}{10400} = \frac{1}{520} + j\frac{4}{10400}$$

$$Y = 10^{-3}(j0.385+1.9231) = 10^{-3} \cdot 1.961 e^{j11°} = \frac{1}{510} e^{j11°}\ mhos$$

(24) $\quad V = \dfrac{I}{Y} = \dfrac{20mA}{1.961 e^{j11°} \times 10^{-3}} = 10.198\, e^{-j11°}\ volts$

$$\text{Estimate and verify}:\ IR = \frac{I}{|Y|} = 20mA \times 510 = 10.2V$$

8 The ω Phasor Method

Problem 804 If impedances are equal in magnitude at radial frequency ω_0 what is the equation for component pairs R and C, R and L, L and C?

Problem 805 At radial frequency ω_0 what is the equation for Q of series pair L and R (Section 1.4.2), and parallel pair R and L?

Problems 806 to 811 Reference figures 806P to 811P. Find an expression for $T(p)=v_2/v_1$, or $T(p)=v_1/i_1$. Find expressions for $T(j\omega)$ for $\omega<<\omega_0$, $\omega=\omega_0$, and $\omega>>\omega_0$

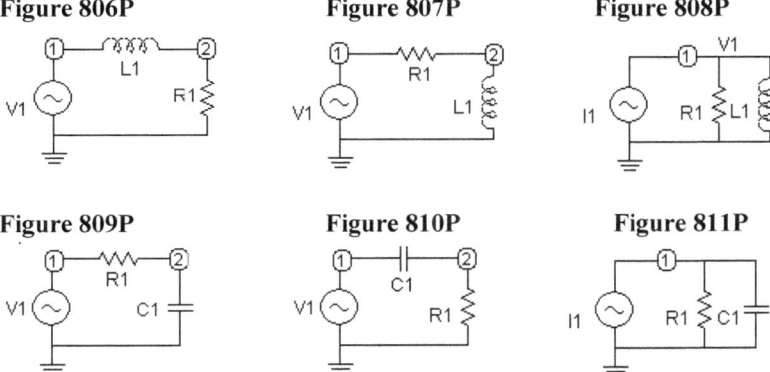

Figure 806P   Figure 807P   Figure 808P

Figure 809P   Figure 810P   Figure 811P

Problems 812 to 814 Reference figures 812P to 814P. Find an expression for $T(p)=v_2/v_1$, or $T(p)=v_3/v_1$. Find expressions for $T(j\omega)$ for $\omega<<\omega_0$, ω, and $\omega>>\omega_0$. Calculate the magnitude and phase in radians of $T(p)$ at frequencies $T(j0)$, $T(j\omega_0/k)$, $T(j\omega_0)$, and $T(j\infty)$.

612 $$k = \frac{R_2}{R_1 + R_2}$$

613 $$\omega_0 = \frac{1}{R_1 C_1} = \frac{1}{R_1 L_1}$$

614 $$\omega_0 = \frac{R_1}{L_1} = \frac{1}{R_2 C_2}$$

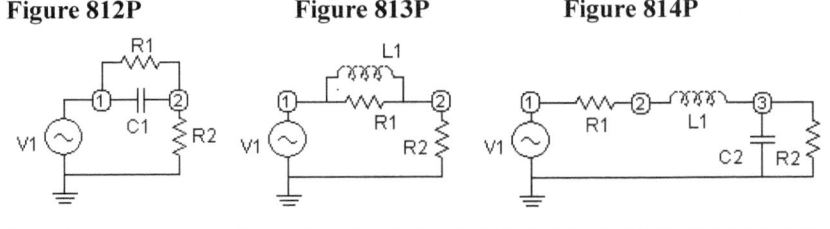

Figure 812P   Figure 813P   Figure 814P

Electric Circuits - Analysis and Design

Summary 8

We explain the phasor idea so that you know about it. Then you can set it aside, because the idea is a special case of the Laplace transform. We chose not to burden you with special cases.

Traditional AC Analysis Traditional AC analysis uses classical methods with sin x and cos x forcing functions. Examples show that the classical methods using sin x and cos x functions for solving electric circuit equations are not efficient.

The ω phasor transform The process solving the equations is simplified when sin x and cos x source functions are replaced by exp(jx). The process is referred to as the ω phasor method. The ω phasor method is a transform of the time domain t to the p=jω domain.

Equation 5 defines F as an ω phasor. P is the phasor transform operator, which is analogous to the Laplace transform operator. The phasor transform of the derivative and integral multiply and divide F by jω..

If $f(t) = F_M \cos(\omega t + \theta)$, then $f(t) = \text{Re}[F_M e^{j(\omega t + \theta)}] = \text{Re}[(F_M e^{j\theta})e^{j\omega t}]$

(5) $P[f(t)] = F_M e^{j\theta} = F$

(6) $P\left[\dfrac{df(t)}{dt}\right] = j\omega F$

(7) $P\left[\displaystyle\int_0^t f(x)dx\right] = \dfrac{1}{j\omega}F$ by a similar process

Impedance in the Steady-State Impedance is a generalization of the resistance concept, and in the steady-state the impedance of R, L, and C is R, jωL, 1/jωC respectively.

Solving problems Go directly to the ω domain. Write the KVL or KCL equations as algebraic equations such as equations 19 and 23.

9 One Mesh RL and RC Circuits

One mesh RL and RC circuits are used everywhere. They are worth knowing about. This is an analysis of RC and RL circuits emphasizing

transfer functions, time constant, impedance, and stored energy.

An inductor is essentially a piece of wire. The wire has resistance that is modeled as an independent resistor. Removing the resistance leaves an ideal inductor that has the steady state impedance $j\omega L$. At zero frequency the impedance goes to zero, because it is just an ideal wire. As frequency increases the impedance increases at +6dB/octave (+20 dB/decade). At high frequencies the impedance $j\omega L$ is infinite (as large as you please). Increased frequency increases inductor voltage ($v=j\omega Li$).

The transfer function $T(\omega)$ from node 1 to node 2 is the ratio v_2/v_1 (Figure 901). When L is connected from node 2 to ground v_2 increases from 0 to v_1 as frequency increases. $T(\omega)$ varies from 0 to 1. When L is in series (Figure 902) v_2 decreases from v_1 to 0 as frequency increases. $T(\omega)$ varies from 1 to 0. The *frequency responses* are produced by the impedance ratios $j\omega L/(R+j\omega L)$, and $R/(R+j\omega L)$.

Figures 901   902   904   905

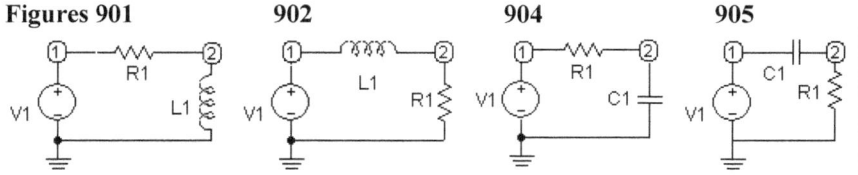

A capacitor is essentially two insulated pieces of metal. The steady state impedance is $1/j\omega C$. As frequency increases the impedance decreases at −6dB/octave or −20dB/decade. At zero frequency the impedance $1/j\omega C$ is infinite. At high frequencies the impedance $1/j\omega C$ goes to zero. Increased frequency increases capacitor current ($i=j\omega Cv$).

When C is a shunt (Figure 904) v_2 decreases from v_1 to 0 as frequency increases. $T(\omega)$ varies from 1 to 0. When C is in series (Figure 905) v_2 increases from 0 to v_1 as frequency increases. $T(\omega)$ varies from 0 to 1. The *frequency responses* are produced by the impedance ratios $(1/j\omega C)/(R+1/j\omega C)$ and $R/(R+1/j\omega C)$.

Electric Circuits - Analysis and Design

9.1 RL Steady State

The elementary parameters of the RL circuit are the time constant L/R seconds, the stored energy $E = \frac{1}{2}Li^2$ joules, and the jωL impedance of an inductor in the steady state. Laplace analysis produces the steady state response to $v_1 = V_M e^{j\omega t}$.

Figure 901 High pass RL circuit   **Figure 902 Low pass RL circuit**

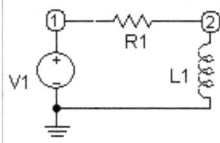 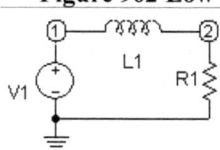

(1a) $\quad i(t) = \dfrac{V_M}{R + j\omega L} e^{j\omega t} - \dfrac{V_M}{R + j\omega L} e^{-\frac{R}{L}t} + i(0) e^{-\frac{R}{L}t}$

(1b) $\quad i(t) = \dfrac{V_M}{R + j\omega L} e^{j\omega t} \quad$ after 5L/R seconds elapse

(1c) $\quad$ time constant $\tau = \dfrac{L}{R} = \dfrac{1}{\omega_0} \quad$ where impedances $\omega_0 L = R$

The high pass and low pass transfer functions:

(2a) $\quad T_{high\_pass}(\omega) = \dfrac{v_2}{v_1} = \dfrac{z_L}{V_m e^{j\omega t}} i(t) = \dfrac{j\omega L}{V_m e^{j\omega t}} \dfrac{V_m e^{j\omega t}}{R + j\omega L} = \dfrac{j\omega L}{R + j\omega L} \quad (Fig\,901)$

(2b) $\quad T_{low\_pass}(\omega) = \dfrac{v_2}{v_1} = \dfrac{z_R}{V_m e^{j\omega t}} i(t) = \dfrac{R}{V_m e^{j\omega t}} \dfrac{V_m e^{j\omega t}}{R + j\omega L} = \dfrac{R}{R + j\omega L} \quad (Fig\,902)$

Manipulate the equations to emphasize the ratio of L and R impedances.

(3a) $\quad T_{high\_pass}(\omega) = \dfrac{1}{1 + \dfrac{R}{j\omega L}} \qquad$ (3b) $\quad T_{low\_Pass}(\omega) = \dfrac{1}{1 + \dfrac{j\omega L}{R}}$

The impedance of any L is $z_L = j\omega L$. The circuit in Figure 901 is a high pass circuit, because z_L increases with frequency as frequency increases above ω_0 so that the transmission approaches magnitude 1 and phase 0°. Z_L decreases with frequency as frequency decreases below ω_0 so that the transmission approaches magnitude 0 and phase 90°. The phase is 45° at ω_0 (Figure 90112). When R and L are exchanged (Figure 902), the behavior above and below ω_0 is exchanged so that we now have a low pass filter (Figures 90211, 90212).

9 One Mesh RL & RC Circuits

Spice Program 9011 - High pass RL

```
Fig9011.ckt high pass RL circuit sine
V1 1 0 AC 1
R1 1 2 1000
L1 2 0 100m
*.PLOT AC VDB(1) VDB(2) -40,10
.AC DEC 200 10 100000
.TEMP 27
.PLOT AC VP(2) 0,100
.end
```

Transmission is reduced by 3dB when the source frequency ω makes the reactance of L equal to R ($\omega_0 L = R$). The -3dB frequency is f_0 The easiest way to find f_0 is to find the 45 degree phase frequency in Figure 90112.

(4) $\quad f_0 = \dfrac{R}{2\pi L} = \dfrac{1000}{2\pi \times 100 \cdot 10^{-3}} = 1590 \; Hz$

Figure 90111 High Pass RL Circuit magnitude of T(ω)

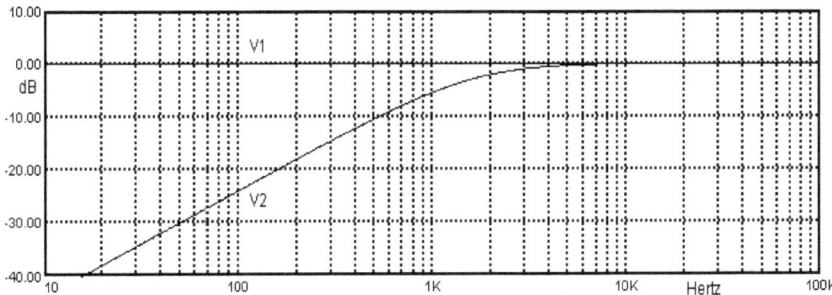

Figure 90112 High Pass RL Circuit phase of T(ω)

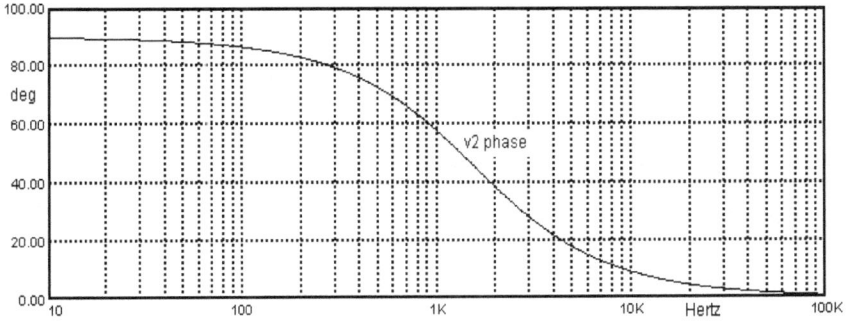

143

Electric Circuits - Analysis and Design

Spice Program 9021 - Low pass RL
```
Fig9021.ckt series RL circuit sine
V1 1 0 AC 1
L1 1 2 100m
R1 2 0 1000
*.PLOT AC VP(2) -100,0
.AC DEC 200 10 100000
.TEMP 27
.PLOT AC VDB(1) VDB(2) -40,10
.end
```

Figure 90211 Low Pass RL Circuit magnitude of T(ω)

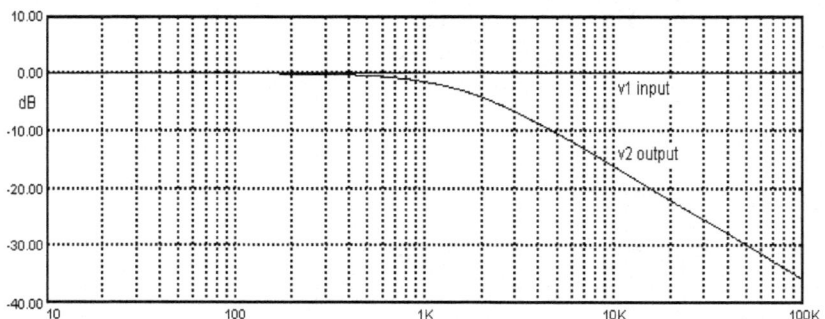

Figure 90212 Low Pass RL Circuit phase of T(ω)

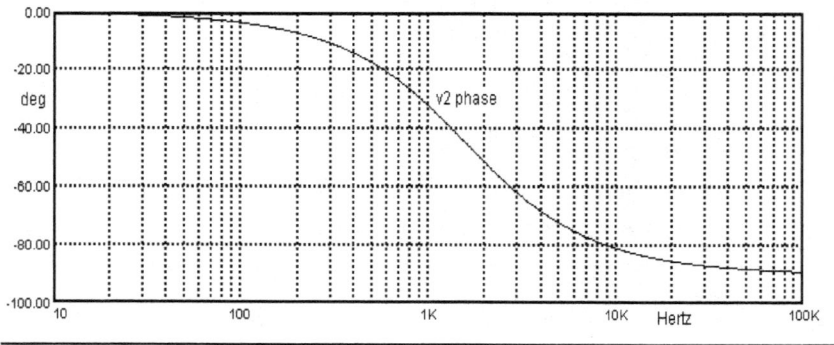

Problem 901 Solve for the Laplace solution equation 1a.

Problem 902 If $\omega_0 L = R$, then find $T_{high}(\infty)$, $T_{high}(\omega_0)$, $T_{high}(\omega_0/20)$. Compare to Figures 90111, 90112.

Problem 903 If $\omega_0 L = R$, then find $T_{low}(0)$, $T_{low}(\omega_0)$, $T_{low}(20\omega_0)$. Compare to Figures 90211, 90212.

9.2 RL Transient State

We examine the series RL circuit transient response to
step functions u(t), pulses u(t)−u(t−T), and ramps (t/T)u(t).

The step at t=0 $v_1(t)=V_M u(t)$ (Figure 907) is in effect an "infinite" dv/dt at time t=0. Since $v_L = L di/dt$ there must be an infinite voltage drop across the inductor. In fact, the inductor voltage is limited to source voltage V_M. This leaves zero voltage across R (Figure 901) that is a way to see that the current starts from zero. As the current increases the di/dt decreases to zero exponentially as time elapses so that in the final analysis all of the voltage V_M is across R (Figure 90121). In other words, the inductor L responds to a step function at t=0 as if L is an open circuit.

Figure 907 Steps u(t), u(t-T)

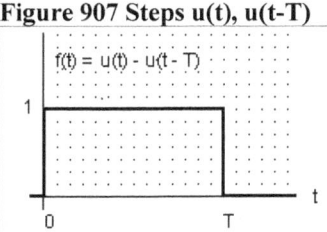

9.2.1 Steps u(t)

The Laplace transform RL analysis (see sidebar) shows that the response to the step function u(t) waveform is a current with the $1-e^{-x}$ waveform. The current starts at 0 and increases to V_M/R, because when the transient is over, L is just a zero ohms wire. The output v_2 has the e^{-x} waveform (Figure 90121). In this RL circuit the pole is at −R/L, and so the exponential is exp(−Rt/L). The frequency-time relationship is this.

Figure 901 Series RL

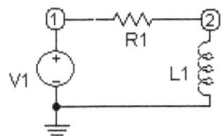

> *Time constant* $\tau = L/R = 1/\omega_0$.

Calculate R and L voltages from step 06 of *Laplace Transform analysis* of the series RL circuit step response.

(5) $v_R(t) = Ri = V_M(1-e^{-\frac{R}{L}t}) = V_M(1-e^{-x})$

(6) $v_L(t) = L\dfrac{di}{dt} = L\dfrac{V_M}{R}\dfrac{R}{L}e^{-\frac{R}{L}t} = V_M e^{-\frac{R}{L}t} = V_M e^{-x}$

(7) verify $v_L(t) + v_R(t) = V_M$

Electric Circuits - Analysis and Design

Equation 5 shows that the resistor voltage has the $1-e^{-x}$ waveform, and equation 6 shows that the inductor voltage has the e^{-x} waveform (see also Figure 90121). Their sum is 1, and the circuit time constant is $\tau = L/R$.

Stored energy The initial energy was stored in the inductor L by current $i(0^-)$. The initial energy stored in the inductor is dissipated in the resistor during the natural response.

(8) $\quad p = v_1 i_1 = i_1^2 R = \left(\dfrac{V_M}{R} e^{-\frac{R}{L}t}\right)^2 R = \dfrac{V_M^2}{R} e^{-\frac{2Rt}{L}}$

(9) $\quad W_R(t) = \int_0^t p\,dx = \int_0^t \dfrac{V_M^2}{R} e^{-\frac{2Rx}{L}}\,dx = -\dfrac{V_M^2}{R}\dfrac{L}{2R} e^{-\frac{2Rx}{L}}\bigg|_0^t = \dfrac{1}{2}L\dfrac{V_M^2}{R^2}(1-e^{-\frac{2Rt}{L}})$

(10) $\quad W_R(\infty) = \dfrac{1}{2}L\dfrac{V_M^2}{R^2} = \dfrac{1}{2}LI_M^2 = W_L(0^-)$

Spice Program 9012
```
Fig9012.ckt   series RL circuit step
* PULSE(Vbase.Vmax Tdelay Trise Tfall Twidth Tperiod )
V1 1 0 Pulse(0 0.9 0.1n 0n 0n 2n 10n)
R1 1 2 1000
L1 2 0 100n
*.PLOT TRAN I(V1) -5M,5M
.TRAN 1e-011 1e-009 0
.TEMP 27
.PLOT TRAN V(1) V(2) -1,1
.end
```

Figure 90121 Series RL circuit step function response, L/R=100pS

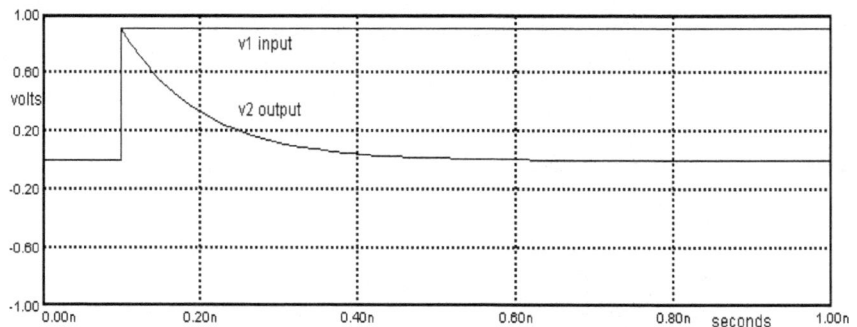

Observe that the v_2 plot replicates equation 5.

146

9 One Mesh RL & RC Circuits

Laplace Transform analysis of the series RL circuit step response

01. $v_s(t) = v_L + v_R = L\dfrac{di(t)}{dt} + Ri(t)$

02. $V_s(p) = [pL + R]I(p)$ [assume $i(0) = 0$]

03. $I(p) = \dfrac{V_s(p)}{pL + R}$ and $Z(p) = pL + R$

04. if $v_s(t) = V_m u(t)$ then $V_s(p) = \dfrac{V_m}{p}$

05. $I(p) = \dfrac{1}{pL+R} V_s(p) = \dfrac{1}{pL+R} \cdot \dfrac{V_M}{p} = \dfrac{V_m}{L} \cdot \dfrac{1}{p+\dfrac{R}{L}} \cdot \dfrac{1}{p}$

06. $i(t) = \dfrac{V_M}{R}(1 - e^{-\dfrac{t}{\tau}})u(t)$ where $\tau = \dfrac{L}{R}$

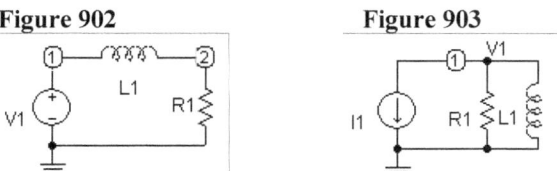

Figure 902   **Figure 903**

Problem 904 Reference Figure 902. Change Spice program Fig9012 to FigP9021 representing Figure 902 and plot the transient response.

Problem 905 Reference Figure 903. Change Spice program Fig9012 to FigP9031 representing Figure 903 and plot the transient response.

Problem 906 Reference Figure 902. If $v_2(t)=5[1-\exp(-t/\tau)]$, $i(t)=0.002 \exp(-2000t)$, and $R=1K\Omega$, then find $v_1(t)$, L_1, τ, $i_{L1}(0)$.

Problem 907 Reference Figure 903. If $i_1(t)=I_m\, u(t)$, $L_1=2\mu H$, $R_1=5K\Omega$, and $i_{L1}(0)=0.02A$, then find τ, $v_1(t)$. and $v_1(0)$.

Electric Circuits - Analysis and Design

9.2.2 Pulses u(t)–u(t–T)

A pulse is formed by using two step functions: u(t) starts the pulse that is terminated by a negative u(t) delayed by T seconds: $i_1(t) = I_M [u(t) - u(t-T)]$. The response is simply the algebraic sum of the responses to the two u(t) step functions. When time delay T is greater than 5 time constants the responses to the two u(t) step functions are, in effect, independent. However, as T decreases from 500pS to 100pS the responses interact (Figure 90311).

Figure 907 u(t)–u(t–T) pulse

Spice Program 9031

```
Fig9031.ckt   parallel RL circuit
* PULSE(Vbase Vmax Tdelay Trise Tfall Twidth Tperiod )
I5 5 0 PULSE(0 -4m 100p 000p 000p 500p 2000p)
L5 5 0 100n  ;T=500ps
R5 5 0 1000

I2 2 0 PULSE(0 -4m 105p 000p 000p 200p 1000p)
L2 2 0 100n  ;T=200ps
R2 2 0 1000

I1 1 0 PULSE(0 -4m 110p 000p 000p 100p 1000p)
L1 1 0 100n  ;T=100ps
R1 1 0 1000
.TRAN 1e-011 1e-009 0
.TEMP 27
.PLOT TRAN V(5) V(2) V(1) -5,5
.end
```

Figure 90311 Parallel RL Circuit Responses, L/R=100pS

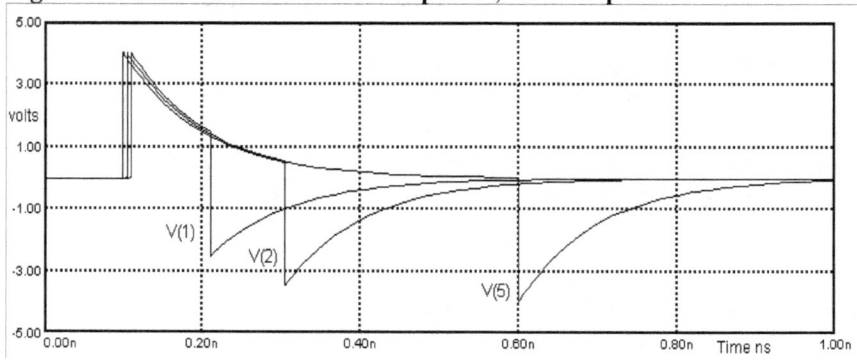

148

9 One Mesh RL & RC Circuits

Laplace Transform analysis of the parallel RL circuit pulse response

(03) $\quad V_1(p) = \dfrac{i}{y} = \dfrac{1}{G + \dfrac{1}{pL}} I_1(p) + \dfrac{1}{G + \dfrac{1}{pL}} \dfrac{i(0^-)}{p} \quad$ Figure 903

(04) $\quad$ If $i_1(t) = I_M [u(t) - u(t-T)]$ then $I_1(p) = \dfrac{I_M}{p}(1 - e^{-pT})$

(05) $\quad V_1(p) = \dfrac{1}{G + \dfrac{1}{pL}} \dfrac{I_M}{p}(1 - e^{-pT}) + \dfrac{1}{G + \dfrac{1}{pL}} \dfrac{i(0^-)}{p}$

$\quad V_1(p) = I_M R \dfrac{1}{p + \dfrac{R}{L}}(1 - e^{-pT}) + i(0^-) R \dfrac{1}{p + \dfrac{R}{L}}$

(06) $\quad v_1(t) = I_m R \left(e^{-\frac{Rt}{L}} u(t) - e^{-\frac{R(t-T)}{L}} u(t-T) \right) + i(0^-) R e^{-\frac{Rt}{L}} u(t)$

Problem 908 Reference Figure 909. Show that the energy stored in the inductor L is dissipated in the resistor R.

Figure 909

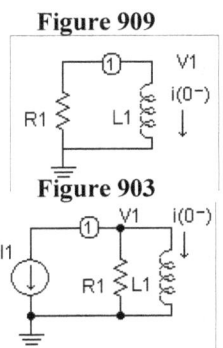

Problem 909 Reference Figure 903. If $i_1(t)=0.007\exp(-10^6 t)$, $v_1(t)=12\exp(-10^6 t)$, then find R_1, L_1, τ (μs), $i(0)$, $\omega_L(0)$.

Figure 903

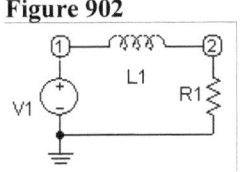

Problem 910 Reference Figure 901 If $R=1200\Omega$, $v_1(t)=5\exp(-4.0 \cdot 10^9 t)$ then find L, τ (ns), $i(0)$, $\omega_L(0)$.

Problem 911 Reference Figure 901. Change Spice program Fig9031 to FigP9011 representing Figure 901 and plot the transient response.

Figure 902

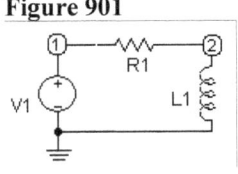

Problem 912 Reference Figure 902. Change Spice program Fig9031 or FigP9011 to FigP9023 representing Figure 902 and plot the transient response.

Figure 901

Problem 913 Reference Figure 901. If $i(t)=0.02[1-\exp(-100t)]$, $v_2(t)=3\exp(-t/\tau)$ and $v_1(t)=10\,u(t)$, then find R_1, L_1, τ.

Electric Circuits - Analysis and Design

9.2.3 Ramps t/T×u(t)

Ramps have many applications. E.g. sweep circuits in oscilloscopes and TVs are ramps. This is one reason why the ramp waveform f(t) = t/T is relevant. Clearly the ramp is not an exponential function. This is another example of the versatility of the Laplace and Spice methods. In practice we use the Spice PULSE source which has T_{rise} and T_{fall} parameters specifying ramps with finite duration.

Figure 902 Series RL

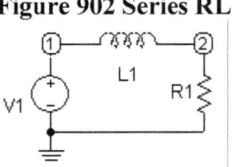

Vxx +node -node   PULSE(V_s V_p T_{delay} T_{rise} T_{fall} T_{width} T_{period})

The series RL (Figure 902) equations when $v_1(t) = V_M$ (t/T) u(t) require a more complex partial fraction expansion (sidebar equation 05), because the transform of a ramp is $1/p^2$.

Laplace Transform analysis of the series RL circuit ramp response

(03) $\quad V_1(p) = (R + pL)I_1(p) - Li(0^-)$

(04) $\quad v_1(t) = V_M \dfrac{t}{T} u(t) \quad \Rightarrow \quad V_1(p) = \dfrac{V_m}{T}\dfrac{1}{p^2}$

(05) $\quad I_1(p) = \dfrac{V_M}{T}\dfrac{1}{R+pL}\dfrac{1}{p^2} + \dfrac{L}{R+pL}i(0^-) = \dfrac{V_M}{TL}\dfrac{1}{p+\dfrac{R}{L}}\dfrac{1}{p^2} + i(0^-)\dfrac{1}{p+\dfrac{R}{L}}$

If $\dfrac{1}{p+\dfrac{R}{L}}\dfrac{1}{p^2} = \dfrac{A}{p+\dfrac{R}{L}} + \dfrac{B}{p} + \dfrac{D}{p^2}$ then

$1 = Ap^2 + Bp(p + \dfrac{R}{L}) + D(p + \dfrac{R}{L}) = (A+B)p^2 + (B\dfrac{R}{L} + D)p + D\dfrac{R}{L}$

$1 = \dfrac{DR}{L}, \qquad 0 = A + B \qquad 0 = B\dfrac{R}{L} + D \quad \text{so that}$

$D = \dfrac{L}{R} \qquad B = -\left(\dfrac{L}{R}\right)^2 \qquad A = \left(\dfrac{L}{R}\right)^2$

(05) $\quad I_1(p) = \dfrac{V_M}{TL}\dfrac{L}{R}\left[\dfrac{1}{p^2} - \dfrac{\dfrac{L}{R}}{p} + \dfrac{\dfrac{L}{R}}{p+\dfrac{R}{L}}\right] + i(0^-)\dfrac{1}{p+\dfrac{R}{L}}$

(06) $\quad i_1(t) = \dfrac{V_M}{RT}\left(t - \dfrac{L}{R} + \dfrac{L}{R}e^{-\dfrac{Rt}{L}}\right)u(t) + i(0^-)e^{-\dfrac{Rt}{L}}u(t)$

150

9 One Mesh RL & RC Circuits

After the transients die out the output is linear in t, but delayed by one L/R time constant.

(11) $\quad i_1(t) = \dfrac{V_M}{RT}\left(t - \dfrac{L}{R}\right)u(t)$

Explanation of the "−L/R" term in the Ramp Response The voltage v_S divides into v_R that is the drop across the resistor R, and the voltage v_L which is the drop across the inductor L. After the transients expire the response current increases linearly. That means di/dt is a constant. Since $v_L = L\,di/dt$ the reduction in v_R is a constant v_L as time elapses. Here are the equations representing these comments

$$\text{from (11)} \quad \frac{di_1(t)}{dt} = \frac{V_M}{RT} \quad \Rightarrow \quad v_L(t) = L\frac{di_1(t)}{dt} = L\frac{V_M}{RT}$$

$$i_R = \frac{v_R}{R} = \frac{1}{R}(v_S - v_L) = \frac{1}{R}\left(\frac{V_M t}{T} - \frac{LV_M}{RT}\right) = \frac{V_M}{R}\frac{1}{T}\left(t - \frac{L}{R}\right)$$

Spice Program 9022

```
Fig9022.ckt   series RL circuit
*PULSE(Vbase Vmax Tdelay Trise Tfall Twidth Tperiod )
V1 1 0 PULSE(0 4 100p 500p 000p 1000p 2000p)
L1 1 2 100n   ; 100nH
R1 2 0 1000   ; 1000 ohms
.TRAN 1e-011 1e-009 0
.TEMP 27
.PLOT TRAN V(1) V(2) 0,5
.end
```

Figure 90221 Series RL Ramp Response. L/R=100pS

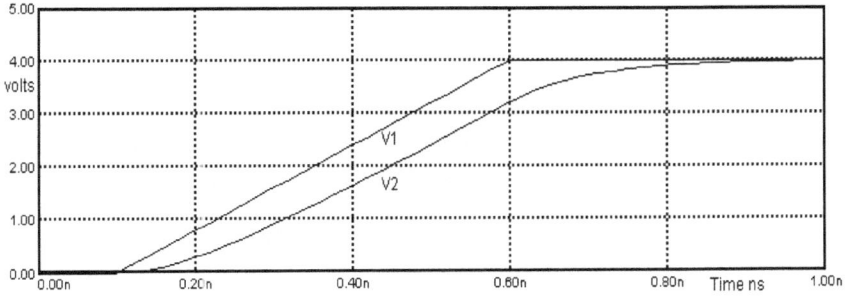

151

Electric Circuits - Analysis and Design

9.3 RC Steady State

The elementary parameters of the RC circuit are the time constant RC seconds, the stored energy $E=\frac{1}{2}Cv^2$, and the capacitor's $1/j\omega C$ impedance in the steady state. Laplace analysis produces the steady state response to $v_1 = V_M e^{j\omega t}$.

Figure 904 Low pass RC Circuit   **Figure 905 High pass RC circuit**

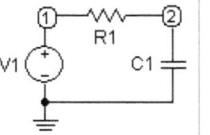

(12a) $\quad i(t) = \dfrac{V_M}{R + \dfrac{1}{j\omega C}} e^{j\omega t} + \dfrac{V_M}{R} \cdot \dfrac{1}{1+j\omega CR} e^{-\frac{t}{RC}} - \dfrac{v(0)}{R} e^{-\frac{t}{RC}}$

(12b) $\quad i(t) = \dfrac{V_M}{R + \dfrac{1}{j\omega C}} e^{j\omega t} = \dfrac{j\omega C}{1+j\omega CR} V_M e^{j\omega t} \quad$ after $5RC$ seconds elapse

(12c) $\quad \tau = RC = \dfrac{1}{\omega_0} \quad$ and $\quad \dfrac{1}{\omega_0 C} = R$

The low pass and high pass transfer functions:

(13a) $T_{low\_Pass}(\omega) = \dfrac{v_2}{v_1} = \dfrac{z_C i(t)}{V_M e^{j\omega t}} = \dfrac{1/j\omega C}{V_M e^{j\omega t}} \dfrac{j\omega C}{1+j\omega CR} V_M e^{j\omega t} = \dfrac{1}{1+j\omega CR} \quad (Fig 904)$

(13b) $T_{high\_pass}(\omega) = \dfrac{v_2}{v_1} = \dfrac{z_R i(t)}{V_M e^{j\omega t}} = \dfrac{R}{V_M e^{j\omega t}} \dfrac{j\omega C}{1+j\omega CR} V_M e^{j\omega t} = \dfrac{j\omega CR}{1+j\omega CR} \quad (Fig 905)$

Manipulate the equations to emphasize the ratio of C and R impedances.

(14a) $\quad T_{low\_pass}(\omega) = \dfrac{1}{1+j\omega CR}$ (14b) $\quad T_{high\_Pass}(\omega) = \dfrac{1}{1+\dfrac{1}{j\omega CR}}$

The impedance of any C is $z_C = 1/j\omega C$. The circuit in Figure 904 is a low pass circuit, because z_C decreases with frequency as frequency increases above ω_0 so that the transmission approaches magnitude 0 and phase $-90°$. Z_C increases with frequency as frequency decreases below ω_0 so that the transmission approaches magnitude 1 and phase $0°$. The phase is $45°$ at ω_0 (Figure 90412). When R and C are exchanged (Figure 905), the behavior above and below ω_0 is exchanged so that we now have a high pass filter (Figures 90511, 90512).

9 One Mesh RL & RC Circuits

Spice Program 9041 Low pass RC Spice Program

```
Fig9041.ckt low pass RC circuit sine
V1 1 0 AC 1
R1 1 2 1000
C1 2 0 100p
.AC DEC 200 10000 1e+009
*.PLOT AC VP(2) -100,0
.TEMP 27
.PLOT AC VDB(1) VDB(2) -40,10
.end
```

Transmission is reduced by 3dB when the source frequency ω makes the reactance of C equal to R ($R=1/\omega_0 C$). The corner frequency is f_0 The easiest way to find f_0 on a plot is to find the 45 degree phase frequency in Figure 90412.

$$(14) \quad f_0 = \frac{1}{2\pi RC} = \frac{1}{2\pi \times 1000 \times 100 \cdot 10^{-12}} = 1.590 \, MHz$$

Figure 90411 Low Pass RC Circuit magnitude of T(ω)

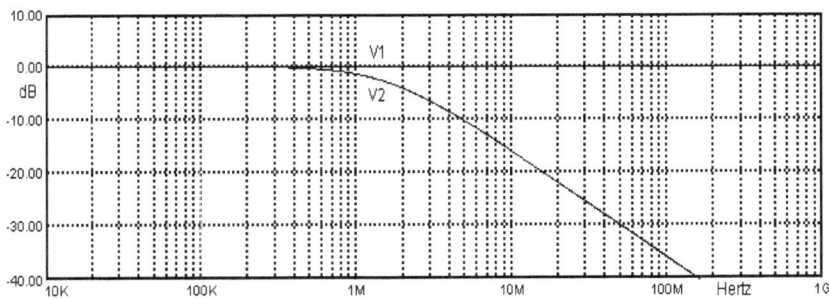

Figure 90412 Low Pass RC Circuit phase of T(ω)

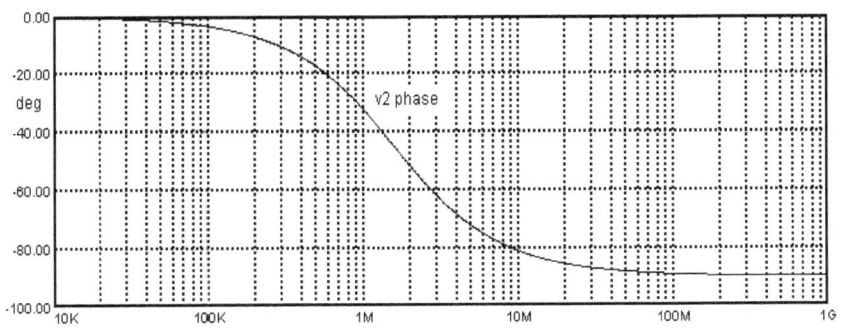

Electric Circuits - Analysis and Design

Spice Program 9051 High pass RC Spice Program
```
Fig9051.ckt series RC circuit sine
V1 1 0 AC 1
C1 1 2 100p
R1 2 0 1000
*.PLOT AC VDB(1) VDB(2) -40,10
.AC DEC 200 10000 1e+009
.TEMP 27
.PLOT AC VP(2) 0,100
.end
```

Figure 90511 High Pass RC Circuit magnitude of T(ω)

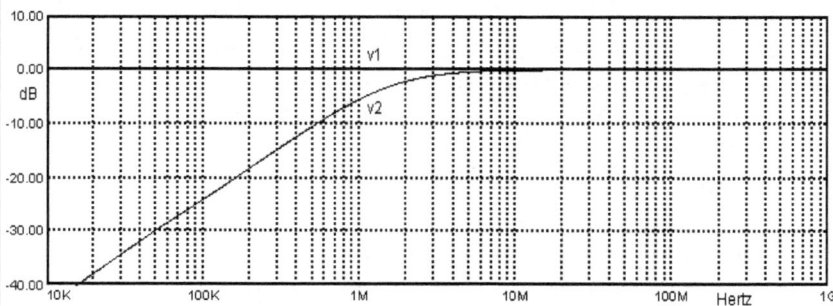

Figure 90512 High Pass RC Circuit phase of T(ω)

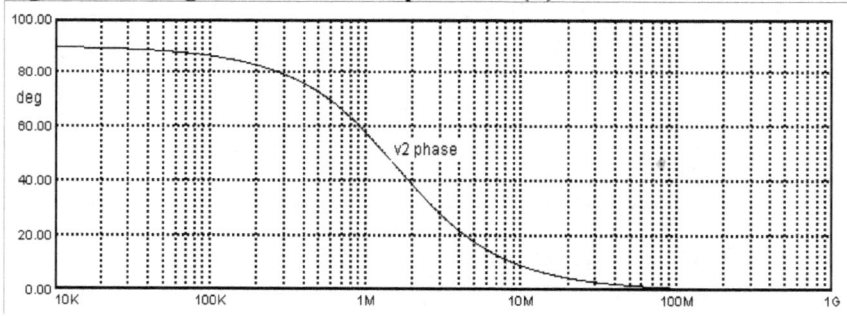

Problem 914 Reference Figure 905. Change Spice program Fig9051 to FigP9051 representing Figure 905 and plot the AC response.

Problem 915 Reference Figure 906. Change Spice program Fig9051 to FigP9061 representing Figure 906 and plot the AC response.

Problem 916 Reference Figure 905. If $v(0^-)=0$, $v_1(t)=9u(t)$, $i(t)=0.003\exp(-2\times 10E6t)$, then find R, C, τ, $v_2(t)$.

154

9.4 RC Transient State

We examine the series RC circuit transient response to
step functions u(t), pulses u(t)−u(t−T), and ramps (t/T)u(t).

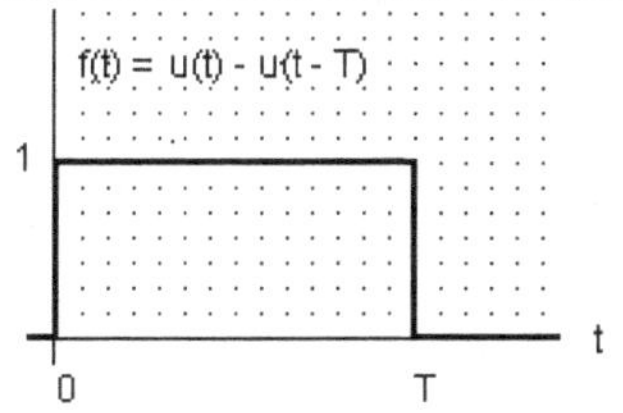

The step at t=0 $v_1(t)=V_M u(t)$ (Figure 907) is in effect an "infinite" dv/dt at time t=0. Since $i_C=C dv/dt$ there must be an infinite current through the capacitor. In fact, the capacitor current is limited to V_M/R. This leaves V_M volts across R, and zero voltage across C that is another way to see that C's voltage starts from zero. As the current decreases the di/dt decreases to zero exponentially as time elapses so that in the final analysis all of the voltage V_M is across C (Figure 90421). In other words, the capacitor C responds to a step function at t=0 as if C is a short circuit.

Figure 907 Steps u(t), u(t-T)

f(t) = u(t) - u(t - T)

9.4.1 Steps u(t)

The Laplace transform RC analysis (see sidebar) shows that the response to the step function u(t) waveform is a current with the e^{-x} waveform. The current starts at V_M/R and decreases to 0, because when the transient is over C is just an open circuit. Output v_2 has the $1-e^{-x}$ waveform. In this RC circuit the pole is at $-1/RC$, and the exponential is $\exp(-t/RC)$. The frequency-time relationship is this.

Figure 904

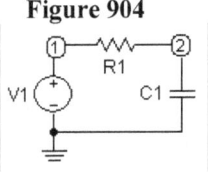

$$\boxed{\text{Time constant } \tau = RC = 1/\omega_0.}$$

Calculate R and C voltages from step 06 of *Laplace Transform analysis* of the series RC circuit step response

$$v_C(t) = \frac{1}{C}\int_0^t i(x)dx = \frac{1}{C}\cdot\frac{V_M}{R}\int_0^t e^{-\frac{x}{\tau}}dx$$

(15) $v_C(t) = \frac{V_M}{RC}(-\tau)\, e^{-\frac{x}{\tau}}\Big|_0^t = V_M\left(1-e^{-\frac{t}{\tau}}\right) = V_M\left(1-e^{-x}\right)$

(16) $v_R(t) = iR = \dfrac{V_M}{R} R e^{-\frac{t}{\tau}} = V_M e^{-\frac{t}{\tau}} = V_M e^{-x}$

(17) $v_R(t) + v_C(t) = V_M e^{-\frac{t}{\tau}} + V_M \left(1 - e^{-\frac{t}{\tau}}\right) = V_M$

The capacitor voltage has the $1-e^{-x}$ waveform, and the resistor voltage has the e^{-x} waveform. Their sum is 1, and the circuit time constant is $\tau = RC$.

Stored energy The initial energy was stored in the capacitor C by voltage $v(0^-)$. The initial energy stored in the capacitor is dissipated in the resistor during the natural response.

(18) $p = vi = i^2 R = \left(\dfrac{V_M}{R} e^{-\frac{t}{RC}}\right)^2 R = \dfrac{V_M^2}{R} e^{-\frac{2t}{RC}}$

(19) $W_R(t) = \int_0^t p\, dx = \int_0^t \dfrac{V_M^2}{R} e^{-\frac{2x}{RC}} dx = -\dfrac{V_M^2}{R} \dfrac{RC}{2} e^{-\frac{2x}{RC}} \Big|_0^t = \dfrac{1}{2} C V_M^2 (1 - e^{-\frac{2t}{RC}})$

(20) $W_R(\infty) = \dfrac{1}{2} C V_M^2 = W_C(0^-)$

Laplace Transform analysis of the series RC circuit step response

(02) $V_S(p) = RI(p) + \dfrac{1}{pC} I(p) = \left(R + \dfrac{1}{pC}\right) I(p)$ assume $i(0) = 0$

(03) $I(p) = \dfrac{1}{R + \dfrac{1}{pC}} V_S(p) = \dfrac{1}{p + \dfrac{1}{RC}} \dfrac{V_m}{R}$

(05) $I(p) = \dfrac{1}{p + \dfrac{1}{RC}} \dfrac{V_M}{R}$

(06) $i(t) = \dfrac{V_M}{R} e^{-\frac{t}{\tau}} u(t)$ where $\tau = RC$

---

9 One Mesh RL & RC Circuits

Spice Program 9042

```
Fig9042.ckt   series RC circuit    step
* PULSE(Vbase Vmax Tdelay Trise Tfall Twidth Tperiod )
V1 1 0 Pulse(0 0.9 1n 0n 0n 2n 10n)
R1 1 2 1000
C1 2 0 100f

.TRAN 1e-012 5e-009 0
.TEMP 27
.PLOT TRAN V(1) V(2) 0,1
.end
```

Figure 90421 Series RC circuit transient response

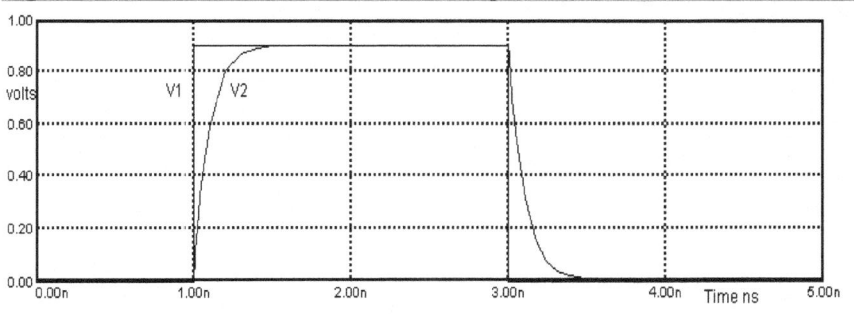

Figure 905   **Figure 906**

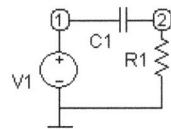 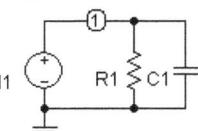

Problem 917 Reference Figure 905. Change Spice program Fig9051 to FigP9051 representing Figure 905 and plot the transient response.

Problem 918 Reference Figure 906. Change Spice program Fig9051 to FigP9062 representing Figure 906 and plot the transient response.

157

9.4.2 Pulses u(t)–u(t–T)

A pulse is formed by using two step functions: u(t) starts the pulse that is terminated by a negative u(t) delayed by T seconds: $i_1(t) = I_M [u(t) - u(t-T)]$. The response is simply the algebraic sum of the responses to the two u(t) step functions.

Figure 907 u(t)–u(t–T) pulse

When time delay T is greater than 5 time constants the responses to the two u(t) step functions are, in effect, independent. However, as T decreases from 500pS to 100pS the responses interact (Figure 90611).

Figure 906 Parallel RC

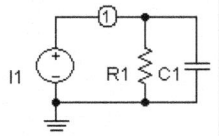

Laplace Transform analysis of the parallel RC circuit pulse response

(03) $\quad V(p) = \dfrac{1}{G+pC} I_S(p) + \dfrac{1}{G+pC} Cv(0^-)$

(04) $\quad$ If $i_S(t) = I_M [u(t) - u(t-T)]$ then $I_S(p) = \dfrac{I_M}{p}\left(1 - e^{-pT}\right)$

(05) $\quad V(p) = \dfrac{1}{G+pC} \dfrac{I_M}{p}\left(1 - e^{-pT}\right) \quad$ assume $v(0^-) = 0$

$\quad V(p) = I_M R \left(\dfrac{1}{p} - \dfrac{1}{p + \dfrac{1}{RC}} \right)\left(1 - e^{-pT}\right)$

(06) $\quad v(t) = I_m R \left[\left(1 - e^{-\frac{t}{RC}}\right) u(t) - \left(1 - e^{-\frac{(t-T)}{RC}}\right) u(t-T) \right]$

9 One Mesh RL & RC Circuits

Spice Program 9061

```
Fig9061.ckt   parallel RC circuit   pulse
* PULSE(Vbase Vmax Tdelay Trise Tfall Twidth Tperiod )
I1 1 0 PULSE(0 -4m 100p 000p 000p 500p 1000p)
C1 1 0 100f    ; 100fF
R1 1 0 1000    ; 1000 ohms

I2 2 0 PULSE(0 -3.9m 100p 000p 000p 200p 1000p)
C2 2 0 100f    ; 100fF
R2 2 0 1000    ; 1000 ohms

I5 5 0 PULSE(0 -3.8m 100p 000p 000p 100p 1000p)
C5 5 0 100f    ; 100fF
R5 5 0 1000    ; 1000 ohms
.TRAN 1e-011 1e-009 0
.TEMP 27
.PLOT TRAN V(1) V(2) V(5) 0,5
`.end
```

Figure 90611 Parallel RC Circuit transient responses

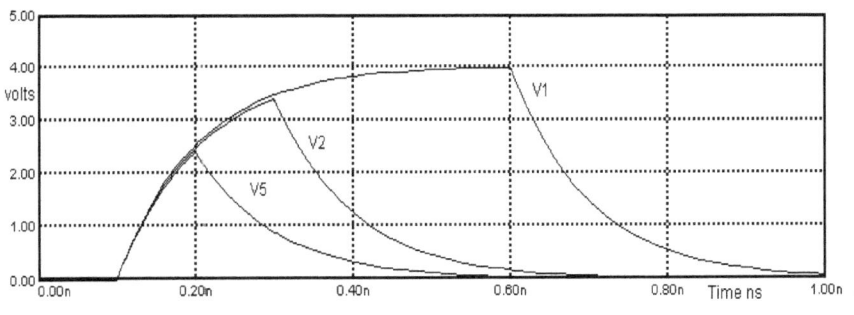

Figure 904

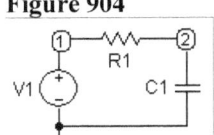

Figure 905 Series RC

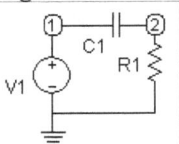

Problem 919 Reference Figure 904. Change Spice program Fig9061 to FigP9043 representing Figure 904 and plot the transient response.

Problem 920 Reference Figure 905. Change Spice program Fig9061 or FigP9043 to FigP9053 representing Figure 905 and plot the transient response.

159

Electric Circuits - Analysis and Design

9.4.3 Ramps t/T×u(t)

The ramp is not an exponential function. The Spice PULSE source has rise and fall parameters that specify ramps with finite duration. The series RC (Figure 905) equations when $v_S(t) = V_M (t/T) u(t)$ require a more complex partial fraction expansion, because the transform of a ramp is V_M/Tp^2. After the transients die out the output is linear in t, but delayed by one RC time constant.

(21) $\quad v_C(t) = \dfrac{1}{C} \displaystyle\int_0^t i(x)dx = \dfrac{1}{C} \dfrac{V_M}{R} \dfrac{RC}{T} \displaystyle\int_0^t \left(1 - e^{-\frac{t}{RC}}\right) dx = \dfrac{V_M}{T}\left[t - RC\left(1 - e^{-\frac{t}{RC}}\right)\right]$

(22) $\quad v_C(t) = V_M \left(\dfrac{t}{T} - \dfrac{RC}{T}\right)$

Laplace Transform analysis of the series RC circuit ramp response

(03) $\quad I(p) = \dfrac{1}{R + \dfrac{1}{pC}} \left(V_S(p) - \dfrac{v(0^-)}{p}\right) \qquad \text{assume } v(0^-) = 0$

(04) $\quad v_S(t) = V_M \dfrac{t}{T} u(t) \quad \Rightarrow \quad V_S(p) = V_M \dfrac{1}{Tp^2}$

(05) $\quad I(p) = \dfrac{1}{R + \dfrac{1}{pC}} \times \dfrac{V_M}{Tp^2}$

Execute a partial fraction expansion.

$I(p) = \dfrac{1}{R + \dfrac{1}{pC}} \dfrac{V_m}{Tp^2} = \dfrac{1}{p + \dfrac{1}{RC}} \dfrac{V_m}{RTp}$

If $\dfrac{1}{p + \dfrac{1}{RC}} \dfrac{1}{p} = \dfrac{A}{p + \dfrac{1}{RC}} + \dfrac{B}{p} \quad$ then $\quad B = RC, \quad A = -RC$

(05) $\quad I(p) = \dfrac{V_m}{RT}\left(-\dfrac{RC}{p + \dfrac{1}{RC}} + \dfrac{RC}{p}\right)$

(06) $\quad i(t) = \dfrac{V_m}{R} \dfrac{RC}{T}\left(1 - e^{-\frac{t}{RC}}\right) u(t)$

9 One Mesh RL & RC Circuits

Spice Program 9052

```
Fig9052.ckt   series RC circuit
*        PULSE(Vbase Vmax Tdelay Trise Tfall Twidth Tperiod )
V3 3 0 PULSE(0    4     100p    500p  000p  1000p  2000p)
R2 3 4 1000   ; 1000 ohms
C2 4 0 100f   ; 100fF
.TRAN 1e-011 1e-009 0
.TEMP 27
.PLOT TRAN V(3) V(4) 0,5
.end
```

Figure 90521 Series RC Ramp transient response

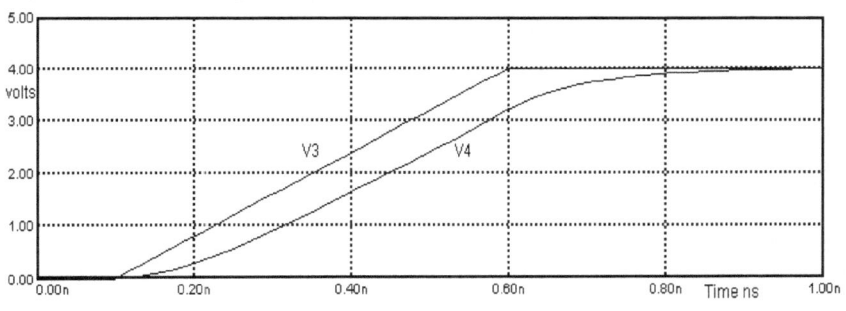

| Figure 904 | Figure 905 | Figure 906 |

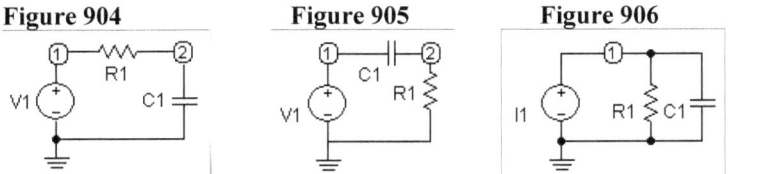

Problem 921 Reference Figure 904. If $v(0^-)=0$, $v_{C1}(t)=V_m[1-\exp(-t/\tau)]$ and $v_1(t)=5u(t)$, $i(t)=10^{-3}\exp(-10^8 t)$, $v_{C1}(0)=2$, then find R_1, C_1, τ.

Problem 922 Reference Figure 905. If $v(0^-)=0$, $v_{R1}(t)=V_m\exp(-t/\tau)$ and $i(t)=0.002\exp(-10^6 t)$ amps, $v_1(t)=8(t)$ volts, $v_{C1}(0)=0$, then find R_1, C_1, τ.

Problem 923 Reference Figure 906. Change Spice program Fig9051 (but retain V3) to FigP9062 representing Figure 906 and plot the transient response.

Electric Circuits - Analysis and Design

9.5 Series RC Differentiator

An ideal voltage *pulse* of time duration T and magnitude V_M is constructed from a sequence of two step functions: V_M u(t) followed by $-V_M$ u(t − T) (Figure 907). After time t = T the algebraic sum of the step functions is zero, and from time 0 to T their sum is one, because u(t − T) is zero during that time. The derivative of this ideal pulse is zero everywhere except at times 0 and T. At times 0 and T its value is $+V_M$ δ(t) and $-V_M$ δ(t − T) where δ(t) is the impulse operator that is the derivative of the step function.

Figure 907 Pulse

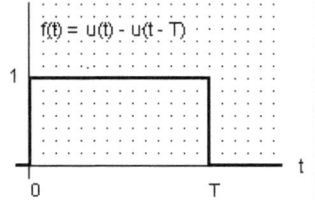

Figure 905 RC

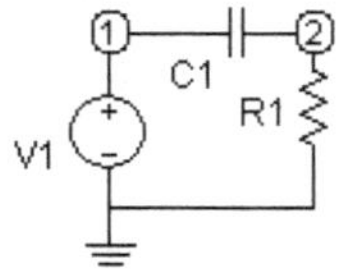

In practice neither the pulse nor the formation of the derivative is ideal. However, a series RC circuit (Figure 905), in effect, forms a "practical" derivative when signal pulse width T >> RC. The capacitor C charges up to V_M volts 5 RC seconds after the onset of the first step function u(t). Significant current flows in the resistor R for only 5 RC seconds. The output voltage duration is 5RC because the voltage equals iR. At time T the second step function u(t − T) switches the source voltage to zero. What ensues is the natural response of the circuit. The initial voltage v(T⁻) is V_M because the capacitor was charged to V_M after the onset of u(t). The current flow is in the opposite direction this time that makes iR negative. This is consistent with a negative step, whose derivative is negative.

In Spice program 9053 the pulse width T = 500 ps, and the time constant τ = RC = 1000Ω×10fF = 10 ps, so that 5 RC = 50 ps = 0.1T.

(23) $\quad \dfrac{v(t)}{V_M} = e^{-\frac{t}{RC}} u(t) - e^{-\frac{t-T}{RC}} u(t-T) \quad \rightarrow \quad \dfrac{v(t)}{V_M} = e^{-\frac{t}{0.1T}} u(t) - e^{-\frac{t-T}{0.1T}} u(t-T)$

Note A step function's dt=0 produces a maximum current V_M/R. A physically real dt is greater than zero that results in a reduced peak value at the output.

9 One Mesh RL & RC Circuits

Spice Program 9053

```
Fig9053.ckt   series RC circuit diff
* PULSE(Vbase Vmax Tdelay Trise Tfall Twidth Tperiod )
V1 1 0 PULSE(0 4 100p 000p 000p 500p 2000p)
C1 1 2 10f ; 10fF
R1 2 0 1000 ; 1000 ohms
.TRAN 1e-011 1e-009 0
.TEMP 27
.PLOT TRAN V(1) V(2) -5,5
.end
```

Figure 90531 RC Differentiator

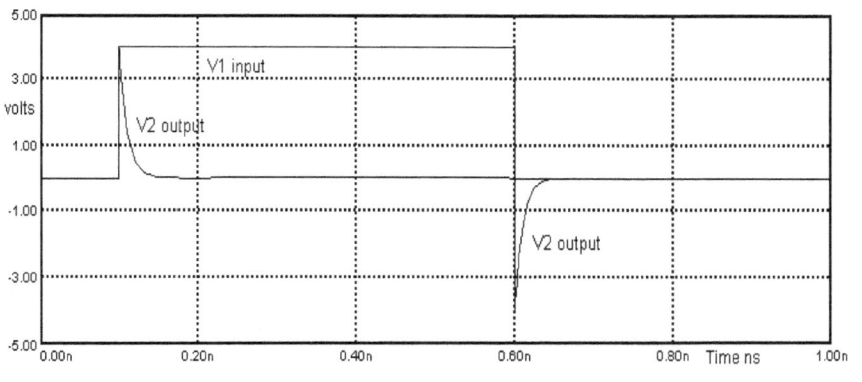

9.6 Square Wave Fidelity in an RC Circuit

This is a good example of how Spice spares us the hard work of mathematical manipulations. An ideal voltage *square wave* of period T and magnitude V_M (Figure 908) is constructed from a sequence of pulses. The average value, the DC value, is $V_M/2$.

Figure 908 Squarewave with period T=2c

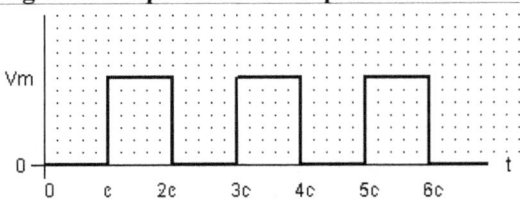

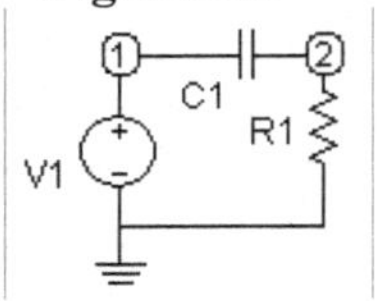

Suppose we need to pass a square wave through an RC circuit (Figure 905) with minimum "distortion". Clearly RC must be >> T for low distortion; i.e. a flat-top square wave. The top "tilts" when the series capacitor acquires charge, and the average output voltage shifts. In other words the DC value is not passed.

Figure 905

Observe the transient shift of the DC value from 1/2 to 0 (Figure 90541). If RC = 2T then the distortion is as shown in Figure 90541. In the real world RC might be equal to 20T.

(23) $\quad \dfrac{v(t)}{V_M} = \sum_{i=0}^{n} \left[e^{-\frac{t-2iT}{RC}} u[t-2iT] - e^{-\frac{t-(2i+1)T}{RC}} u[t-(2i+1)T] \right]$

(24) $\quad \dfrac{v(t)}{V_M} = \left[e^{-\frac{t}{RC}} u[t] - e^{-\frac{t-T}{RC}} u[t-T] \right] + \left[e^{-\frac{t-2T}{RC}} u[t-2T] - e^{-\frac{t-3T}{RC}} u[t-3T] \right] + ...$

(25) $At\ time\ T \quad v_C(T) = V_M(1 - e^{-\frac{T}{RC}}) = V_m(1 - 1 + \dfrac{T}{RC} - ...) = V_M \dfrac{T}{RC}$

(26) $If \quad \dfrac{v_C}{V_M} = \dfrac{1}{100} \quad then\ RC = 100T \quad for\ 1\%\ distortion$

9 One Mesh RL & RC Circuits

Mathematics dominates the electrical engineering in problems such as these. A practical procedure uses Spice to solve these problems.

If $RC = T$ then in a half period $\quad e^{-\frac{T}{2RC}} = e^{-0.5} = 0.6$

(27) $V_C = V_M \left(1 - e^{-\frac{T}{2RC}} \right) = V_M (1 - 0.6) = 0.4 V_M$

Spice Program 9054

```
Fig9054.ckt   series RC circuit
* PULSE(Vbase Vmax Tdelay Trise Tfall Twidth Tperiod )
V1 1 0 PULSE(0 4 000p 000p 000p 1000p 2000p)
C1 1 2 2000f
R1 2 0 1000

.TRAN 1e-011 1e-008 0
.TEMP 27
.PLOT TRAN V(1) V(2) -5,5
.end
```

Figure 90541 Square Wave Fidelity

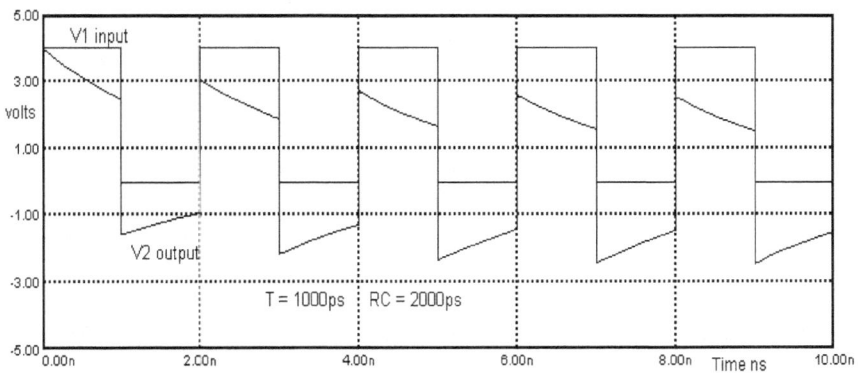

Electric Circuits - Analysis and Design

10 The Bode Method

The Bode method is an important, tremendously useful, graphical method that converts abstract ratios of polynomials encountered in electric circuit mathematics into comprehensible visual displays. For example, the Bode analysis in Section 6.3 page 95 is a practical example of the Bode Method.

The amplitude and phase of any immittance[1] or circuit response provides important and valuable information in the analysis and design of circuits. For example, TV (and radio) transmitters must restrict their output energy to allocated bands of frequencies in order avoid interference with each other. The analog part of the telephone system uses sophisticated filters and amplifiers to maintain the integrity of its multitude of information channels.

The magnitude response shows how the sinusoidal responses to sinusoidal inputs vary in magnitude over a frequency range of interest. The same may be said for the phase response. The significance of this information becomes clear as we proceed. In the majority of applications, response plots accurate to within 1% or 2% are sufficient. However, calculating the points to be plotted is a very laborious process. Fortunately there is an efficient solution when you want to do hand plots. Hendrick Bode[2] (1905-1982) presented (circa 1935) a straightforward, and widely adopted, method for making graphical displays of the magnitude and phase of electric circuit functions. The method uses a straight line approximation scheme that is implemented by logarithmic scales.

The use of the Laplace transform to transform lumped parameter electric circuit integro-differential equations into algebraic equations showed that the general form of electric circuit functions is a ratio of two polynomials $N(p)/D(p)$. Mathematicians call these functions rational algebraic functions of p. A polynomial of degree n has n roots. The roots of the numerator $N(p)$ are called zeros, whereas the roots of the denominator $D(p)$ are called poles. The poles and zeros (7.9 page 125) determine circuit behavior.

> *A polynomial of degree n has n roots, which means it can be written as a product of n factors.*

---

[1] Bode's word for impedance or admittance
[2] H. W. Bode, Network Analysis and Feedback Amplifier Design, ISBN 1124 038 647

10 Bode Method

Logarithms - a brief review
Let $y = 10^x$, $y_1 = 10^{x_1}$, $y_2 = 10^{x_2}$

If $y = y_1 y_2$, then $\log y = \log(y_1 y_2) = \log(10^{x_1} 10^{x_2}) = \log(10^{(x_1+x_2)})$
$= (x_1 + x_2)\log 10 = x_1 + x_2 = \log y_1 + \log y_2$

If $y = \dfrac{y_1}{y_2}$, then $\log y = \log\left(\dfrac{y_1}{y_2}\right) = \log(10^{x_1} 10^{-x_2}) = \log(10^{x_1 - x_2})$
$= (x_1 - x_2)\log 10 = x_1 - x_2 = \log y_1 - \log y_2$

Logarithm of a complex number
If $z = x + jy = re^{j\theta}$,

then $\log z = \log(re^{j\theta}) = \log r + \log e^{j\theta} = \log r + j\theta \log e$

In practice the factor log e is omitted so that we can plot phase in degrees (log e only represents a scale change).

Logarithm of a circuit function T(p) Any lumped parameter circuit function is the ratio of two polynomials in p. For example log T is the sum of the logs of the zero factors such as (p+z₁) minus the sum of the logs of the pole factors such as (p+p₁). For example

(3) $\quad T(p) = K\dfrac{N(p)}{D(p)} = \dfrac{n_2 p^2 + n_1 p}{d_2 p^2 + d_1 p + d_0} = \dfrac{n_2}{d_2} \dfrac{p(p+z_1)}{(p+p_1)(p+p_2)}$

(4) $\quad \log T(p) = \log\dfrac{n_2 z_1}{d_2 p_1 p_2} + \log\left(\dfrac{p}{1}\right) + \log\left(1 + \dfrac{p}{z_1}\right) - \log\left(1 + \dfrac{p}{p_1}\right) - \log\left(1 + \dfrac{p}{p_2}\right)$

Factors of T(p) The factors of T(p) are one or more of four kinds
a) K that is a constant
b) p that has root 0 at the origin
c) (p + a) that has root −a on the negative real axis
d) (p + a + jb)(p + a − jb) that have complex conjugate roots −a±jb.

Decibels A decibel (dB) is defined as 20 times the log of a ratio's magnitude.
(5) $\quad T_{dB} = 20\log|T(p)|$

10.1 Corner Frequency

A graphical display of the steady-state amplitude and phase is especially useful. We have said that in the majority of applications, response plots accurate to within 1% or 2% are sufficient. However, the laborious process calculating the points to be plotted is avoided by doing hand plots using the Bode straight line approximation method presented here. Use Spice for detailed plots (such as those in Chapter 6).

Bode uses frequency response asymptotes as straight line approximations.

The transfer function of the RC circuit (equation 6 and Figure 1000) is calculated by Spice program 10001. Bode's "corner frequency" is defined as $\omega_0 = 1/R_1 C_1$, where resistance R equals reactance X_C (Figure 100011). We calculate T(p) for various frequencies to determine the asymptotes.

Figure 1000

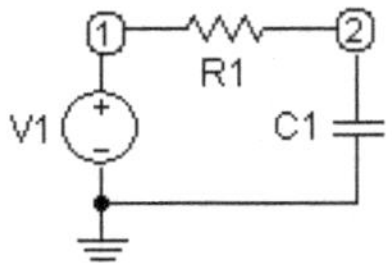

$$(6) \quad T(p) = \frac{v_2}{v_1} = \frac{\frac{1}{pC_1}}{R_1 + \frac{1}{pC_1}} = \frac{1}{1 + pC_1 R_1} = \frac{1}{1 + j\omega C_1 R_1} = \frac{1}{1 + j\frac{\omega}{\omega_0}} \quad \left(R_1 = \frac{1}{\omega_0 C_1} \right)$$

(7a) $T(0) = 1 e^{-j0°}$

(7b) $T(0.1\omega_0) = \frac{1}{1 + j0.1} = \frac{1}{1.005} e^{-j5.7°}$

(7c) $T(\omega_0) = \frac{1}{1 + j1} = \frac{1}{\sqrt{2}} e^{-j45°}$

(7d) $T(10\omega_0) = \frac{1}{1 + j10} = \frac{1}{10.05} e^{-j84.2°}$

(7e) $T(20\omega_0) = \frac{1}{1 + j20} = \frac{1}{20.02} e^{-j87.1°}$

(7f) $T(100\omega_0) = \frac{1}{1 + j100} = \frac{1}{100.005} e^{-j89.4°}$

Magnitude Equations 7a and 7b show that the asymptote from zero frequency has slope 0 (a horizontal line).

If T(0)=1 is defined as 0dB, then equation 7c shows that $T(\omega_0)$ magnitude is −3dB (20 log $1/\sqrt{2}$).

Equations 7d and 7e show that the asymptote past ω_0 is −6dB/octave (20log 10.05/20.02).

Equations 7d and 7f show that the asymptote past ω_0 is −20dB/decade (20log 10.05/100.005). Also see Figure 100011.

10 Bode Method

Phase The 0dB/octave asymptote from zero frequency has associated phase 0°. The −6dB/octave asymptote past ω_0 has associated phase −90° (Figure 100012). The phase is −45° at ω_0.

> The R and C impedances are equal by definition at corner ω_0. This is why the asymptotes cross at ω_0, and why we say ω_0 is a corner frequency (Figure 100011).

Spice program 10001

```
Fig10001.ckt   factor 1/(p+a)
V1 1 0 AC 1 0   ; volts
R1 1 2 1000
C1 2 0 0.0159155u
*.PLOT AC VDB(2) -40,10
.AC DEC 200 1C 1e+007
.PLOT AC VP(2) -100,0
.TEMP 27
.end
```

Figure 100011 Magnitude of T(p)=v_2/v_1

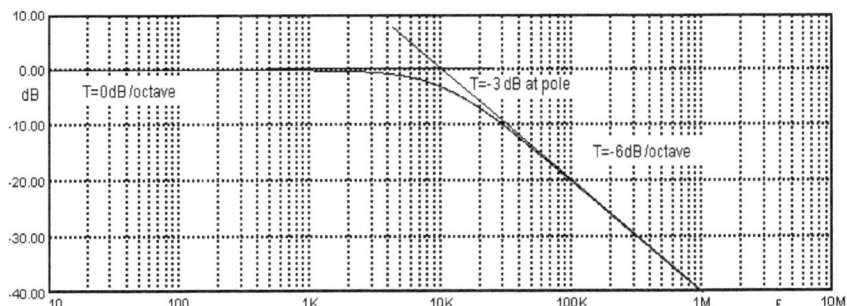

Figure 100012 Phase of T(p) =v_2/v_1

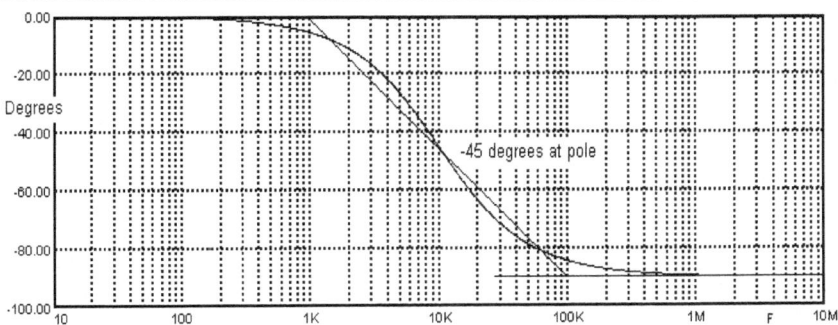

Electric Circuits - Analysis and Design

10.2 Factor K

The Bode diagram of a real constant term K is a horizontal line, because K is independent of complex frequency variable p. If the constant K=5 the line intersects the ordinate at +14dB (20 log 5), and phase is 0°. A K=1/5 constant intersects the ordinate at −14dB, , and phase is 180°. K can be positive or negative. The phase of positive and negative constants K is zero and −180 degrees respectively.

Spice program 10011

```
Fig10011.ckt  constant K
*source voltages
V1 1 0 AC 5 0    ; K=5
V2 2 0 AC 0.2 -180 ; K=-1/5
*.PLOT AC VDB(1) VDB(2) -20,20
.AC DEC 200 1 1000
.TEMP 27
.PLOT AC VP(1) VP(2) -200,50
.end
```

Figure 100111 Bode diagram of a constant K - magnitude

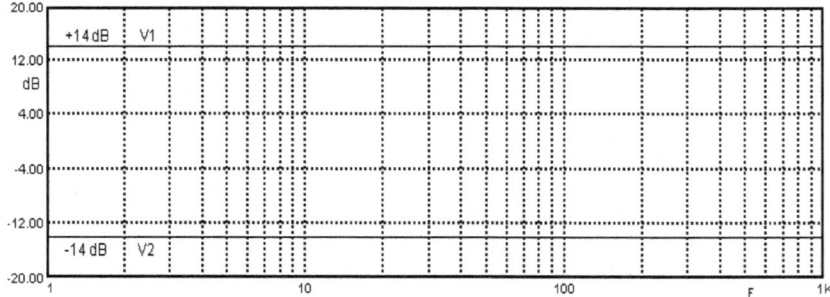

Figure 100112 Bode diagram of a constant K - phase

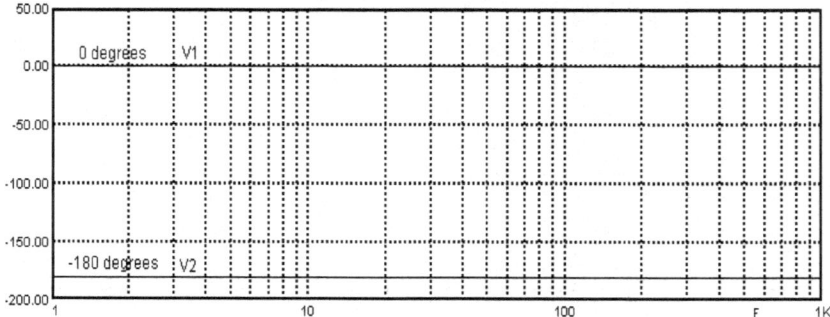

10.3 Factors p and 1/p

If p is a factor of the denominator of T(p), then its exponent is −1. On the real frequency axis $p = j\omega$ so that the magnitude $1/\omega$ is divided by 2 when the frequency doubles. Since $20 \log ½ = −6dB$, we say the factor $1/p$ has a slope of −6dB per octave on the real frequency axis (Figure 100211). The normalized factor ω_0/p crosses the 0dB line when $|\omega_0/p| = 1$ (20log1=0dB).

On the other hand, if the factor p is in the numerator its exponent is +1, and its magnitude is multiplied by 2 when the frequency doubles. Then the slope is +6dB/octave (Figure 100211).

The magnitude ω increases in value 10 times when the frequency increases 10 times. The slope is +20dB/decade that is the same slope as +6dB/octave, and a −20dB/decade slope is the same as −6dB/octave slope.

Note that we ignore the 20Log e factor in all phase terms, because we want to plot phase in degrees or radians (Figure 100212). This does not change the basic result, because $20\text{Log}_{10}e$ only represents a change of the scale.

(8) $T_{zero} = T_z(j\omega) = 20\log\dfrac{p}{\omega_0} = 20\log\left(\dfrac{\omega}{\omega_0}e^{j\frac{\pi}{2}}\right) = 20\log\dfrac{\omega}{\omega_0} + j\dfrac{\pi}{2}20\log(e)$

$T_{zero} \approx 20\log\dfrac{\omega}{\omega_0} + j\dfrac{\pi}{2}$

(9) $T_{pole} = T_p(j\omega) = 20\log\dfrac{\omega_0}{p} = 20\log\left(\dfrac{\omega_0}{\omega}e^{-j\frac{\pi}{2}}\right) \approx 20\log\dfrac{\omega_0}{\omega} - j\dfrac{\pi}{2}$

Magnitude and phase are calculated for real frequencies ($p=j\omega$) (Spice program 10021)

(10) $T_p(p) = \dfrac{\omega_0}{p} \qquad T_p(j\omega) = \dfrac{\omega_0}{j\omega} \qquad T_z(p) = \dfrac{p}{\omega_0} \qquad T_z(j\omega) = j\dfrac{\omega}{\omega_0}$

  Magnitude Phase Magnitude Phase

$|T_p(j\omega)| = \dfrac{\omega_0}{\omega} \qquad -\dfrac{\pi}{2} = -90° \qquad |T_z(j\omega)| = \dfrac{\omega}{\omega_0} \qquad \dfrac{\pi}{2} = 90°$

Electric Circuits - Analysis and Design

Spice program 10021

```
Fig10021.ckt  factors p and 1/p

*source currents
I1 1 0 AC 1m 0   ; 1mA
I2 2 0 AC 1m 0   ; 1mA

* names/nodes/values
C1 1 0 159.15p
R1 1 0 1E14   ; "infinite" R required by Spice
L2 2 0 159.15u

*.PLOT AC VDB(1) VDB(2) -25,25
.AC DEC 200 100000 1e+007
.TEMP 27
.PLOT AC VP(1) VP(2) -100,100
.end
```

Figure 100211 Bode diagram of p and 1/p - magnitude

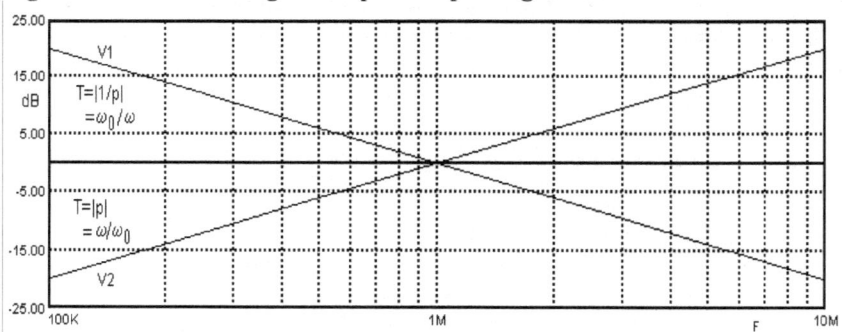

Figure 100212 Bode diagram of p and 1/p - phase

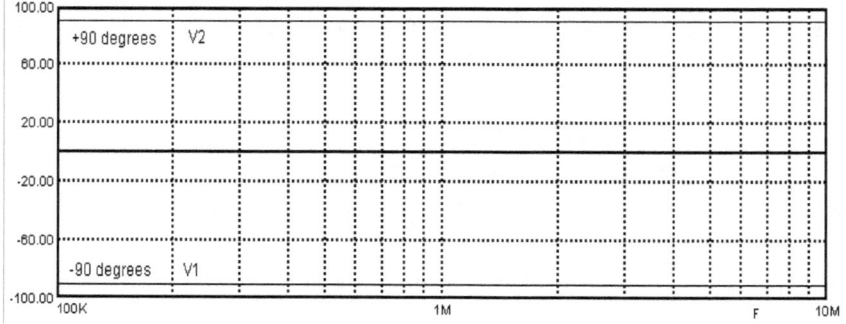

10 Bode Method

10.4 Factors p+ω₀ and 1/(p+ω₀)

A factor (p+ω₀) that is in the numerator of a rational algebraic function is referred to as a zero of that function. The factor is zero when p = −ω₀. The magnitude of a function tends to zero as p approaches the value of any zero. A factor in the denominator it is referred to as a pole. The magnitude of a function tends to infinity as p approaches the value of any pole.

(11) $T_z = 20\log\left(\dfrac{p+\omega_0}{\omega_0}\right) = 20\log\left(1+\dfrac{\omega^2}{\omega_0^2}\right)^{\frac{1}{2}} + j\,arctg\dfrac{\omega}{\omega_0}$

(12) $T_p = 20\log\left(\dfrac{\omega_0}{p+\omega_0}\right) = -20\log\left(\dfrac{p+\omega_0}{\omega_0}\right) = -20\log\left(1+\dfrac{\omega^2}{\omega_0^2}\right)^{\frac{1}{2}} - j\,arctg\dfrac{\omega}{\omega_0}$

Magnitude and phase are calculated for real frequencies (p=jω)
(Spice Programs 10031 and 10041).

$T_p(p) = \dfrac{\omega_0}{p+\omega_0}$ $\quad T_p(j\omega) = \dfrac{\omega_0}{\omega_0+j\omega}$ $\quad T_z(p) = \dfrac{p+\omega_0}{\omega_0}$ $\quad T_z(j\omega) = \dfrac{\omega_0+j\omega}{\omega_0}$

(13) *Magnitude*   *Phase*   *Magnitude*   *Phase*

$|T_p(j\omega)| = \dfrac{1}{\left(1+\dfrac{\omega^2}{\omega_0^2}\right)^{\frac{1}{2}}}$ $\quad -arctg\dfrac{\omega}{\omega_0}$ $\quad |T_z(j\omega)| = \left(1+\dfrac{\omega^2}{\omega_0^2}\right)^{\frac{1}{2}}$ $\quad arctg\dfrac{\omega}{\omega_0}$

If |p/ω₀| << 1 then the logarithm of |1+p/ω₀| is approximated by 20 log 1=0. The magnitude plot is a horizontal line at the 0dB level. If |p/ω₀| >> 1 and the factor is a zero of the function, the logarithm of |1+p/ω₀| is, in effect, 20 log ω/ω₀ whose magnitude has a +6dB/octave slope (Figure 100411).

If |p/ω₀| >> 1 and the factor is a pole, the logarithm of |1/(1+p/ω₀)| is approximated by −20log ω/ω₀ whose magnitude has a −6dB/octave slope (Figure 100311). These straight asymptotic lines provide a simple way to plot (on the real frequency axis) very good approximations for 1+p/ω₀ pole and zero factors.

> The deviations of exact plots from the magnitude and phase asymptotes are listed in Table 1001 p178.

Spice program 10031 pole at 10KHz

```
Fig10031.ckt   factor 1/(p+w₀)
*asymptote source voltages
V1 1 0 AC 100 0  ; volts
* names/nodes/values
R1 1 2 1000
C1 2 0 .0159u
*.PLOT AC VP(1) VP(2) -100,100
.AC DEC 200 10 1e+007
.TEMP 27
.PLOT AC VDB(2) -40,10
.end
```

Bode plots are drawings of straight line asymptotes only.

Note: For comparison purposes asymptotes are added to Figures 100311 and 100312. The straight lines are the Bode plots.

Figure 100311 Bode diagram of factor $\omega_0/(p+\omega_0)$ - magnitude

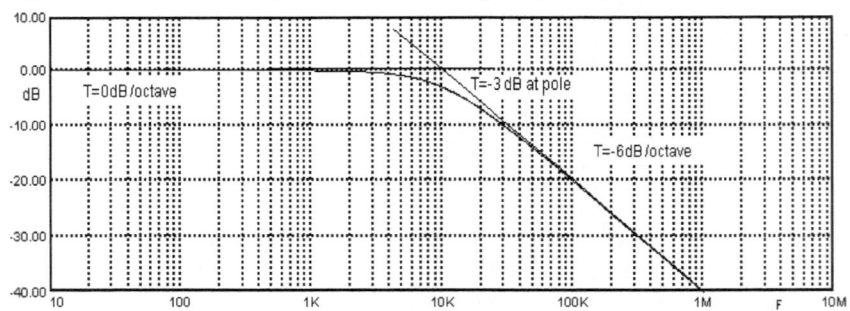

Figure 100312 Bode diagram of factor $\omega_0/(p+\omega_0)$ - phase

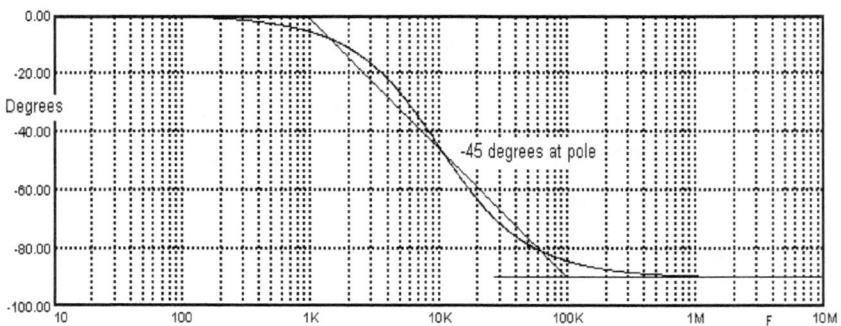

10 Bode Method

Spice program 10041 zero at 10KHz

```
Fig10041.ckt   (p+w₀)/w₀
*asymptote source voltages
I1 1 0 AC 1m 180   ; volts

* names/nodes/values
R1 1 2 1000
L1 2 0 31m

.AC DEC 200 10 1e+007.TEMP 27
.PLOT AC VP(1) 0,100
.PRINT AC VDB(1)
.PRINT AC VP(1)
.end
```

Figure 100411 Bode diagram of factor $(p+\omega_0)/\omega_0$ - magnitude

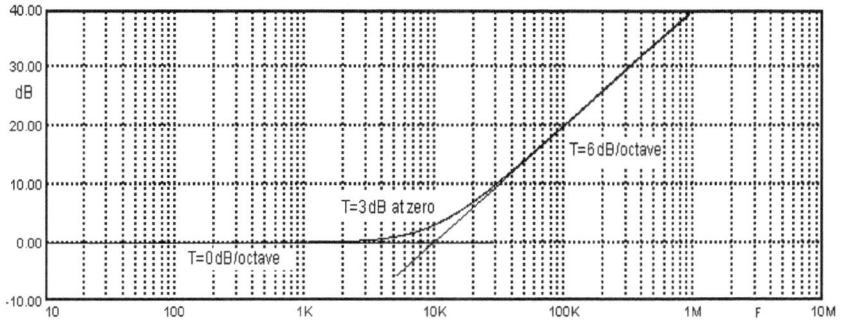

Figure 100412 Bode diagram of factor $(p+\omega_0)/\omega_0$ - phase

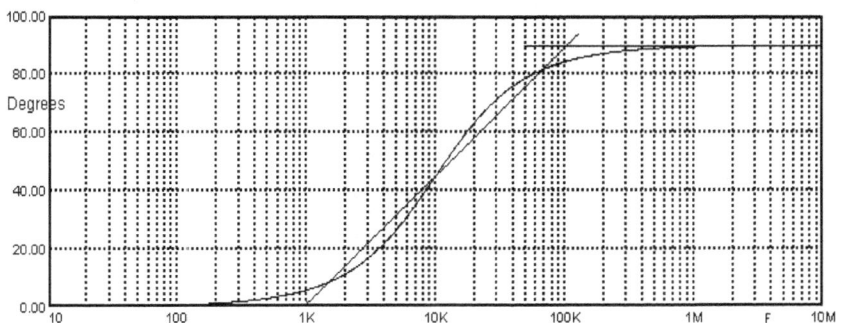

10.5 Factors p+a+jb and (p+a−jb)

The $(p+\omega_0)$ factor, with one parameter ω_0, has only one set of deviations from asymptotes to use when making exact plots (Table 1001 page 178). The complex conjugate factors $(p+a+jb)(p+a-jb)$ have many sets of deviations from the asymptotes (Table 1002 page 179) because there are two parameters a and b. The equations relate parameters a and b to Q and ω_0. The deviation set used in any specific case depends upon the Q of L and C components (pages 33, 40). $Q_{series}=\omega L/R_S$ or $1/\omega CR_S$. $Q_{parallel}=\omega CR_P$ or $R_P/\omega L$. (A damping parameter $d = 1/2Q$ is used in some writings).

The complex factors' magnitude and phase are shown in Table 1002 and Figures 100511, 100512.

if $p_1 = a + jb$ and $p_2 = a - jb$, then

$$p_1 p_2 = (a+jb)(a-jb) = a^2 + b^2 = \omega_0^2$$

$$p_1 + p_2 = a + jb + a - jb = 2a$$

(14) $(p+p_1)(p+p_2) = p^2 + (p_1+p_2)p + p_1 p_2 = p^2 + 2ap + \omega_0^2$

$$(p_1+p)\times(p_2+p) = p_1\left(1+\frac{p}{p_1}\right)\times p_2\left(1+\frac{p}{p_2}\right) = p_1 p_2 \left(1+\frac{p}{p_1}\right)\left(1+\frac{p}{p_2}\right)$$

(15) $T(p) = \left(1+\dfrac{p}{p_1}\right)\left(1+\dfrac{p}{p_2}\right) = 1 + \left(\dfrac{1}{p_1}+\dfrac{1}{p_2}\right)p + \dfrac{p^2}{p_1 p_2} = 1 + \dfrac{2a}{\omega_0}\cdot\dfrac{p}{\omega_0} + \dfrac{p^2}{\omega_0^2}$

(16) if $\dfrac{2a}{\omega_0} = \dfrac{1}{Q}$, then $T(p) = \dfrac{p^2}{\omega_0^2} + \dfrac{1}{Q}\cdot\dfrac{p}{\omega_0} + 1$

The asymptotes stand out clearly when Q=½ because T then takes the form $(1+p/\omega_0)^2$. The asymptotes for the logarithm of this squared factor are a horizontal (0dB/octave) line at the 0dB level when $\omega/\omega_0 \ll 1$. If the factor is a zero of T, then for $\omega/\omega_0 \gg 1$ T is approximated by $20\log (p/\omega_0)^2$, or $40 \log (p/\omega_0)$ that has a +12dB/octave slope. For $\omega/\omega_0 \gg 1$, if the factor is a pole of T, T is about $-40\log p/\omega_0$ which has a −12dB/octave slope (Spice Program 10051).

When Q>½ the response has a peak at $f = f_0$.

176

10 Bode Method

Spice program 10051

```
Fig10051.ckt   factors  (p^2+2ap+w0^2)
*asymptote source voltages
V1  1  0 AC 1 0  ; volts
V11 11 0 AC 1 0  ; volts

* names/nodes/values
R1  1  2  1200
C1  3  0  0.01591u
L1  2  3  15.91m
R11 11 12 100
C11 13  0  0.01591u
L11 12 13 15.91m
*.PLOT AC VDB(3) VDB(13) -30,20
.AC DEC 200 100 1e+006
.TEMP 27
.PLOT AC VP(3) VP(13) -200,0
.end
```

Figure 100511 Bode diagram of factor $1/(p^2+\omega_0 p/Q+\omega_0^2)$ - magnitude

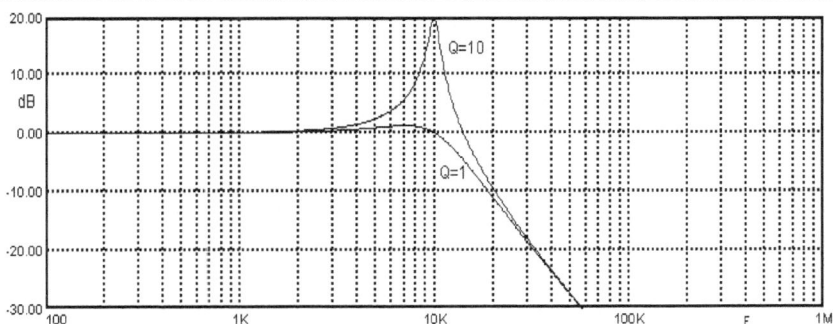

Figure 100512 Bode diagram of factor $1/(p^2+\omega_0 p/Q+\omega_0^2)$ - phase

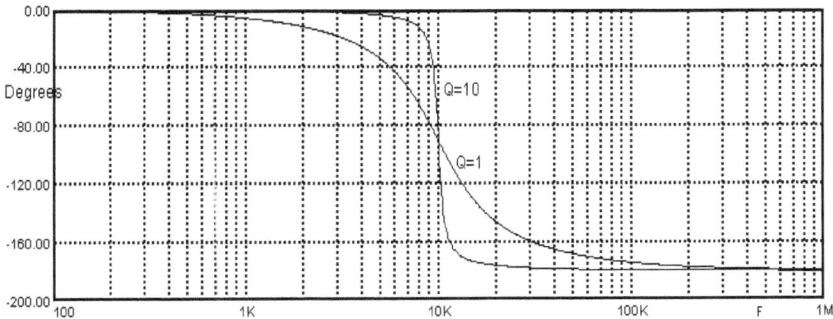

Table 1001 Magnitude and phase of $20\text{Log}(p+\omega_0)$

Magnitude of $20\log[1+j(\omega/\omega_0)]$ dB Accurate magnitude plots, on the real frequency axis, for $1+p/\omega_0$ factors are made by drawing the asymptotes as reference lines (e.g. see Figure 100411), and marking off the deviations Δ listed here from the asymptotes.

$\dfrac{\omega}{\omega_0}$	Asymptote	Exact	Δ
–	0	0	0
1/4	0	0.3	0.3
1/2	0	1	1
1	0	3	3
2	6	7	1
4	12	12.3	0.3
8	18	18	0

Phase of $20\log[1 + j(\omega/\omega_0)]$ dB Accurate phase plots for $(1+p/\omega_0)$ factors are made by drawing the asymptotes as reference lines (e.g. see Figures 100312, 100412), and marking off the deviations Δ listed here from the asymptotes.

$\dfrac{\omega}{\omega_0}$	Asymptote	Exact	Δ	$\dfrac{\omega}{\omega_0}$	Asymptote	Exact	Δ
0.01	0	0	0	1	45	45	0
1/64	0	0.9	+0.9	2	58.5	63.4	+4.9
1/32	0	1.8	+1.8	4	72	76	+4
1/16	0	3.6	+3.6	8	85.6	82.9	−2.7
0.1	0	5.7	+5.7	10	90	84.3	−5.7
1/8	4.4	7.1	+2.7	16	90	86.4	−3.6
1/4	18	14	−4	32	90	88.2	−1.8
1/2	31.5	26.6	−4.9	64	90	89.1	−0.9
1	45	45	0	100	90	90	0

10 Bode Method

Table 1002 Magnitude and phase of $-20\text{Log}(1 + \lambda/Q + \lambda^2)$

Normalized frequency $\lambda = p/\omega_0$

Magnitude of $-20\text{Log}(1 + \lambda/Q + \lambda^2)$

dB magnitude for various values of Q.

ω/ω_0	$Q = 10$	5	2.50	1.00	0.50
1/10	0	0	0	0	0
1/4	0.56	0.55	0.51	0.26	-0.53
1/2	2.5	2.4	2.2	0.9	-1.9
4/5	8.7	8.1	6.3	1.1	-4.3
1	20	14	8	0	-6
5/4	4.9	4.2	2.5	-2.7	-8.2
2	-9.6	-9.6	-9.8	-11.1	-14
4	-23.5	-23.5	-23.6	-23.8	-24.6
10	-40	-40	-40	-40	-40

Phase of $-20\text{Log}(1 + \lambda/Q + \lambda^2)$

Degrees deviation from 0° for various values of Q.

ω/ω_0	$Q = 10$	5	2.50	1.00	0.50
1/10	$+0.6$	$+1.2$	$+2.3$	$+5.8$	$+11.4$
1/4	$+1.5$	$+3.1$	$+6.1$	$+14.9$	$+28.1$
1/2	$+3.8$	$+7.6$	$+14.9$	$+33.7$	$+53.7$
4/5	$+12.5$	$+24$	$+41.5$	$+65.8$	$+77.3$
1	$+90$	$+90$	$+90$	$+90$	$+90$

Degrees deviation from $-180°$ for various values of Q.

5/4	$+12.5$	$+24$	$+41.6$	$+65.8$	$+77.3$
2	$+3.8$	$+7.6$	$+14.9$	$+33.7$	$+53.1$
4	$+1.5$	$+3.1$	$+6.1$	$+14.9$	$+28.1$
10	$+0.6$	$+1.2$	$+2.3$	$+5.8$	$+11.4$

10.6 Bode Plots

Cascade of 2 Circuits

(17) $T_1(p) = \dfrac{p+\omega_0}{p+4\omega_0} = \dfrac{1}{4} \times \dfrac{p+\omega_0}{\omega_0} \times \dfrac{4\omega_0}{p+4\omega_0} = \dfrac{1}{4} \times \left(1 + \dfrac{p}{\omega_0}\right) \times \dfrac{1}{1 + \dfrac{p}{4\omega_0}}$

$T_1(p)$ is the product of three terms: a ¼ constant, a zero at ω_0, and a pole at $4\omega_0$. $T(p)$ is the transfer function from node 1 to node 2 when $R_1 = 3R_2$ (Figure 1006). The three terms are plotted on dB scales to emphasize the ±6dB/octave slopes (Figure 1007).

Figure 1006

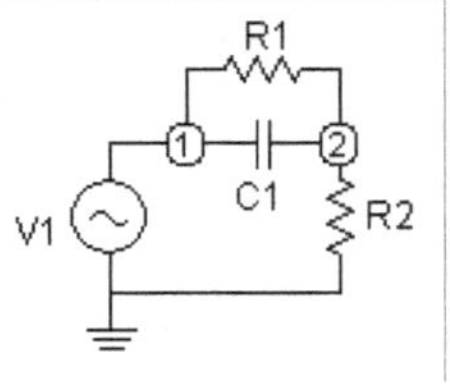

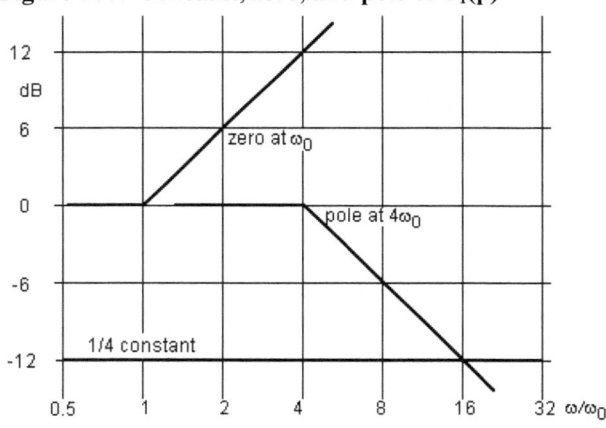

Figure 1007 Constant, zero, and pole of $T_1(p)$

The sum of the three terms is $T_1(p)$, because dB add. Observe how the +1 zero added to the −1 pole produces a 0 slope from $4\omega_0$ onward (Figure 1008).

Figure 1008 Sum of terms of $T_1(p)$

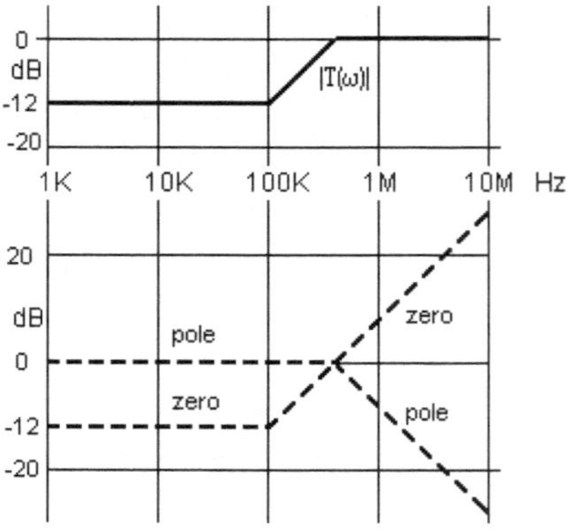

180

Cascade of 3 Circuits

(18) $T_2(p) = 2\dfrac{\omega_0^2}{(p+\omega_0)^2} \times \dfrac{p+2\omega_0}{p+4\omega_0} = \dfrac{1}{\left(1+\dfrac{p}{\omega_0}\right)^2} \times \left(1+\dfrac{p}{2\omega_0}\right) \times \dfrac{1}{\left(1+\dfrac{p}{4\omega_0}\right)}$

$T_2(p)$ is the sum of three terms:
1) a pole of order 2 at ω_0 with -2 slope, 2) zero at $2\omega_0$ with $+1$ slope and
3) a pole at $4\omega_0$. with -1 slope.

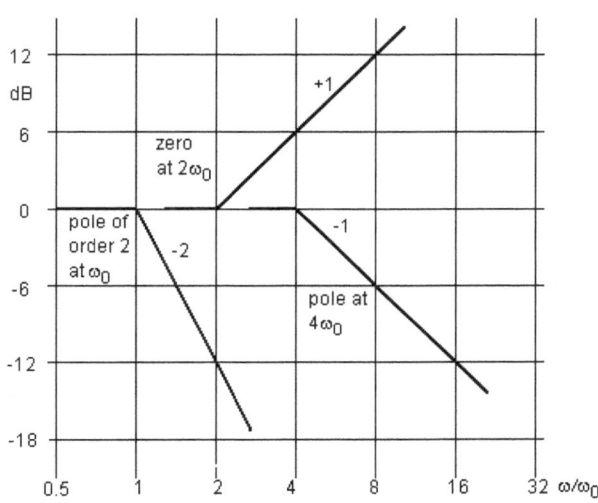

Figure 1009 Order 2 pole, zero, and pole of $T_2(p)$

A practical way to sum the terms is to work by octaves here.

0.5 to 1 $0+0+0=0$

1 to 2 $-2+0+0=-2$

2 to 4 $-2+1+0=-1$

4 to 8 $-2+1-1=-2$

and -2 onward.

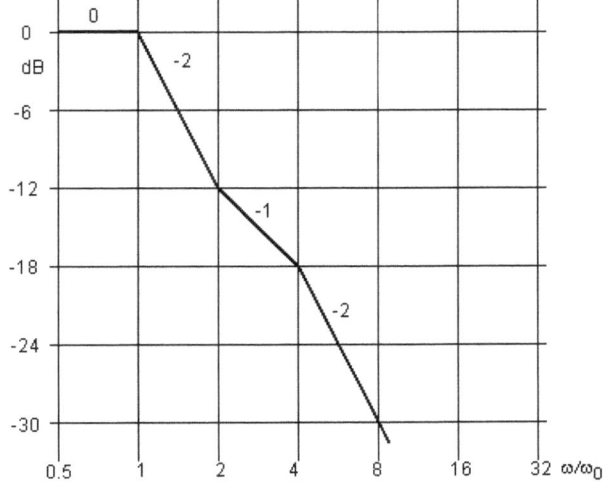

Figure 1010 Sum of terms of $T_2(p)$

Electric Circuits - Analysis and Design

Problems 10

1000 A good grasp of dB is useful to have. Consider ratios n=1, 2, 4, 5, 8, 9, 10.

Calculate log n and 20log n. Mark off a linear scale in tenths from 0 to 1. Label this log n. Above the scale enter the corresponding values of n. Below the scale enter the corresponding values of 20 log n. Observe that the dB scale is linear. Know that 6, 12, and 18 dB are ratios of 2, 4, 8. Know that 14 and 19 dB are ratios of 5 and 9.

How many dB for the ratios $\sqrt{2}$, $1/\sqrt{2}$? Relate dB for the ratios n and 1/n?

Useful definitions.

$$\omega_0 L_1 = R_1 \qquad \frac{1}{\omega_0 C_1} = R_1 \qquad \frac{1}{\omega_0 C_1} = \omega_0 L_1 \qquad \lambda = \frac{p}{\omega_0}$$

Example of a conversion from p to λ.

$$T(p) = \frac{v_2}{v_1} = \frac{\omega_0}{\omega_0 + p} = \frac{1}{1+\lambda} \qquad \text{where } \lambda = \frac{p}{\omega_0}$$

1001P to 1018P

Make magnitude and phase Bode plots of T(p)=output/input. Can you draw the circuit schematic representing T(p)? Hint convert back to p. Be careful and note whether its a voltage or current source.

1001P RL
$$T(p) = \frac{v_2}{v_1} = \frac{1}{1+\lambda}$$

1002P RL
$$T(p) = \frac{v_2}{v_1} = \frac{\lambda}{1+\lambda}$$

1003P RL
$$T(p) = \frac{v_2}{i_1} = R\frac{\lambda}{1+\lambda}$$

1004P RC
$$T(p) = \frac{v_2}{v_1} = \frac{1}{1+\lambda}$$

1005P RC
$$T(p) = \frac{v_2}{v_1} = \frac{\lambda}{1+\lambda}$$

1006P RC
$$T(p) = \frac{v_2}{i_1} = \frac{R}{1+\lambda}$$

1007P LC
$$T(p) = \frac{v_2}{v_1} = \frac{\lambda^2}{1+\lambda^2}$$

1008P LC
$$T(p) = \frac{v_2}{i_1} = \sqrt{\frac{L}{C}} \cdot \frac{\lambda}{1+\lambda^2}$$

1009P RLC
$$T(p) = \frac{v_2}{v_1} = \frac{1}{Q}\frac{\lambda}{1+\frac{1}{Q}\lambda+\lambda^2}$$

1010P RLC
$$T(p) = \frac{v_2}{v_1} = \frac{1}{1+\frac{1}{Q}\lambda+\lambda^2}$$

1011P RLC
$$T(p) = \frac{v_2}{v_1} = \frac{\lambda^2}{1+\frac{1}{Q}\lambda+\lambda^2}$$

1012P RRC
$$T(p) = \frac{v_2}{i_1} = k\frac{1+\lambda}{1+k\lambda} \qquad \omega_0 = \frac{1}{R_1 C_1}$$

1013P RRL
$$T(p) = \frac{v_2}{i_1} = \frac{1+\lambda}{1+\lambda/k} \qquad \omega_0 = \frac{R_1}{L_1}$$

1014P RLGC
$$T(p) = \frac{V_2}{V_1} = \frac{1}{1+yz} = \frac{1}{1+R_1 G_2(1+\lambda)^2} \qquad \omega_0 = \frac{R_1}{L_1} = \frac{G_2}{C_2}$$

1015P Product (cascade) of 2 Circuits
$$T_1(p)T_2(p) = \frac{\omega_0^2}{(p+\omega_0)^2} \cdot \frac{p+4\omega_0}{p+8\omega_0}$$

1016P Product (cascade) of 3 Circuits
$$T_1(p)T_2(p)T_3(p) = \frac{\omega_0}{p+\omega_0} \cdot \frac{\omega_0}{p+4\omega_0} \cdot \frac{p+16\omega_0}{p+32\omega_0}$$

1017P Product (cascade) of 2 Circuits
$$T_1(p)T_2(p) = \frac{p}{p+8\omega_0} \cdot \frac{p+2\omega_0}{p+\omega_0}$$

1018P Product (cascade) of 3 Circuits
$$T_1(p)T_2(p)T_3(p) = \frac{\omega_0}{p+\omega_0} \cdot \frac{\omega_0}{p+10\omega_0} \cdot \frac{p+20\omega_0}{p+80\omega_0}$$

Electric Circuits - Analysis and Design

11 Reactance Chart Method

The reactance chart is a designer's assistant, as well as a useful aid for checking circuit analyses. A designer needs to make estimates, and to select "ballpark" values for components appropriate to the problem at hand. The visual display provided by reactance charts saves a great deal of effort. You do not draw on the chart, you simply examine it. For example, you look at a chart to see that the 1000 ohm resistance line crosses the 0.01μF reactance line at 15.9KHz (Figure 11012). You see that the reactance of 0.159μF equals (resonates with) the reactance of 1.59mH at about 10KHz (Figure 11052). The numbers 1.59, 15.9, etc. appear, because $1/2\pi = 0.159$.

A reactance chart is a set of straight line plots of impedance magnitude of R, L, and C components as a function of frequency. The chart is also applicable to voltages and currents, because they are proportional to immittances[1]. In fact the chart is a useful adjunct to design or analysis of any electric circuit. Two benefits accrue from having logarithmic scales on the y and x axes: (1) many decades of magnitude and frequency can fit on one page, and (2) all plots are straight lines.

The magnitude of a capacitor's impedance $|Z_C|$ is referred to as the capacitor's *reactance* that is written as X_C, and plots as $1/\omega C$ that is not a straight line on a linear scale. However, log X_C plots as a straight line with negative slope on log-log scales. Inductive reactance X_L, and resistance R plots are straight lines on linear scales or log-log scales.

$z_C = \dfrac{1}{pC}$ when $p = j\omega$ $\Rightarrow$ $z_C = \dfrac{1}{j\omega C}$ and $|z_C| = X_C = \dfrac{1}{\omega C} = \dfrac{1}{2\pi f C}$

(1) $\log |z_C| = -\log \omega C$

$z_L = pL$ when $p = j\omega$ $\Rightarrow$ $z_L = j\omega L$ and $|z_L| = X_L = \omega L = 2\pi f L$

(2) $\log |z_L| = \log \omega L$

E.g. a 100μμF capacitor line crosses the 1KΩ line at $f_0 = 1.59$MHz. A 100μH inductor line crosses the 1KΩ line at $f_1 = 1.59$MHz, and so forth.

(3) $f_0 = 1.59 MHz$, $\dfrac{1}{\omega_0 C} = \dfrac{1}{2\pi \cdot 1.59 \cdot 10^6 \cdot 100 \cdot 10^{-12}} = 1K\Omega$, $\omega_0 L = 1K\Omega$

---

[1] Bode's word for impedance or admittance

11 Reactance Chart Method

11.1 Impedance Plots - $|Z(\omega)|$

11.1.1 RC in series

The impedance of a series RC circuit approaches R as frequency goes to infinity ($Z_C \rightarrow 0$). When the frequency goes to zero the impedance approaches $1/\omega C$ that is now much greater than R. The key to simple plots is plotting asymptotic values log R and log $1/\omega C$ (Figures 11011, 11012). Deviation of the actual value of $|Z|$ from the asymptotes is shown in Figure 11011. The deviations are the same as Bode plot deviations (Table 1001 page 178). Knowing values of R and C you can quickly find $|Z|$ at any frequency with small percentage error. Note: $10nF = 10^4 \ \mu\mu f = 0.01 \mu F$.

(4) $\quad Z(\omega) = R + \dfrac{1}{j\omega C} \quad \Rightarrow \quad \lim\limits_{\omega \to \infty} |Z| = R \quad and \quad \lim\limits_{\omega \to 0} |Z| = \dfrac{1}{\omega C}$

Figure 11011 Magnitude of Z as a function of frequency

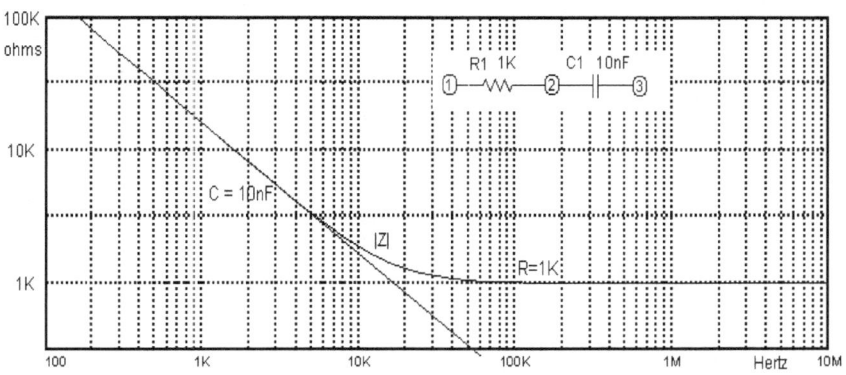

Figure 11012 R and C Asymptotes on a reactance chart

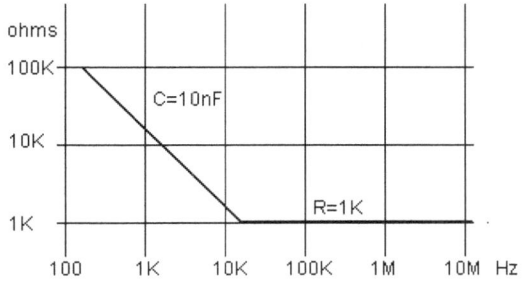

Electric Circuits - Analysis and Design

11.1.2 RC in parallel

The impedance of a parallel RC circuit approaches $1/\omega C$ as frequency goes to infinity ($Z_C \to 0$). When the frequency goes to zero the impedance approaches R that is now much less than $1/\omega C$. The key to simple plots is to plot asymptotic values log R and log $1/\omega C$ (Figures 11021, 11022). Deviation of the actual value of |Z| from the asymptotes is shown in Figure 11021. For any given value of R and the corner frequency f_0, you can quickly find C with small percentage error.

$$(5) \quad Z(\omega) = \frac{R \cdot \frac{1}{j\omega C}}{R + \frac{1}{j\omega C}} \quad \Rightarrow \quad \lim_{\omega \to \infty} |Z| = \frac{1}{\omega C} \quad \text{and} \quad \lim_{\omega \to 0} |Z| = R$$

Figure 11021 Magnitude of Z as a function of frequency

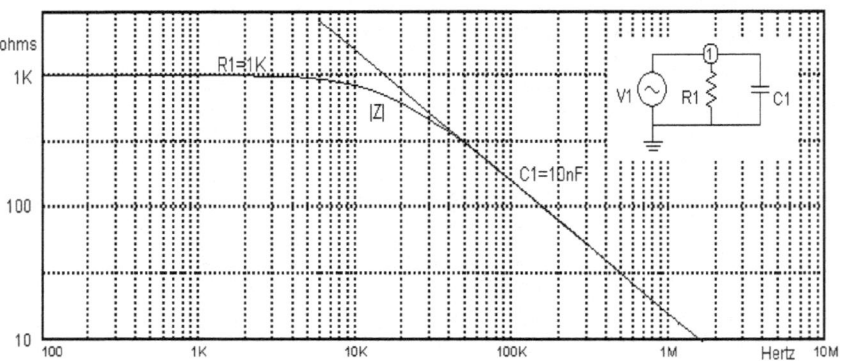

Figure 11022 R and C Asymptotes on a reactance chart

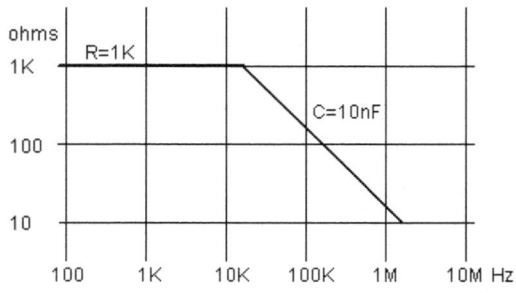

Note: $10nF = 10^4 \mu\mu f$

186

11 Reactance Chart Method

11.1.3 RL in series

The impedance of a series RL circuit approaches ωL as frequency goes to infinity ($Z_L \to \infty$). When the frequency goes to zero the impedance approaches R that is now much greater than ωL. The key to simple plots is to plot asymptotic values log R and log ωL (Figures 11031, 11032). Deviation of the actual value of |Z| from the asymptotes is shown in Figure 11031. For a given value of R (L) and the corner frequency f_0 you can quickly find L (R) with small percentage error.

(6) $Z(\omega) = R + j\omega L$ ⇒  $\lim_{\omega \to \infty} |Z| = \omega L$ and $\lim_{\omega \to 0} |Z| = R$

Figure 11031 Magnitude of Z as a function of frequency

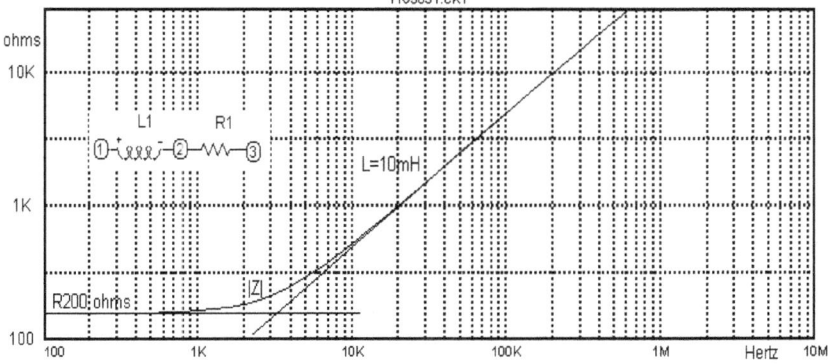

Figure 11032 R and L Asymptotes on a reactance chart

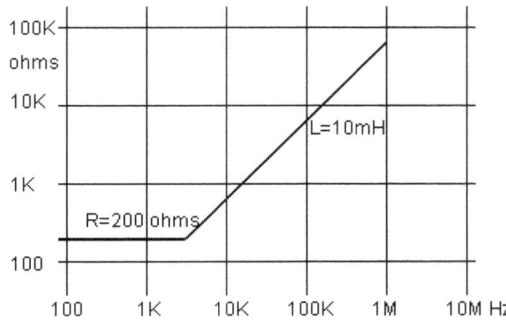

Electric Circuits - Analysis and Design

11.1.4 RL in parallel

The impedance of a parallel RL circuit approaches R as frequency goes to infinity ($Z_L \to \infty$). When the frequency goes to zero the impedance approaches ωL that is now much less than R. The key to simple plots is to plot asymptotic values log R and log ωL (Figures 11041, 11042). Deviation of the actual value of |Z| from the asymptotes is shown in Figure 11041. For any values of R and L you can quickly find |Z| at any frequency with small percentage error. Or, given R (L) and f_0 find L (R).

(7) $\quad Z(\omega) = \dfrac{R \cdot j\omega L}{R + j\omega L} \quad \Rightarrow \quad \lim\limits_{\omega \to \infty} |Z| = R \quad and \quad \lim\limits_{\omega \to 0} |Z| = \omega L$

Figure 11041 Magnitude of Z as a function of frequency

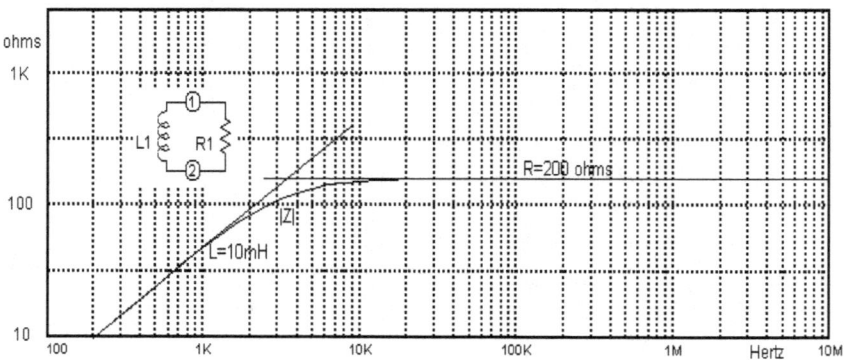

Figure 11042 R and L Asymptotes on a reactance chart

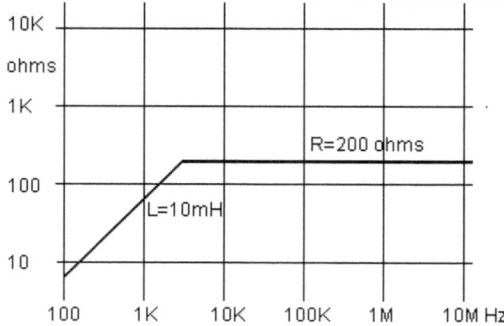

11.1.5 RLC in series

The impedance of a series RLC circuit approaches ωL as frequency goes to infinity ($Z_L \to \infty$). When the frequency goes to zero the impedance approaches 1/ωC that is now much greater than R and ωL. The key to simple plots is to plot asymptotic values log ωC and log ωL (Figures 11051, 11052). Deviation of the actual value of |Z| from the asymptotes is shown in Figure 11051. For various combinations of 2 or 3 of the R, L, C components, and f_0 you can quickly find the rest with small percentage error.

(8) $Z(\omega) = R + j\omega L + \dfrac{1}{j\omega C} \Rightarrow \lim_{\omega \to \infty}|Z| = \omega L \text{ and } \lim_{\omega \to 0}|Z| = \dfrac{1}{\omega C}$

and $Z(\omega_0) = R \quad \omega_0^2 = \dfrac{1}{LC} \quad Q = \dfrac{\omega_0 L}{R} \quad Z_0 = \sqrt{\dfrac{L}{C}}$

Figure 11051 Magnitude of Z as a function of frequency

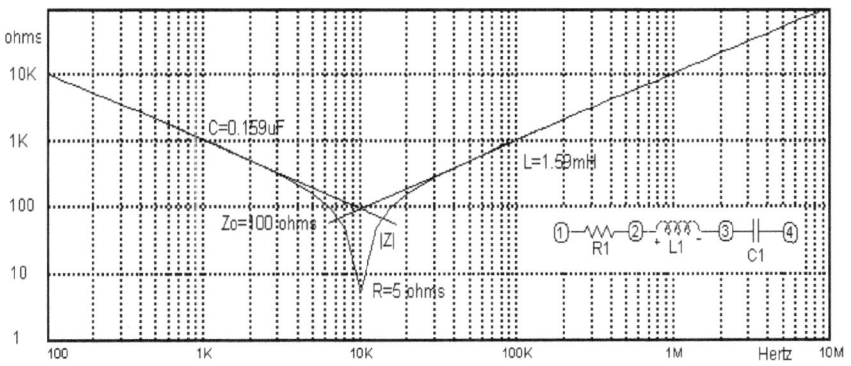

Figure 11052 L and C Asymptotes on a reactance chart

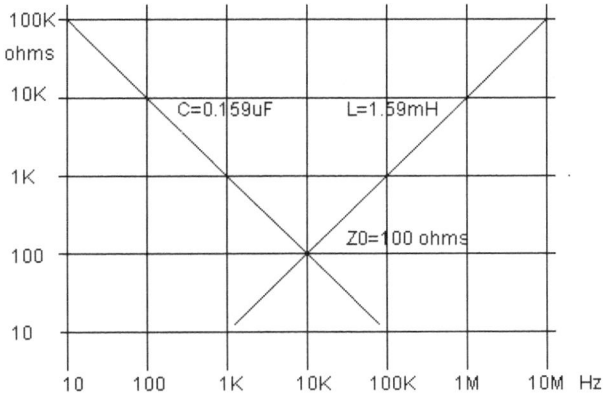

189

Electric Circuits - Analysis and Design

11.2 Transfer Function Plots - |T(ω)|

11.2.1 Series C, Shunt R High Pass Filter

The capacitive reactance decreases as 1/f as frequency is increased, and so the capacitive reactance ultimately decreases to zero as frequency increases without limit. Consequently the circuit transmission increases at a rate of 6dB/octave and equals 1 in the high frequency limit.

(9) $T(\omega) = \dfrac{R}{R + \dfrac{1}{j\omega C}} \Rightarrow \lim_{\omega \to \infty}|T| = 1 \text{ and } \lim_{\omega \to 0}|T| = \omega RC = \dfrac{\omega}{\omega_0} = 0$

$\omega_0 = \dfrac{1}{RC}, \text{ and } T(\omega_0) = \dfrac{1}{1 + \dfrac{1}{j\omega_0 RC}} = \dfrac{1}{1 - j1} = \dfrac{1}{\sqrt{2}} e^{j\pi/4}$ (−3dB)

Figure 11061 Magnitude of T as a function of frequency

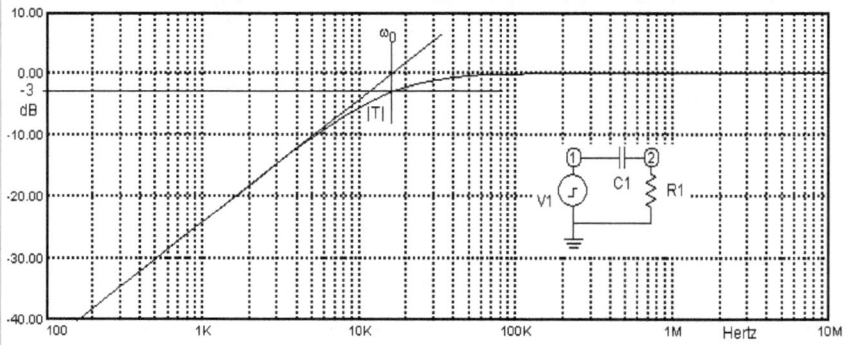

11.2.2 RC phase compensation network

Fig 1107 RC Network

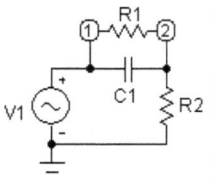

At zero frequency the capacitive reactance is infinite and transmission T(0)=1/4 when $R_1/R_2=3$. At high frequencies the capacitive reactance is essentially zero, and so transmission T(∞)=1. The transition of T from 1/4 to 1 is implemented by a zero/pole combination. See Sections 6.3 page 95 and 10.6 page 180.

(10) $T(p) = \dfrac{V_2}{V_1} = \dfrac{g_1 + pC_1}{g_1 + g_2 + pC_1} = \dfrac{p + \omega_z}{p + \omega_p}$ where $\omega_z = \dfrac{g_1}{C_1}$ & $\omega_p = \dfrac{g_1 + g_2}{C_1}$

11 Reactance Chart Method

$C_1 = 530.5\,pF, \quad R_1 = 3000\Omega, \quad R_2 = 1000\Omega$

(11) $\quad \dfrac{\omega_p}{\omega_z} = \dfrac{g_1 + g_2}{g_1} = 1 + \dfrac{R_1}{R_2} = 4$

(12) $\quad f_z = \dfrac{g_1}{2\pi C_1} = \dfrac{1}{2\pi R_1 C_1} = 100\,KHz \text{ and } f_p = 400\,KHz$

Figure 110711 Transmission - actual magnitude

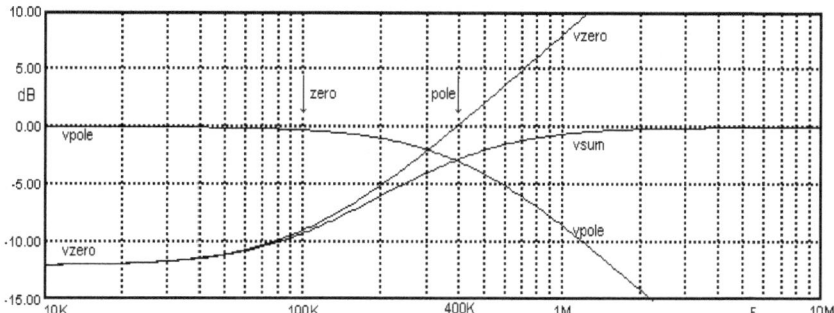

Figure 110712 Transmission - actual phase

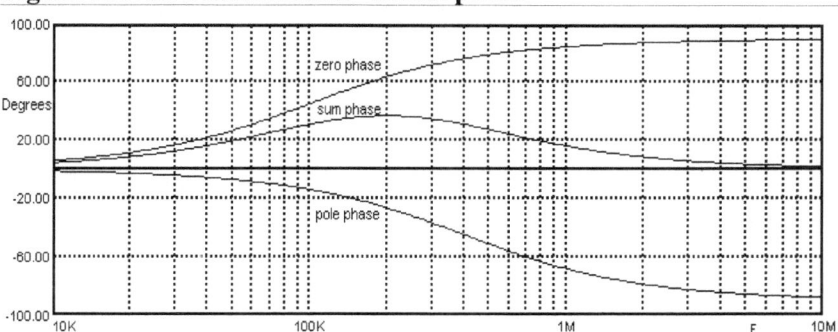

Figure 110713 Transmission - magnitude reactance chart

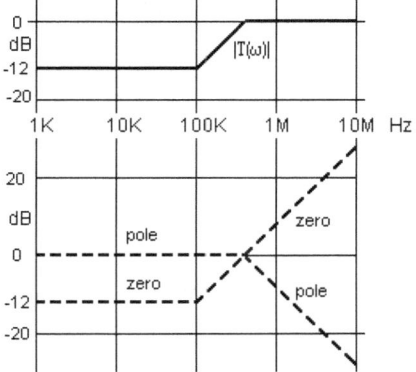

191

Electric Circuits - Analysis and Design

12 How to write AC, DC and TRAN Spice Programs

A Spice program requires a *title statement* as the first line, a *.end statement* (dot end) as the last line, and a *.temp* statement to specify temperature.

Required program lines Between the first and last lines you insert, in any order, a list of *data statements* that describe the components of the circuit to be simulated, and a list of *control statements* that describe the circuit analysis to be performed. Any line can be empty/blank.

Comments An asterisk (*) in the first column indicates that the line is a comment line. A semicolon (;) anywhere in a line means the rest of the line is a comment. Comment lines may be placed anywhere in a Spice program.

```
Fig2011.ckt  ;title statement must be on the first line

*The R3 line is a data statement
R3 5 7 8.2K      ;8.2K resistor connected to nodes 5 and 7

*The dot plot line is a DC control statement
.PLOT DC V(1) V(2) 0,-50

.TEMP 27
.end;         ;dot end must be on last line - ends the program
```

Our numbering scheme We write a Spice program on a word processor such as Notebook or Wordpad or some version of Spice (all Spice programs are text files). When we want to evaluate the performance of the circuit in Figure 201 for example, we save the text as text file Fig2011.ckt. The first line of a program includes the name Fig2011.ckt. Plots of results from program Fig2011.ckt are labeled Fig20111, Fig20112, etc.

A second program for Figure 201 is given the name Fig2012.ckt. Plots are labeled Fig20121, Fig20122, and so forth. In this way we know how circuit figures, Spice programs, Spice files, and plots are related.

> *A Spice program is a text file created on any word processor.*

12 How to write AC, DC, and TRAN Spice Programs

Remark You will see Spice programs written by others with complicated symbols that, in our opinion, make the program VERY hard to read.

We copy symbols from the simple symbols on our circuit schematics.

Note: We only use the "spice-text" feature of the commercial Spice programs that are really complex wrap arounds to the basic Berkeley Spice. Click on File, New and select spice-text. Type into the blank screen page. This is how we write most Spice programs. (Your version of spice may be different.)

Hint: save your time and energy - copy relevant lines from other programs.

In any Spice program we need to know how to write expressions for voltages and currents in a dot PLOT line. Expressions such as V(3) for node voltage 3, I(I8) for current source I8, and IC(qp3) for qp3's collector current. Here is how we find the answers in our version of spice.

Assume you have just written a program. When, for example, transient analysis is invoked (Alt A T) the *Transient Analysis Limits* dialog box appears. We soon discovered to left click (*not* right click as instructed) on *Y expression* in the dialog box. A list appears. Select *variables* to see *device currents* for example. Select *device currents* to see how the currents in the current program are written. And so forth

Electric Circuits - Analysis and Design

12.1 DC Spice program Fig4012.ckt

Write the program   (Spice is NOT case sensitive)
```
Fig4012.ckt
```
First line The first line of every program is assumed by Spice to be a title statement. The title statement can include any words.

```
Is 1 0 DC 0
```
Current source All sources have to be included in a Spice program. The current source i_S is connected to nodes 1 and 0. The i_S signal is a direct current with 0 ampere magnitude. See dot DC below.

```
R1 1 0 10K    ;the circuit - see figure 401b
R2 1 3 5.6K
Vy 3 2 DC 0   ;iy ammeter
R3 2 0 3.9K
R4 2 4 2.7K
Vx 4 0 DC 0   ;ix ammeter
```
Circuit components The 10K resistor is assigned the unique name R1, value 10K, and is wired to nodes 1 and 0. The other resistors are represented in the same way. The zero voltage sources v_y and v_x have zero resistance so that circuit performance is not changed. Spice calculates currents flowing through all voltage sources. This is why v_y and v_x are *ammeters*, and how circuit currents i_y and i_x are calculated by Spice. Node 0 is always reserved for ground.

```
.DC LIN Is 0 10m 1m
```
Control statements Dot DC (.DC or .dc) is a control statement. The .DC line controls the I_S current source. Dot DC sweeps I_S from 0 to 10mA in LIN (linear) 1mA steps.

```
.PLOT DC V(1) V(2) 0,-50
.PRINT DC I(Vy) I(Vx) ;produces numeric output data
.PRINT DC V(1) V(2)
```
After dot plot (.plot), and before the variables, you enter dc (without the dot), because a dot dc statement created the data. Dot plot DC executes the dot DC LIN statement, calculating voltages V(1) V(2) with a vertical scale of 0 to −50V The dot print statements produce *numeric output* data.
```
.TEMP 27
```
Temperature Dot temp (.temp) defines temperature as 27 degrees C.
```
.end
```
Last line The last line of every program is .end (dot end).

12 How to write AC, DC, and TRAN Spice Programs

Spice Program 4012 Calculates DC Response

```
Fig4012.ckt    ;title which is mandatory first line
      ;spaces are ignored
Is 1 0 DC 0    ;current source
R1 1 0 10K     ;the circuit
R2 1 3 5.6K
Vy 3 2 DC 0    ;iy ammeter
R3 2 0 3.9K
R4 2 4 2.7K
Vx 4 0 DC 0    ;ix ammeter

.DC Is 0 10m 1m    ;range of the current source
.PLOT DC V(1) V(2) 0,-50 ;what to plot
.PRINT DC I(Vy) I(Vx) ;create numeric file
.PRINT DC V(1) V(2)
.TEMP 27
.end       ;mandatory last line


Numeric output
Is=10mA V1=41.845V V2=9.278V I3=5.815mA I4=3.436mA
```

Figure 401a   **Figure 401b**

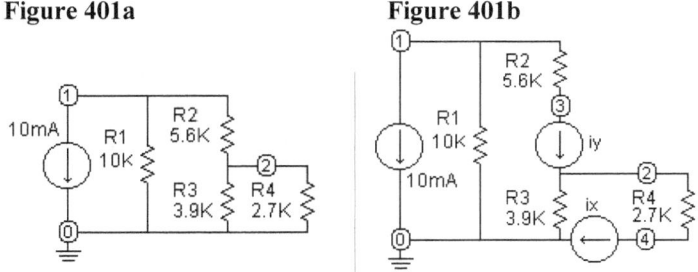

Figure 40121 Voltages out - current in

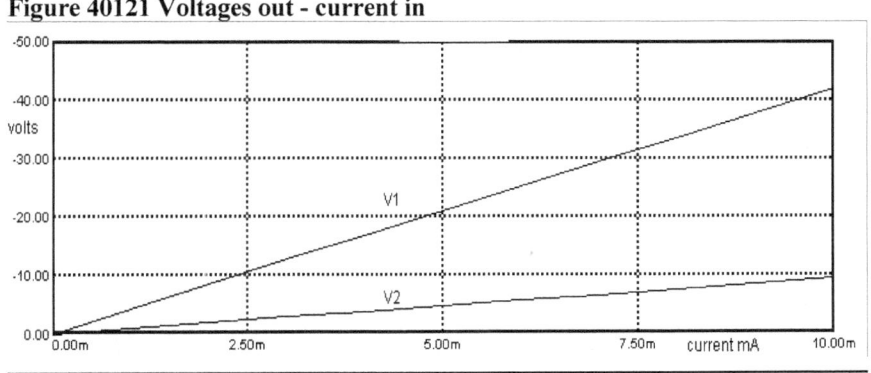

195

12.2 AC Spice program Fig4021.ckt

```
Fig4021.ckt       factor 1/(p+a)
```
First line The first line of every program is assumed by Spice to be a title statement. The title statement can include any words.

```
V1 1 0 AC 1 0   ; volts
```
Voltage source Signal generator v_1 is connected to nodes 1 and 0. The v_1 signal is a AC sinewave with 1 volt magnitude and 0 phase. V1 is defined as a one volt sinewave for *all* frequencies

```
R1 1 2 1000
C1 2 0 .0159155u
```
Circuit components Resistor R_1 is connected to nodes 1 and 2. Capacitor C_1 is connected to nodes 2 and 0 (Figure 402)

```
.AC DEC 200 10 1e+007
```
AC control statement Dot AC DEC defines a decades frequency range from 10 Hz to 10^7 Hz with points to be calculated every 200Hz.

```
.PLOT AC VDB(2) -40,10  ;magnitude dB
*.PLOT AC VP(2) -100,0  ;phase degrees
.PRINT AC V(2)    ;produces numeric output data
.PRINT AC VDB(2)
```
Plot the data dot plot AC executes the dot AC statement calculating the v_2 magnitude in dB, or phase in degrees, from 10Hz to 10^7Hz. The asterisk that defines a line as a comment deactivates the VP(2) phase plot.

```
.TEMP 27
```
Temperature Dot temp (.temp) defines temperature as 27 degrees C.
```
.end
```
Last line The last line of every program is .end (dot end).

V_1 is the Spice voltage source connected from node 1 to node 0, and AC 1 0 means the source is sinusoidal with 1volt peak amplitude and 0 degrees phase. The dot AC (.AC) line defines a frequency range of 10Hz to 10MHz incremented by 200Hz.

Spice starts by selecting 10Hz, and calculating the v_2/v_1 ratio (i.e. $|T(j2\pi 10)|$). Then Spice increments the frequency by 200 Hz to 210Hz, and calculates $|T(j2\pi\ 210)|$. The process is repeated every 200Hz up to 10MHz. The dot plot lines produce figures 40211 and 40212.

12 How to write AC, DC, and TRAN Spice Programs

The corner frequency is at 10KHz. The magnitude decreases 20dB over the decade 100KHz to 1MHz. The slope is –20dB/decade, –6dB/octave. The phase at 10KHz corner is –45 degrees.

Spice Program 4021 Calculates Frequency Response

Figure 402

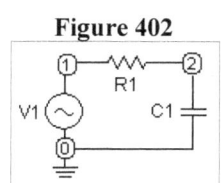

```
Fig4021.ckt    factor 1/(p+a)
V1 1 0 AC 1 0    ; volts
R1 1 2 1000
C1 2 0 .0159155u
*.PLOT AC VDB(2) -40,10;magnitude dB
.AC DEC 200 10 1e+007
.PLOT AC VP(2) -100,0 ;phase degrees
.PRINT AC V(2)
.PRINT AC VDB(2)
.TEMP 27
.end
```

Figure 40211 Magnitude of T(p)=V_2/V_1

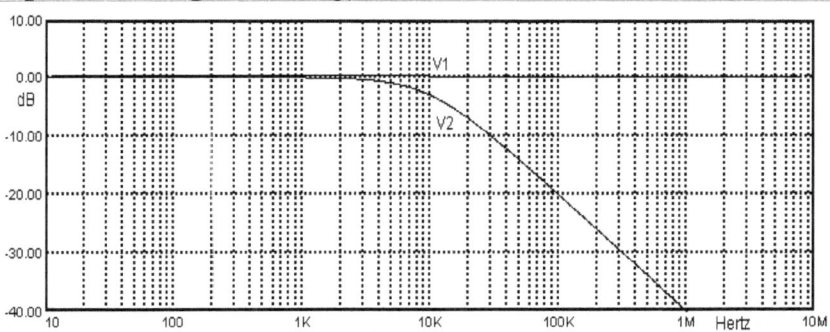

Figure 40212 Phase of T(p) =V_2/V_1

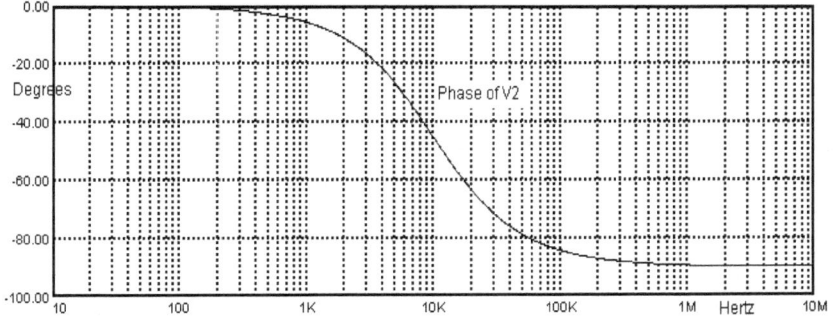

197

Electric Circuits - Analysis and Design

12.3 TRAN Spice program Fig5051.ckt

Spice program for the Series RLC Circuit

```
Fig5051.ckt   series RLC
```
First line The first line of every program is assumed by Spice to be a title statement. The title statement can include any words.

```
V 1 0 PULSE(0    2.5 0     0     0    1000u   2000u)
              V_lo V_hi  t_delay t_rise t_fall t_width t_period
```
Voltage source Pulse generator v_1 is connected to nodes 1 and 0. The pulse is a step function in effect, because the 1000μs width and 2000μs period greatly exceed the 10μs plot time.

```
R1 1 2 10  ; Q=10 under damped - see figure 505
L1 2 3 15.9u
C1 3 0 1590p
```
Circuit components Resistor R_1 is connected to nodes 1 and 2. Inductor L_1 is connected to nodes 2 and 3. Capacitor C_1 is connected to nodes 3 and 0. Three circuits are in the program. Each has a different R to produce a different Q.

```
.TRAN 1e-009 1e-005 0 1e-008
```
TRAN control statement Dot TRAN defines a time range from 0 to 10μs (`1e-005`) that is incremented every 1ns (`1e-009`).

```
.PLOT TRAN V(3) V(13) V(23) 0,5
.PRINT TRAN V(3) V(13) V(23) ;produces numeric output data
```
Plot the data dot plot TRAN executes the dot TRAN statement calculating the v_3, v_{13}, v_{23} from 0s to 10μs every 1ns.

```
.TEMP 27
```
Temperature Dot temp (.temp) defines temperature as 27 degrees C.

.end

Last line The last line of every program is .end (dot end).

12 How to write AC, DC, and TRAN Spice Programs

Spice Program 5051 Calculates Transient Response
```
Fig5051.ckt series RLC

* The trick to viewing responses for various Q's is to enter
* one circuit for each Q value.

V 1 0 PULSE(0 2.5 0 0 0 1000u 2000u)
*in effect pulse is a step function, because plot time is
10µs

R1  1  2  10    ; Q=10 under
L1  2  3  15.9u
C1  3  0  1590p

R12  1  12  100   ; Q=1 close to critical
L12  12  13  15.9u
C12  13   0  1590p

R21  1  22  1000 ; Q=0.1 over
L21  22  23  15.9u
C21  23   0  1590p

.TRAN 1e-009 1e-005 0 1e-008     ;1E-5=10us
.TEMP 27
.PLOT TRAN V(3) V(13) V(23) 0,5
.PRINT TRAN V(3) V(13) V(23) ;produces numeric output data
.end
```

Figure 505 RLC

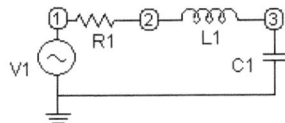

Figure 50512 Series RLC Transient Response

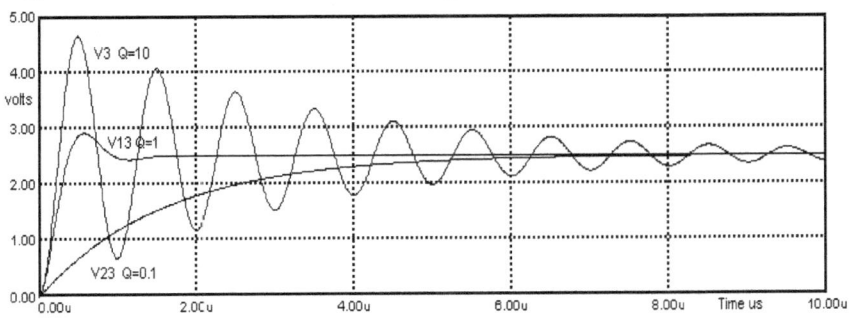

Electric Circuits - Analysis and Design

Appendix

A1 Partial Fraction Expansions

The terms of partial fraction expansions of F(p) are transforms of exp(−at), because each term of a partial fraction expansion has a pole of order one or higher. This is one reason why partial fractions are important.

pole of order one   pole of order n+1

(1) $f(t) = e^{-at} \Leftrightarrow F(p) = \dfrac{1}{p+a}$ (2) $f(t) = \dfrac{t^n}{n!} e^{-at} \Leftrightarrow F(p) = \dfrac{1}{(p+a)^{n+1}}$

The sum of the inverse Laplace transforms of each term of the partial fraction equals the inverse Laplace transform of F(p). For example,

if $F(p) = \dfrac{17}{2j}\left(\dfrac{1}{p - j\omega_0} - \dfrac{1}{p + j\omega_0}\right) + \dfrac{1}{p + \dfrac{R}{L}} + \dfrac{8}{p}$ then

(3) $f(t) = \left[\dfrac{17}{2j}\left(e^{j\omega_0 t} - e^{-j\omega_0 t}\right) + e^{-\dfrac{R}{L}t} + 8\right] u(t) = \left[17\sin\omega_0 t + e^{-\dfrac{R}{L}t} + 8\right] u(t)$

Any complex frequency domain function arising from the linear, lumped parameter electric circuits is the ratio of two polynomials in the complex frequency variable p. We can always divide the denominator into the numerator so that the remainder F(p) is a proper rational fraction that means the degree of N(p) is less than the degree of D(p). For example:

if $G(p) = \dfrac{p^4 + 5p^3 + 8p^2 + 3p + 6}{p^3 + 4p^2 + 3p} = p + 1 + \dfrac{p^2 + 6}{p^3 + 4p^2 + 3p}$

(4) let $F(p) = \dfrac{p^2 + 6}{p^3 + 4p^2 + 3p} = \dfrac{N(p)}{D(p)}$ (a proper fraction)

Note: $f(t) = \dfrac{d\delta(t)}{dt} \Leftrightarrow p$ and $f(t) = \delta(t) \Leftrightarrow 1$

$\delta(t)$ is the impulse function where $\delta(t) = du(t)/dt$.

There are two problems to solve:
1. Find the roots of D(p) (we assume you know how to do this).
2. Convert F(p) into a sum of terms with known inverse Laplace transforms.

Appendix

1. Method for any roots - equate coefficients This method is based on the fact coefficients of similar terms of two polynomials are equal.

(5) $\quad F(p) = \dfrac{p^2 + 6}{p^3 + 4p^2 + 3p}$

(6) $\quad \dfrac{p^2 + 6}{p(p+1)(p+3)} = \dfrac{k_1}{p} + \dfrac{k_2}{p+1} + \dfrac{k_3}{p+3}$

cross multiply:

(7) $\quad p^2 + 6 = k_1(p+1)(p+3) + k_2 p(p+3) + k_3 p(p+1)$

Do not simplify by multiplying out. It is easier to substitute selected values of p.

if $p = 0$ *then*
$0 + 6 = k_1(0+1)(0+3) + k_2 0(0+3) + k_3 0(0+1)$
$\quad 6 = 3k_1 \Rightarrow \quad$ (8) $\quad k_1 = 2$

if $p = -1$ *then*
$1 + 6 = k_1(-1+1)(-1+3) + k_2(-1)(-1+3) + k_3(-1)(-1+1)$
$\quad 7 = -2k_2 \Rightarrow \quad$ (9) $\quad k_2 = -\dfrac{7}{2}$

if $p = -3$ *then*
$9 + 6 = k_1(-3+1)(-3+3) + k_2(-3)(-3+3) + k_3(-3)(-3+1)$
$\quad 15 = 6k_3 \Rightarrow \quad$ (10) $\quad k_3 = \dfrac{5}{2}$

higher order roots Each higher order root of order n requires a sum of terms. One term for each power of p from 1 to n.

(11) $\quad F(p) = \dfrac{5p^3 - 6p - 3}{p^3(p+1)^2}$

(12) $\quad \dfrac{5p^3 - 6p - 3}{p^3(p+1)^2} = \dfrac{k_1}{p} + \dfrac{k_2}{p^2} + \dfrac{k_3}{p^3} + \dfrac{k_4}{p+1} + \dfrac{k_5}{(p+1)^2} \quad$ *now cross multiply*

(13) $\quad 5p^3 - 6p - 3 = k_1 p^2(p+1)^2 + k_2 p(p+1)^2 + k_3(p+1)^2 + k_4 p^3(p+1) + k_5 p^3$

if $p = 0$, *then* $-3 = k_3$, *and if* $p = -1$, *then* $-5 + 6 - 3 = -k_5$

If we equate coefficients of terms, then for

$p^4 : 0 = k_1 + k_4 \quad p^3 : 5 = 2k_1 + k_2 + k_4 + k_5 \quad p : -6 = k_2 + 2k_3$

(14) $\quad k_3 = -3,\ k_5 = 2,\ k_2 = 0,\ k_1 = 3,\ k_4 = -3$

Electric Circuits - Analysis and Design

quadratic factors Quadratic factors require numerators to be linear in p.

(15) $\quad F(p) = \dfrac{16}{p(p^2+4)^2}$

(16) $\quad \dfrac{16}{p(p^2+4)^2} = \dfrac{k_1}{p} + \dfrac{k_2 p + k_3}{p^2+4} + \dfrac{k_4 p + k_5}{(p^2+4)^2}$

Now cross multiply

$$16 = k_1(p^2+4)^2 + (k_2 p + k_3)p(p^2+4) + (k_4 p + k_5)p$$

(17) $\quad 16 = k_1(p^4 + 8p^2 + 16) + k_2(p^4 + 4p^2) + k_3(p^3 + 4p) + k_4 p^2 + k_5 p$

if $p = 0$, then $16 = k_1 16 \quad k_1 = 1$

If we equate coefficients of terms, then for

p^4) $0 = k_1 + k_2 \quad p^3$) $0 = k_3 \quad p^2$) $0 = 8k_1 + 4k_2 + k_4 \quad p$) $0 = 4k_3 + k_5$

(18) $\quad k_1 = 1,\ k_2 = -1,\ k_3 = 0,\ k_4 = -4,\ k_5 = 0$

2. Method for real roots of order 1 This is a formal equivalent to cross-multiplying. The process is straightforward when roots are real and distinct.

(19) $\quad F(p) = \dfrac{14p + 11}{(p+1)(p+3)}$

(20) $\quad \dfrac{14p+11}{(p+1)(p+3)} = \dfrac{k_1}{p+1} + \dfrac{k_2}{p+3}$

(21) $\quad k_1 = (p+1)F(p)\big|_{p=-1} = \dfrac{14(-1)+11}{(-1+3)} = -\dfrac{3}{2}$

(22) $\quad k_2 = (p+3)F(p)\big|_{p=-3} = \dfrac{14(-3)+11}{(-3+1)} = \dfrac{31}{2}$

(23) $\quad F(p) = -\dfrac{3}{2}\dfrac{1}{p+1} + \dfrac{31}{2}\dfrac{1}{p+3} \Rightarrow f(t) = -\dfrac{3}{2}e^{-t} + \dfrac{31}{2}e^{-3t}$

Appendix

A2 Complex Numbers

The words complex and imaginary are potentially misleading, because complex numbers are not complicated and imaginary operators are not part of someone's imagination. Both words are labels: they are technical terms used to designate a class of numbers. A complex number z is represented by an ordered pair of real numbers x and y written as (x, y).

Multiplication by -1 and $\sqrt{-1}$ A number can be represented as a distance on a number line. We define steps to the right as positive so that distance AB=+4. Multiply +4 by -1 to get -4 that is the distance AC. Multiply AC by -1 to get back to AB. Clearly multiplication by -1 in effect *rotates* AB and AC by 180°.

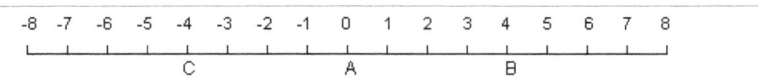

If +4 is multiplied by $\sqrt{-1}$ the result is $4\sqrt{-1}$. Multiply $4\sqrt{-1}$ by $\sqrt{-1}$ to get -4. Hence multiplication by $\sqrt{-1}$ two times rotates AB by 180°. And so multiplication by $\sqrt{-1}$ implements a 90° rotation of AB.

The world has agreed that numbers such as $4\sqrt{-1}$ are *imaginary* numbers. To save writing $\sqrt{-1}$ is replaced by *i* in the mathematical literature. However EE's use i to designate current. That is why they use j for $\sqrt{-1}$.

Complex numbers The ordered pair (x_1, y_1) is a point in the (x, jy) plane that can be reached by starting from the origin, marching along the x-axis for a distance x_1, rotating $\pi/2$ radians, and marching parallel to the jy-axis for distance y_1 (Figure CN1a).

Working with ordered pairs (x, y) does not have much appeal, which is why the world adopted the well known alternative z=x+jy that is easier to work with.

In other words: taking our clue from the rotation operation we use j as a $\pi/2$ rotation operator. Then we say jy_1 is a vector we add to vector x_1 so that $z_1=x_1+jy_1$. This replaces the ordered pair (x_1, y_1). We say z is a complex number whose real part is x and whose imaginary part is y. Keep in mind that x and y are real numbers.

203

Electric Circuits - Analysis and Design

Figure CN1 Complex numbers in Cartesian and polar coordinates

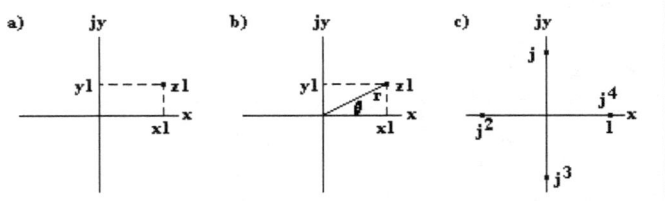

Polar coordinates: If r is the distance from the origin to the point z, then $x = r \cos \theta$, and $y = r \sin \theta$ (Figure CN1b). See Euler relation below.

(1) $\quad z = x + jy = r\cos\theta + jr\sin\theta = re^{j\theta}$

(2) $\quad \tan\theta = \dfrac{y}{x} \quad \text{so that} \quad \theta = \tan^{-1}\dfrac{y}{x}$

Multiples of j Representing j as a $\pi/2$ rotation yields the same results as the $\sqrt{-1}$ representation (Figure CN1c, Euler).

(3) $\quad j = e^{j\frac{\pi}{2}} = \cos\dfrac{\pi}{2} + j\sin\dfrac{\pi}{2} = 0 + j1 = j$

(4) $\quad j^2 = e^{j\frac{\pi}{2}2} = e^{j\pi} = \cos\pi + j\sin\pi = -1 + j0 = -1$

(5) $\quad j^3 = e^{j\frac{\pi}{2}3} = e^{j\frac{3\pi}{2}} = \cos\dfrac{3\pi}{2} + j\sin\dfrac{3\pi}{2} = -0 - j1 = -j$

(6) $\quad j^4 = e^{j\frac{\pi}{2}4} = e^{j2\pi} = \cos 2\pi + j\sin 2\pi = 1 + j0 = 1$

Addition The sum of complex numbers is found by adding the two x's and then the y's.

$$z_1 + z_2 = (x_1 + jy_1) + (x_2 + jy_2)$$
(7) $\quad z_1 + z_2 = (x_1 + x_2) + j(y_1 + y_2)$

Multiplication Find the product $z_1 z_2$. To find it multiply z_1 and z_2, while treating j as another real number. Then substitute -1 for j^2.

(8) $\quad z_1 z_2 = (x_1 + jy_1)(x_2 + jy_2)$

$\qquad = x_1 x_2 + x_1 jy_2 + jy_1 x_2 + jy_1 jy_2$

$\qquad = x_1 x_2 + jy_1 jy_2 + jy_2 x_1 + jy_1 x_2$

$\qquad = x_1 x_2 + j^2 y_1 y_2 + j(x_2 y_1 + x_1 y_2)$

$\qquad = (x_1 x_2 - y_1 y_2) + j(x_2 y_1 + x_1 y_2)$

Appendix

Subtraction Subtraction is defined as addition of positive and negative complex numbers.

(9) $\quad z_1 - z_2 = z_1 + [-z_2] = (x_1 + y_1) + (-x_2 - jy_2)$
$\qquad = (x_1 - x_2) + j(y_1 - y_2)$

Division Division is facilitated by the complex conjugate concept, where j is replaced by $-j$.

$$\text{If } z = x + jy, \text{ then } \bar{z} = x - jy$$

$$z\bar{z} = (x + jy)(x - jy) = x^2 - j^2 y^2 + jxy - jyx$$

(10) $\quad z\bar{z} = x^2 + y^2 = r^2 = |z|^2 = |z| \times |z|$

$$\frac{z_1}{z_2} = \frac{z_1}{z_2} \times \frac{\bar{z}_2}{\bar{z}_2} = \frac{(x_1 + jy_1)(x_2 - jy_2)}{r_2^2} = \frac{x_1 x_2 - j^2 y_1 y_2 - jx_1 y_2 + jy_1 x_2}{r_2^2}$$

(11) $\quad \dfrac{z_1}{z_2} = \dfrac{x_1 x_2 + y_1 y_2}{r_2^2} + j\dfrac{x_2 y_1 - x_1 y_2}{r_2^2}$

Euler Relation (Figure CN1b)

$$\text{If } r = 1 \text{ then } z = \cos\theta + j\sin\theta$$

$$\frac{dz}{d\theta} = -\sin\theta + j\cos\theta = j(\cos\theta + j\sin\theta) = jz$$

(12) $\quad \dfrac{dz}{z} = jd\theta$

Integrating $\quad \ln z = j\theta + constant$

If $\theta = 0$ then $z = 1$ so that $\ln 1 = j0 + constant$

However, $\ln 1 = 0$ so that $constant = 0$

$\therefore \ln z = j\theta \quad \Rightarrow \quad z = e^{j\theta}$

(13) $\quad e^{j\theta} = \cos\theta + j\sin\theta$

Electric Circuits - Analysis and Design

A3 Oliver Heaviside's Method

You have studied circuit laws, and the node and mesh analytical methods in the context of resistor R circuits. When we add to the circuits inductors L and capacitors C the mathematics escalates, because algebraic methods must be extended to include the calculus.

Current i flows through a resistor from one terminal to the other. Electrons pass through the atomic structure of the resistance material creating voltage v between terminals. The current-voltage relationship, the vi constraint, is Ohm's Law $v=iR$.

A capacitor is like a battery that can be charged and discharged. However the battery voltage is fixed, whereas the capacitor voltage $v=q/C$. The voltage is proportional to charge q, not current i. Current pours charge into a capacitor that stores that charge. The vi constraint is $i=Cdv/dt$ that is derived from $q=Cv$. In an inductor *change* of current is what matters. Current creates flux, and change of flux creates voltage that opposes the applied voltage. The vi constraint is the differential equation $v=Ldi/dt$. In resistor R circuits *time* is not explicit, because there are no transients in resistor R circuits. On the other hand the differential equations for L and C vi constraints introduce time explicitly that brings us to the calculus. We quote Thompson and Gardner[1].

"These dreadful symbols are
(1) *d* which merely means 'a little bit of'.
Thus *dx* means a little bit of *x*, or *du* means a little bit of *u*. Ordinary mathematicians think it more polite to say 'an element of', instead of 'a little bit of'. Just as you please. But you will find that these little bits (or elements) may be considered to be infinitely small.
(2) $\int$ which is merely a long S, and may be called (if you like) 'the sum of'. Thus $\int dx$ means the sum of all the little bits of *x*; or $\int dt$ means the sum of all the little bits of *t*. Ordinary mathematicians call this symbol 'the integral of'. Now any fool can see that *x* is made up of a lot of little bits, each of which is called *dx*. if you add them up all together you get the sums of the *dx's* (which is the same thing as the whole of *x*). The word integral simply means 'the whole'."

---

[1] Thompson and Gardner, "Calculus Made Easy", page 39, ISBN 0312 185 480
Note: The short Prolog is great reading.

Appendix

Oliver Heaviside was a superb electrical engineer, and a mathematical whiz. His work shows what clear thinking can produce! He showed that the differential symbol dx and the integral symbol $\int$ obey all the laws of algebra. He defined the letter p as an *operator* representing differentiation.

(1) p is a *differentiation operator*

Integration is the reverse process of differentiation, and Heaviside must have asked what cancels p? Well, in algebra 1/p cancels p ($p \times 1/p = 1$). So Heaviside came to the conclusion

(2) $\dfrac{1}{p}$ is an *integration operator*

For L and C the vi constraints become algebraic equations in p. In other words pL and 1/pC are *operational impedances*.

(3a) $v = L\dfrac{di}{dt} \rightarrow v = Lpi = pLi$ (3b) $i = \dfrac{1}{L}\int v dt \rightarrow i = \dfrac{v}{pL}$

(4a) $i = C\dfrac{dv}{dt} \rightarrow i = Cpv = pCv$ (4b) $v = \dfrac{1}{C}\int i dt \rightarrow v = \dfrac{i}{pC}$

RL circuit Consider the series RL circuit.

(5) $v(t) = Ri(t) + L\dfrac{di}{dt} = Ri(t) + Lpi(t) = (R + pL)i(t)$

(6) $i(t) = \dfrac{v(t)}{R + pL} = \dfrac{v(t)}{pL} \cdot \dfrac{1}{1 + \dfrac{R}{pL}} = \dfrac{v(t)}{R} \cdot \dfrac{1}{p\tau} \cdot \dfrac{1}{1 + \dfrac{1}{p\tau}}$ where $\tau = \dfrac{L}{R}$

Heaviside's symbol [1] is his notation for a step function (Figure 501), whose modern notation is u(t).

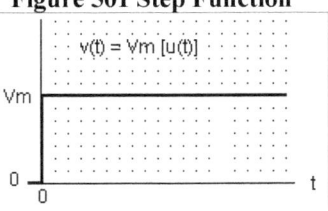

Figure 501 Step Function

(7) $v(t) = V_M[1] = V_M u(t)$

Heaviside used this series expansion of integrations in his transient response analysis. From 6 we get

207

Electric Circuits - Analysis and Design

(8) $\dfrac{1}{1+y} = 1 - y + y^2 - y^3 + \ldots$ $\qquad y = \dfrac{1}{p\tau}$

(9) $i(t) = \dfrac{V_M}{R} \cdot \dfrac{1}{p\tau} \cdot \left[1 - \left(\dfrac{1}{p\tau}\right) + \left(\dfrac{1}{p\tau}\right)^2 - \left(\dfrac{1}{p\tau}\right)^3 + \ldots \right] u(t)$

Each term is one or more 1/p integrations. For example

(10) $\left(\dfrac{1}{p\tau}\right)^3 u(t) = \left(\dfrac{1}{p\tau}\right)\left(\dfrac{1}{p\tau}\right)\left(\dfrac{1}{p\tau}\right) \cdot u(t) = \left(\dfrac{1}{p\tau}\right)\left(\dfrac{1}{p\tau}\right)\left(\dfrac{t}{\tau}\right) \cdot u(t)$

$= \left(\dfrac{1}{p\tau}\right) \cdot \dfrac{1}{1 \cdot 2}\left(\dfrac{t}{\tau}\right)^2 u(t) = \dfrac{1}{1 \cdot 2 \cdot 3}\left(\dfrac{t}{\tau}\right)^3 u(t) = \dfrac{x^3}{3!} \cdot u(t)$

Integrate term by term within the brackets.

(11) $i(t) = \dfrac{V_M}{R} \cdot \dfrac{1}{p\tau} \cdot \left[1 - \dfrac{x^1}{1!} + \dfrac{x^2}{2!} - \dfrac{x^3}{3!} + \ldots \right] u(t)$

The leading 1/p factor integrates all terms within the brackets.

(12) $i(t) = \dfrac{V_M}{R} \left[\dfrac{x^1}{1!} - \dfrac{x^2}{2!} + \dfrac{x^3}{3!} - \dfrac{x^4}{4!} + \ldots + (-1)^{n-1} \dfrac{x^n}{n!} + \ldots \right] u(t)$

It is not obvious that this series equals $1 - e^{-x}$, but it does. Verify it.

(13) $i(t) = \dfrac{V_M}{R}\left[1 - e^{-x}\right]u(t) = \dfrac{V_M}{R}\left[1 - e^{-\frac{t}{\tau}}\right]u(t)$

RC circuit

(14) $v(t) = Ri(t) + \dfrac{1}{C}\int i(x)\,dx = Ri(t) + \dfrac{1}{pC}i(t) = \left(R + \dfrac{1}{pC}\right)i(t)$

(15) $i(t) = \dfrac{v(t)}{R + \dfrac{1}{pC}} = v(t) \cdot \dfrac{pC}{1 + pCR} = \dfrac{v(t)}{R} \cdot p\tau \cdot \dfrac{1}{1 + p\tau}$ where $\tau = RC$

The 1/(1+pτ) term produces $1-e^{-x}$, and the p term differentiates it.

(16) $i(t) = \dfrac{v(t)}{R} \cdot p\tau \cdot \left(1 - e^{-x}\right) = \dfrac{v(t)}{R} \cdot \tau \cdot \left(0 + \dfrac{1}{\tau}e^{-\frac{t}{\tau}}\right) = \dfrac{v(t)}{R} e^{-\frac{t}{\tau}}$

Appendix

RLC circuit

(17) $v(t) = Ri(t) + L\dfrac{di}{dt} + \dfrac{1}{C}\displaystyle\int_0^t i(x)\,dx = Ri(t) + pLi(t) + \dfrac{1}{pC}i(t)$

$= \left(R + pL + \dfrac{1}{pC}\right)i(t)$

(18) $i(t) = \dfrac{v(t)}{R + pL + \dfrac{1}{pC}} = \dfrac{v(t)}{L} \cdot \dfrac{p}{p^2 + \dfrac{R}{L}p + \dfrac{1}{LC}}$

if $R = 3,\ L = 1,\ C = 1/2$, then

(19) $\dfrac{i(t)}{v(t)} = \dfrac{p}{p^2 + 3p + 2} = \dfrac{p}{(p+1)(p+2)} = \dfrac{2}{p+2} - \dfrac{1}{p+1}$

The $1/(1+p\tau)$ terms produce $1-e^{-x}$, where τ equals ½ and 1.

(20) $\dfrac{i(t)}{v(t)} = (1 - e^{-2t}) - (1 - e^{-t}) = e^{-t} - e^{-2t}$

> Important observation: the 1's cancel.

if $R = 2,\ L = 1,\ C = 1/2$, then

(21) $\dfrac{i(t)}{v(t)} = \dfrac{p}{p^2 + 2p + 2} = \dfrac{p}{(p+1+i)(p+1-i)} = \dfrac{1}{2i}\left[\dfrac{1}{p+1-i} - \dfrac{1}{p+1+i}\right]$

The $1/(1+p\tau)$ terms produce $1-e^{-t/\tau}$, where $1/\tau$ equals $-1+i$ and $-1-i$.

(20) $\dfrac{i(t)}{v(t)} = \dfrac{1}{2i}\left(1 - e^{(-1-i)t}\right) - \dfrac{1}{2i}\left(1 - e^{(-1+i)t}\right) = \dfrac{1}{2i}\left(e^{(-1+i)t} - e^{(-1-i)t}\right)$

$= e^{-t}\left(\dfrac{e^{it} - e^{-it}}{2i}\right) = e^{-t}\sin t$

> The *Laplace Transform* in effect simplifies Heaviside's procedures, and introduces initial values as a bonus!

Electric Circuits - Analysis and Design

A4 Easy Method for Evaluating Laplace Transforms

The method applies basic transforms proven previously.

Definition of the Laplace transform F(p) of any function f(t)

(101) $F(p) = \mathscr{L}[f(t)] = \int_0^\infty f(t)e^{-pt}\,dt$

(102) $\mathscr{L}[f'(t)] = pF(p) - f(0)$

(103) $\mathscr{L}[-t \cdot f(t)] = F'(p)$

(104) $\mathscr{L}[e^{at} f(t)] = F(p-a)$

(105) $\mathscr{L}[f(t-T)] = e^{pT} F(p)$

if $\mathscr{L}[f_1(t)] = F_1(p)$ and $\mathscr{L}[f_2(t)] = F_2(p)$ then

(106) $\mathscr{L}[a_1 f_1(t) + a_2 f_2(t)] = a_1 F_1(p) + a_2 F_2(p)$

(107) $\mathscr{L}[a_1 f_1(t)] = a_1 F_1(p)$

Examples

Laplace transform of 1.
    if $\mathscr{L}[1] = F(p)$ then
by 102 $\mathscr{L}[0] = pF(p) - f(0) = pF(p) - 1$
clearly $\mathscr{L}[0] = 0$ so that

$$0 = pF(p) - 1 \quad \rightarrow \quad F(p) = \frac{1}{p}$$

Laplace transform of t.
    if $\mathscr{L}[t] = F(p)$ then
by 102 $\mathscr{L}[1] = pF(p) - f(0) = pF(p) - 0$
also $\mathscr{L}[1] = \frac{1}{p}$ so that

$$\frac{1}{p} = pF(p) - 0 \quad \rightarrow \quad F(p) = \frac{1}{p^2}$$

Appendix

Laplace transform of e^{-at}.

$$\mathcal{L}[1] = F(p) = \frac{1}{p} \quad \text{and}$$

by 104 $\quad \mathcal{L}[e^{at} \cdot 1] = F(p-a) \quad$ so that

$$\mathcal{L}[e^{at} \cdot 1] = \frac{1}{p-a}$$

Laplace transform of $\sin \omega t$ and $\cos \omega t$.

Let $\mathcal{L}[\sin \omega t] = F(p)$ then

by 102 $\quad \mathcal{L}[\omega \cos \omega t] = pF(p) - 0$

by 102 $\quad \mathcal{L}[-\omega^2 \sin \omega t] = p^2 F(p) - \omega$

by 107 $\quad \mathcal{L}[-\omega^2 \sin \omega t] = -\omega^2 F(p)$

so $\quad p^2 F(p) - \omega = -\omega^2 F(p) \quad \to \quad F(p) = \dfrac{\omega}{p^2 + \omega^2} = \mathcal{L}[\sin \omega t]$

and $\quad \mathcal{L}[\cos \omega t] = \dfrac{p}{\omega} F(p) = \dfrac{p}{p^2 + \omega^2}$

Laplace transform of $e^{-at} \sin \omega t$ and $e^{-at} \cos \omega t$.

$$\mathcal{L}[\sin \omega t] = F(p) = \frac{\omega}{p^2 + \omega^2}$$

by 104 $\quad \mathcal{L}[e^{-at} \sin \omega t] = F(p+a) = \dfrac{\omega}{(p+a)^2 + \omega^2}$

$$\mathcal{L}[e^{-at} \cos \omega t] = \frac{p+a}{(p+a)^2 + \omega^2}$$

Electric Circuits - Analysis and Design

A5 Topology Applied to Electric Circuits

Topology is a branch of mathematics concerned with properties of figures that are invariant with continuous deformations of the figures. Think of figures drawn on a sheet of rubber or a blob of putty, or made of soft copper wire. Topology is applicable because the properties of electric circuits (assembled from elements with lumped constant parameters) do not change when physical layout is changed (Figure A51). The primary data of an electric circuit are the numbers of branches B, nodes N, and subcircuits S. Subcircuits are circuits without metallic connections to each other. However, the numbers of independent nodes N–S and meshes M are the significant data about any electric circuit. We will show that

$$N-S = \text{nodes } N - \text{subcircuits } S$$

is the number of independent node equations one must formulate to solve any electric circuit problem using the node voltage method.

Figure A51 Electric circuit topological forms

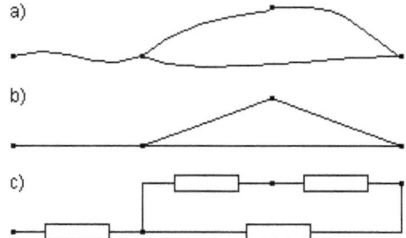

Whereas independent meshes = branches – independent nodes.

(1) $\quad M = B - (N - S)$

M is the number of independent mesh equations one must formulate to solve any electric circuit problem using the mesh method.

Observe that the number of independent node equations (N–S), plus the number of independent mesh equations M equals B the number of branches.

$B = (N-S) + M$

Appendix

Nodes

A node is a terminal that is common to one, or two, or several circuit elements (Figure A52). Two or more terminals connected by zero resistance wires (short circuits) count as one node (e.g. the thick lines in Figure A52c).

Figure A52 Nodes in electric circuits

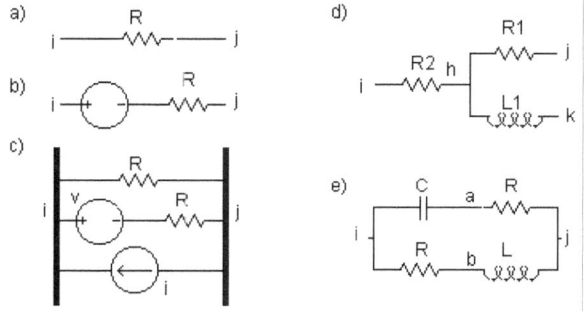

If we were to list all possible node pairs in a circuit we would first choose one out of N nodes for one member of a pair, and then choose one out of the N–1 remaining nodes for the second member of a pair. However, since node pair (i j) is the same as node pair (j i) the node pair total is reduced by ½. Therefore a circuit with N nodes has ½N(N–1) possible node pairs. The number of node pairs actually used depends on the circuit. When there is no component connected from node i to node j we say node pair (i j) is not used.

A practical convention is to ground one node so that its voltage is zero. This leaves N–1 node voltages with respect to ground that need to be determined. In a connected circuit, where S=1, there are N–1 independent voltages that are basic to the circuit. When ½N(N–1) is greater than N–1 the basic N–1 node pairs may be chosen in many ways. For example, if N = 4, then N–1 = 3 and ½N(N–1) = 6. Therefore up to 6 circuit node pairs may have components connected. If an electric circuit has S magnetically coupled subcircuits with $N_1, N_2, ..., N_S$ nodes respectively, the total number of independent nodes N is calculated as follows:

(2) $(N_1 - 1) + (N_2 - 1) + ... + (N_S - 1) = N - S$

Electric Circuits - Analysis and Design

Network Graphs

Network geometry is concerned solely with how the terminals of circuit elements are connected together. In a network schematic emphasizing network geometry, each element is represented by a line with a circle at each end representing a terminal (Figure A53b). Circles coalesce into nodes when terminals are connected. The resulting schematic is a graph of the network. Numbers are used to identify the branches.

A graph has separate parts when magnetic coupling exists (Figure A54). So, three ideas are associated with a graph: branches B, nodes N, and separate parts S. Clearly the graph only shows the network's geometric features. If the separate parts are joined at one node (Figure A54c) so that the branch currents and voltages do not change, then we only need to discuss graphs with one separate part.

Note: In Figure A53 N=5, N−1=4 and ½N(N−1)=10 possible node pairs. Observe that Figure A53 "uses" only 6 out of 10 node pairs. A more extreme example is Figure A55 where N=10, N−1=9 and ½N(N−1)=45 possible node pairs. In this example only 14 of 45 node pairs are "used." *The low usage results from no components crossing over another component. The circuit is flat like a fish net. Most electric circuits are "fish nets."*

Figure A53 Network and one of its graphs

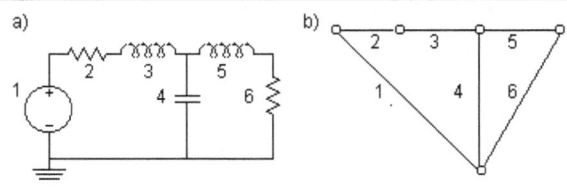

Figure A54 Network with separate parts and one of its graphs

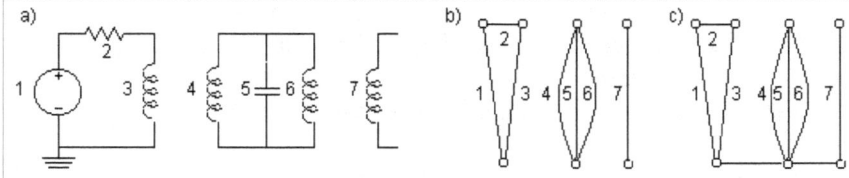

Appendix

Trees and Meshes

In general a network graph has closed paths (Figure A55a). If certain branches are removed there remains a tree of branches connecting all of the nodes (Figure A55b). The tree has no closed paths. In general many trees can be created from a network graph, because removing different sets of branches leaves different trees connecting all of the nodes.

> *Tree: any minimum set of branches in the original graph that connects all the nodes so that there are no closed paths.*

> *Link: any branch removed from a network graph in the process of forming a tree.*

If there are 10 N nodes then 9 N–1 branches connect the 10 N nodes to form a tree (Figure A55b), and 5 M links complete the graph. 10 N nodes and 5 M links mean there are B=M+N–1=5+10–1=14 branches in the original graph.

Figure A55 Network and one of its trees

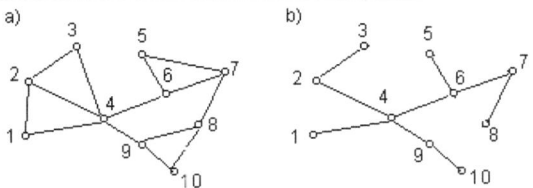

Network Variables Removal of the 5 M link branches forms a tree. Consequently all branch currents in a tree are zero. In other words: somehow if 5 M link currents become zero then all currents in the network are zero. Therefore the 9 N–1 tree branch currents depend upon the 5 M link currents. That is to say, only the 5 M link currents are independent. We can conclude that there are only 5 M independent current variables: i.e. *only 5 M mesh current equations need to be formulated.*

The original network's branch voltages divide into two groups: 9 N–1 tree branch voltages and 5 M link branch voltages. If the 9 N–1 tree branch voltages are set to zero by shorting 10 N nodes together, then the 5 M link branch voltages are forced to zero. Therefore the 5 M link branch voltages depend upon the 9 N–1 tree branch voltages so that only the 9 N–1 node voltages are independent. However, there are 10 N nodes. The Nth node voltage can be set to zero (connect it to earth). This is why *only 9 N–1 node voltage equations need to be formulated.*

Electric Circuits - Analysis and Design

Mesh currents   Each link current flows in a separate closed path (remember the tree). Immediately we know how to select M independent closed path currents that are referred to as the M mesh currents.

Node pairs   Any node voltage can be set to zero (connect it to earth, i.e. ground it) and every other node can be paired with it creating N – 1 node pairs. So, we can select N – 1 independent voltages by grounding any one of the N nodes.

> *If any node is connected to earth so that its potential with respect to earth is zero, then that node is said to be grounded.*

Figure A56 General branch forms - series and parallel

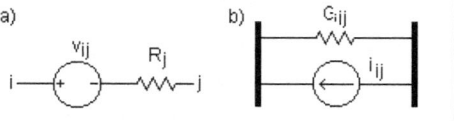

Review

Suppose we have a circuit with B branches and N nodes that we disassemble so that there are B branches on the table. Any circuit tree is constructed by using N–1 of these branches to create a skeleton of the original circuit with N nodes. We say skeleton, because we note that there are no closed loops that can conduct current in this tree of N–1 branches and N nodes. This means we have B–(N–1) branches left over. Now we connect the remaining branches, the links M, to the skeleton thereby recreating the original circuit. We observe that a mesh is created each time we add a link. Clearly there are N–1 independent nodes and M = B–(N–1) independent meshes. Examples: 5=14–(10–1), and when many branches are in parallel 12=14–(3–1). Draw the graphs.

> If M = B – (N–1) is smaller than N–1, then mesh analysis is probably preferred. If N–1 is smaller, then node analysis is probably preferred. We say probably, because other factors can influence the choice of method.

Appendix

A6 Delta-Wye Transformation

A delta-wye transformation of wyes or deltas convert hard to analyze bridge circuits into series-parallel circuits. For example the R_a, R_b, R_c Y is transformed into the R_1, R_2, R_3 delta (Figure A61).

Figure A61 Wye to Delta transformation

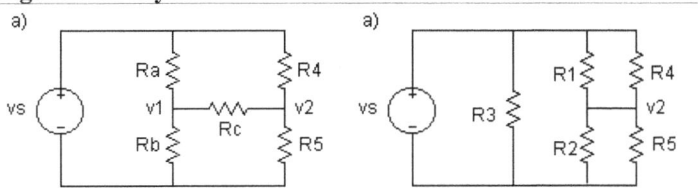

Mesh and node analytic methods facilitate a straightforward derivation of the delta-wye transformation.

Analysis of a T circuit (Figure A62a) that is a wye circuit, produces two mesh equations. Two voltage sources V_1 and V_2 generate two mesh currents I_1 and I_2. The solution to the equations produces two equations where mesh currents I_1 and I_2 are functions of the two source voltages V_1 and V_2.

Analysis of a Pi circuit (Figure A62b) that is a delta circuit, produces two node equations. Two current sources I_1 and I_2 generate two node voltages V_1 and V_2. The solution to the node equations produces two equations where node voltages V_1 and V_2 are functions of the two source currents I_1 and I_2.

Figure A62 T and pi circuits

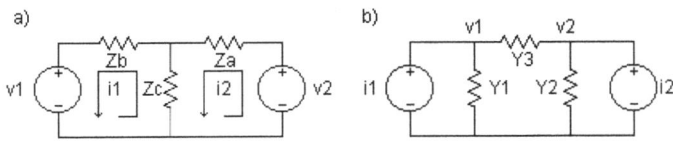

Equivalence implies that currents and voltages of the two circuits are equal:

$$V_{\text{T mesh i1}} = V_{\text{pi source i1}}, \quad V_{\text{T mesh i2}} = V_{\text{pi source i2}}$$
$$V_{\text{T source v1}} = V_{\text{pi node v1}}, \quad V_{\text{T source v2}} = V_{\text{pi node v2}}$$

217

Electric Circuits - Analysis and Design

Mesh analysis of a T wye circuit (Figure A62a) produces mesh equations T1 where voltages generate currents, and solutions T2 where currents are functions of voltages.

(T1) $V_1 = (z_b + z_c)I_1 - z_c I_2$ (T2) $I_1 = \dfrac{z_a + z_c}{\Delta_Z} V_1 - \dfrac{z_c}{\Delta_Z} V_2$
$V_2 = -z_c I_1 + (z_a + z_c)I_2$
$\Delta_Z = z_a z_b + z_b z_c + z_c z_a$ $I_2 = -\dfrac{z_c}{\Delta_Z} V_1 + \dfrac{z_b + z_c}{\Delta_Z} V_2$

Node analysis of a delta circuit (Figure A62b) produces node equations P1 where currents generate voltages, and solutions P2 where voltages are functions of currents.

(P1) $I_1 = (y_1 + y_3)V_1 - y_3 V_2$ (P2) $V_1 = \dfrac{y_2 + y_3}{\Delta_y} I_1 + \dfrac{y_3}{\Delta_y} I_2$
$I_2 = -y_3 V_1 + (y_2 + y_3)V_2$
$\Delta_y = y_1 y_2 + y_2 y_3 + y_3 y_1$ $V_2 = \dfrac{y_3}{\Delta_y} I_1 + \dfrac{y_1 + y_3}{\Delta_y} I_2$

The *wye to delta*, T to P, transformation equations result from equating coefficients of the node analysis I_1 & I_2 current equations (P1) and mesh analysis I_1 & I_2 current equations (T2).

(P1 = T2) $y_1 + y_3 = \dfrac{z_a + z_c}{\Delta_Z}$, $y_3 = \dfrac{z_c}{\Delta_Z}$, $y_2 + y_3 = \dfrac{z_b + z_c}{\Delta_Z}$

$y_1 = \dfrac{1}{z_1} = \dfrac{z_a}{\Delta_Z} = \dfrac{z_a}{z_a z_b + z_b z_c + z_c z_a}$

$y_2 = \dfrac{1}{z_2} = \dfrac{z_b}{\Delta_Z} = \dfrac{z_b}{z_a z_b + z_b z_c + z_c z_a}$

$y_3 = \dfrac{1}{z_3} = \dfrac{z_c}{\Delta_Z} = \dfrac{z_c}{z_a z_b + z_b z_c + z_c z_a}$

The *delta to wye*. P to T, transformation equations result from equating coefficients of the mesh analysis V_1 & V_2 voltage equations (T1), and the node analysis V_1 & V_2 voltage equations (P2).

(T1 = P2) $z_b + z_c = \dfrac{y_2 + y_3}{\Delta_y}$ $z_c = \dfrac{y_3}{\Delta_y}$ $z_a + z_c = \dfrac{y_1 + y_3}{\Delta_y}$

$z_a = \dfrac{y_1}{\Delta_y} = \dfrac{y_1}{y_1 y_2 + y_2 y_3 + y_3 y_1} = \dfrac{z_2 z_3}{z_3 + z_1 + z_2}$

$z_b = \dfrac{y_2}{\Delta_y} = \dfrac{y_2}{y_1 y_2 + y_2 y_3 + y_3 y_1} = \dfrac{z_1 z_3}{z_3 + z_1 + z_2}$

$z_c = \dfrac{y_3}{\Delta_y} = \dfrac{y_3}{y_1 y_2 + y_2 y_3 + y_3 y_1} = \dfrac{z_1 z_2}{z_3 + z_1 + z_2}$

Appendix

A7 Equivalent circuits

Equivalent circuits are important, because analysis or synthesis of a circuit is usually simplified when equivalents are used. An entire circuit or just a "piece" of a circuit can be replaced by an equivalent circuit. For example, one resistor replaces three resistors in series. Source conversion is one application of this idea: voltage source to current source or vice-versa (Figure A71). Here are the equations that establish equivalency. The key to this analysis is calculation of voltage and current at the two terminals:

1) The terminals have no load connected. Find the open circuit voltage v_{OC}.
2) The terminals are short circuited. Find the short circuit current i_{SC}.

(1) Figure A71a $\quad v_{oc} = v_T \quad$ and $\quad i_{sc} = \dfrac{v_T}{R_T}$

$$v_{ab} = v_T - i_L R_T$$

(2) Figure A71b $\quad v_{oc} = i_N R_N \quad$ and $\quad i_{sc} = i_N$

$$i_L = i_N - \dfrac{v_{cd}}{R_N} \quad \Rightarrow \quad v_{cd} = i_N R_N - i_L R_N$$

if $v_{ab} = v_{cd}$, then $v_T = i_N R_N$ and $R_T = R_N$

Figure A71 Equivalent sources

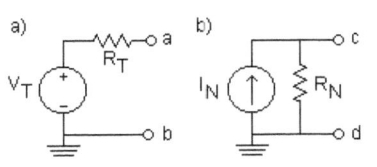

The theorems of Charles Leon Thévenin (1857-1926) and Edward Lawry Norton (1898-1983) show us how to find two-terminal circuits (such as those in Figure A71) equivalent to any circuit when any two nodes in that circuit are selected. Selection of a different node-pair in the same circuit results in a different equivalent circuit. The significance of the equivalent circuit is the ease with which we can calculate the effect of any load on the selected terminals. A node-pair is loaded when two terminals of any "load" circuit are connected to the two selected terminals.

There are several methods you can use to calculate the equivalent circuit between any two terminals. We prefer method 3 which is developed in upcoming paragraphs.

Method 1	When there are no dependent sources, deactivate all sources. Then calculate R_T. Calculate $v_T = v_{OC}$ per method 2.
Method 2	With sources activated find the open circuit voltage v_{OC} and the short circuit current i_{SC}. Then calculate $R_T = v_{OC}/i_{SC}$. And, $v_T = v_{OC}$
Method 3	Do not change the circuit. Add a source voltage v_Z across the two designated terminals, and use the mesh method to solve for i_Z. Or, add a source current i_Y, and use the node voltage method to solve for v_Y.

A7.1 Method 1 Sources Deactivated

Find the Thévenin and Norton equivalents of a circuit that does not have dependent sources (Figure A73).

Calculate R_T with the sources deactivated (Figure A72), because R_T is independent of the voltage and current sources.

The resistance of an ideal voltage source is zero. A deactivated voltage source is replaced by a wire that is a short circuit. The resistance of an ideal current source is "infinite." A deactivated current source is replaced by "nothing" which is an open circuit (Figure A72).

Figure A72 Circuit with deactivated sources

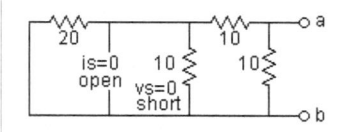

An ad hoc solution is reasonable for this elementary circuit. Starting from the left, 20Ω in parallel with 10Ω equals 20/3Ω. Add this to the 10Ω in series for a total equal to 50/3Ω. The 50/3Ω is in parallel with the 10Ω between terminals ab. Therefore R_{AB} that equals the product over the sum is (500/3)/(80/3) = 6.25 = 25/4Ω. This is R_T.

Appendix

A7.2 Method 2 Sources Activated

Calculation of open circuit voltage (v_{OC}) or short circuit current (i_{SC}) requires activated sources. The mesh and nodal methods provide straightforward solutions.

With sources activated (Figure A73) calculate the open circuit voltage v_{OC} by using Kirchhoff's current law at the two nodes (see sidebar).

(3) at node v_1 $\quad 0 = g_1(v_1 - 0) + i_S + g_4(v_1 - v_S) + g_2(v_1 - v_{oc})$

at node v_c $\quad 0 = g_2(v_{oc} - v_1) + g_3(v_{oc} - 0)$

so that $\quad v_{oc} = (-i_S + g_4 v_S) \dfrac{g_2}{(g_1 + g_2 + g_4)g_3 + (g_1 + g_4)g_2}$

$$v_{oc} = -\dfrac{5}{2} i_S + \dfrac{1}{4} v_S \quad \text{for values shown in Figure A73}$$

Study Figure A73. If $i_S = 0$ the open circuit voltage is the response to v_S. If $v_S = 0$ the open circuit voltage is the response to i_S. The sum is the response to both sources. This is an example of the superposition theorem. In fact any analysis is simplified if you activate one source at a time, calculate the responses, and then add. Next we calculate the short circuit current i_{sc} with sources still activated.

Figure A73 Circuit with sources

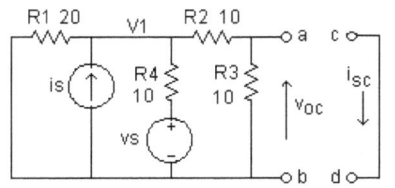

(4) at node v_1

$$-i_S + g_4 v_S = (g_1 + g_4 + g_2)v_1$$
$$i_{sc} = g_2 v_1$$

so that $\quad i_{sc} = g_2 v_1 = g_2 \dfrac{-i_S + g_4 v_S}{g_1 + g_2 + g_4} = (-i_S + g_4 v_S) \dfrac{g_2}{g_1 + g_2 + g_4}$

$$i_{sc} = -\dfrac{10}{25} i_S + \dfrac{1}{25} v_S \quad \text{for values shown in Figure A73}$$

Next calculate the equivalent two-terminal resistance R_T between the two terminals (v_{oc} is calculated on the next page).

(5) $\quad R_T = \dfrac{v_{oc}}{i_{sc}} = \dfrac{-\dfrac{5}{2} i_S + \dfrac{1}{4} v_S}{-\dfrac{10}{25} i_S + \dfrac{1}{25} v_S} = \dfrac{\dfrac{1}{4}}{\dfrac{1}{25}} \times \dfrac{-\dfrac{20}{2} i_S + v_S}{-10 i_S + v_S} = \dfrac{25}{4} \Omega$

Electric Circuits - Analysis and Design

Node method --Method 2 Solution for Figure A73 Sources Activated

1. Select a reference node and ground it to make its value zero volts. Assign unknown voltages to the other nodes

2. Assign a current through every branch, and write a KCL (Kirchhoff's current law) equation for each of N - 1 independent nodes.
$$0 = I_1 + i_S + I_4 + I_2$$
$$0 = I_2 + I_3$$

Use component v–i constraints to relate branch currents to node voltages. Then replace branch currents in KCL equations.
$$0 = g_1 V_1 + i_S + g_4(V_1 - v_S) + g_2(V_1 - v_{oc})$$
$$0 = g_2(v_{oc} - V_1) + g_3 v_{oc}$$

Put equations in standard form.
$$-i_S + g_4 v_S = (g_1 + g_2 + g_4)V_1 - g_2 v_{oc}$$
$$0 = -g_2 V_1 + (g_2 + g_3)v_{oc}$$

3. Use Cramer's rule to solve for the node voltages.
$$\Delta_G = (g_1 + g_2 + g_4)(g_2 + g_3) - (-g_2)(-g_2)$$
$$= (g_1 + g_2 + g_4)g_2 + (g_1 + g_2 + g_4)g_3 - g_2 g_2$$
$$= (g_1 + g_2 + g_4)g_3 + (g_1 + g_4)g_2$$
$$= g_1 g_2 + g_1 g_3 + g_2 g_3 + g_2 g_4 + g_3 g_4$$

$$v_{oc} = -\frac{\Delta_{12}}{\Delta_G}(-i_S + g_4 v_S) = (-i_S + g_4 v_S)\frac{g_2}{\Delta_G}$$

$$v_{oc} = (-i_S + g_4 v_S)\frac{g_2}{(g_1 + g_4 + g_2)g_3 + (g_1 + g_4)g_2}$$

$$v_{oc} = (-i_S + \frac{1}{10}v_S)\frac{\frac{1}{10}}{\left(\frac{1}{20} + \frac{1}{10} + \frac{1}{10}\right)\frac{1}{10} + \left(\frac{1}{20} + \frac{1}{10}\right)\frac{1}{10}}$$

$$= (-i_S + \frac{1}{10}v_S)\frac{\frac{1}{10}}{\left(\frac{5}{20}\right) + \left(\frac{3}{20}\right)} = (-i_S + \frac{1}{10}v_S)\frac{20}{8} = -\frac{5}{2}i_S + \frac{1}{4}v_S$$

Appendix

A7.3 Method 3

We need a context. First, connect a current source I_Y to the terminals a and b of the Thevenin equivalent circuit (Figure A74a). Then write the mesh equation. Second, repeat the process with a voltage source V_Z (Figure A74b).

(6) $V_y = R_T I_y + V_T$

(7) $I_z = \dfrac{1}{R_N} V_z - I_N$

Figure A74 Circuit with added source

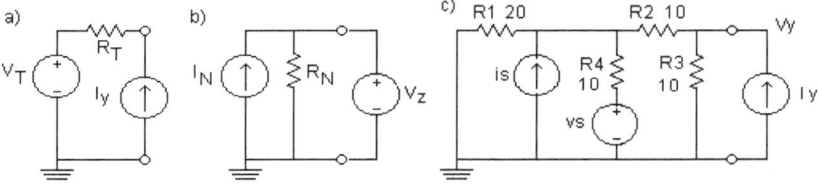

The idea behind this method is this: add a source I_Y or V_Z, solve for V_Y or I_Z, recast the solution in the formats in equations 6 and 7. Identify the equivalent parameters V_T and R_T, or I_N and R_N.

Here is one example. Add a current source I_Y to the terminals a and b of a circuit (Figure A74c). Formulate the node equations, and solve for V_Y.

(8) $(-i_S + g_4 v_S) = (g_1 + g_2 + g_4) v_1 - g_2 v_y$

$\quad i_y = -g_2 v_1 + (g_2 + g_3) v_y$

$\quad \Delta_g = (g_1 + g_2 + g_4)(g_2 + g_3) - g_2 g_2$

$\qquad = (g_1 + g_4)(g_2 + g_3) + g_2 g_3$

$\quad v_y = \dfrac{g_1 + g_2 + g_4}{\Delta_G} i_y + \dfrac{g_2}{\Delta_G}(-i_S + g_4 v_S)$

$\quad v_y = R_T i_y + V_T$

You can readily show that these results are the same as the results of method 2.

Electric Circuits - Analysis and Design

Use method 3 on a circuit with dependent sources (Figure A75).

Add current source I_y.

Find the Thevenin equivalent circuit for Figure A75
Step 1 Connect a current source I_Y to terminals a and b.
Step 2 Formulate the node equations.
Step 3 Solve for node voltage $V_Y = V_2$ in the form $V_Y = cI_Y + d$.
Step 4 Thevenin parameters: $Z_T = c$ and $V_T = d$.

Figure A75 Example for finding Thevenin equivalent circuit

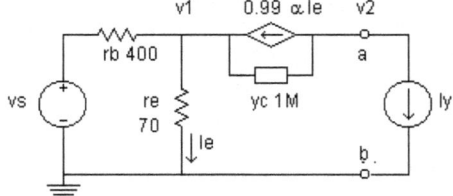

(9a) node1 $\quad g_b V_S = (g_b + g_e + y_c)V_1 - y_c V_2 + \alpha I_e$
(9b) node2 $\quad I_y = \quad\quad\quad -y_c V_1 + y_c V_2 - \alpha I_e$

$\quad\quad\quad\quad\quad \alpha I_e = -\alpha g_e V_1$

(10a) node1 $\quad g_b V_S = [g_b + g_e(1-\alpha) + y_c]V_1 - y_c V_2$
(10b) node1 $\quad I_y = \quad\quad\quad (-y_c + \alpha g_e)V_1 + y_c V_2$

$\quad\quad\quad\quad\quad \Delta_y = y_c(g_b + g_e)$

(11) $\quad V_2 = \dfrac{g_b + g_e(1-\alpha) + y_c}{\Delta_y} I_y + \dfrac{g_b(y_c - \alpha g_e)}{\Delta_y} V_S$

(12) $\quad V_2 = Z_T I_y + V_T$

Appendix

Add voltage source V_z.

Find the Norton equivalent circuit for Figure A76
Step 1 Connect a voltage source V_Z to terminals a and b.
Step 2 Formulate the mesh equations.
Step 3 Solve for mesh current I_Z in the form $I_Z = eV_Z + f$.
Step 4 Norton parameter: $Y_N = e$ and $I_N = f$.

Figure A76 Example for finding Norton equivalent circuit

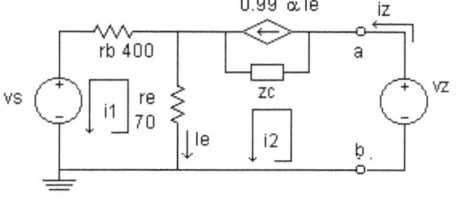

(13a) mesh1 $V_S = (r_b + r_e)I_1 \quad\quad - r_e I_2$
(13b) mesh2 $V_z = \quad\quad -r_e I_1 + (r_e + z_c)I_2 - z_c \alpha I_e$
$\quad\quad\quad\quad I_e = I_2 - I_1$

(14a) mesh1 $V_S = \quad (r_b + r_e)I_1 \quad\quad - r_e I_2$
(14b) mesh2 $V_z = (-r_e + z_c \alpha)I_1 + [r_e + z_c(1-\alpha)]I_2$

$$\Delta_z = r_b r_e + z_c[r_e + r_b(1-\alpha)]$$

(15) $\quad\quad I_2 = \dfrac{r_b + r_e}{\Delta_z} V_z - \dfrac{(-r_e + z_c \alpha)}{\Delta_z} V_S$

(16) $\quad\quad I_z = Y_n V_z - I_n$

Electric Circuits - Analysis and Design

A8 Signal Sources

The battery is the constant voltage source $v_S(t) = V_M$. A battery's $v_S(t)$ waveform versus time is a horizontal line that represents its constant voltage over time. Batteries, or their equivalents, are used in all types of portable electronic apparatus, because a constant voltage power supply is required to energize electronic circuits. A constant voltage source is also known as a DC voltage source.

Figure A81 Waveforms of forcing functions

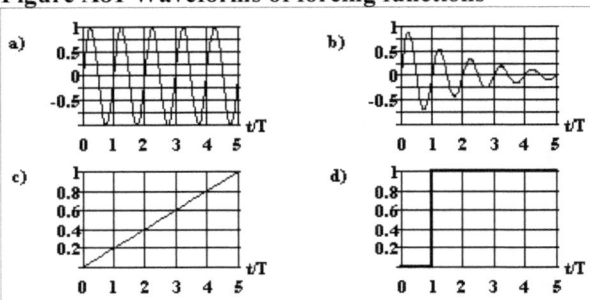

Most electric power systems throughout the world produce electrical energy in the form of a sinewave with a constant frequency of 60Hz (USA) or 50Hz (World). One Hz (Hertz) is a frequency of one cycle per second. The voltage available for your USA home appliances and home lighting is a sinewave (Figure A81a) with constant frequency and constant maximum amplitude of $V_M = 120\sqrt{2}$ volts.

(1) *sine waveform*: $V_S(t) = V_M \sin(\omega t + \theta)$ (*Figure A81a*)

High fidelity sound systems' forcing functions are complex voice and music sounds that are converted by electromechanical devices, such as microphones, into electrical signal sources. The signals are primarily sums of sinewaves (Figure A81a) and damped sinewaves (Figure A81b). Audio signal frequencies can range from about 20Hz to over 100,000Hz.

(2) *damped sine wave*: $V_S(t) = V_M e^{-\sigma t} \sin(\omega t + \theta)$ $\sigma > 0$ (*Figure A81b*)

Appendix

Many parts of telephone and television systems include circuits whose forcing functions are also sinewave sources with various frequencies like those contained in your speech. Other parts of these systems use pulse sources to process digital data. In both cases the system behavior over some frequency range is specified and important. This is why we need to understand how circuits behave over broad ranges of frequency to reproduce voice and digital signals faithfully. Contrast these requirements over a frequency range to the single frequency world of electric power systems. However power system transients are difficult transmission line problems.

Other parts of telephone and television systems use non-linear signals where the emphasis is on waveforms in the time domain. For example, the electron beam in a TV or computer display tube is swept back and forth by a sawtooth-like waveform. The sawtooth's ramp f(t) = t/T (Figure A81c).

Digital signals consist of abrupt transitions, or steps (Figure A81d), from 0 to 1.8 volts (for example) and vice-versa.

(3) $\quad ramp: V_S(t) = V_M \dfrac{t}{T}$

(4) $step\ function: V_S(t) = V_M u(t-1)$

Step Functions eliminate a switch The circuit (Figure A82) has a switch arm that will be moved from position B to position A at time t = 0. Energy is stored in the circuit while the switch is in position B. The stored energy is represented by the inductor current $i(0^-)$ = V(0)/R (in a non transient state the inductor is just a wire). We are interested in this current at time $t = 0^-$, because 0^- is the instant of time just prior to moving the switch to position A.

Figure A82 RL circuit

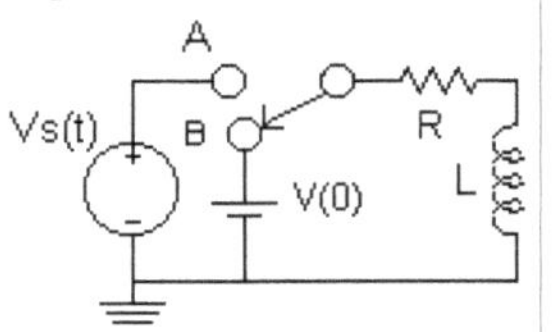

Suddenly, we need to know how to analyze a circuit with a switch. We analyze circuits with switches by eliminating them. This practical process is executed by multiplying the signal source by the step function u(t), because, by definition, u(t) switches from 0 to 1 at time t = 0. In other words u(t) is the mathematical equivalent of the switch. Note: u(t−1) is shown in Figure A81d.

227

Electric Circuits - Analysis and Design

A9 Impedance, Admittance, and Immittance

We have learned that in a circuit assembled with resistors, the ratio of voltage to current, V/I, is a resistance whose dimension is ohms.

What are the ratios of voltage to current for inductors and capacitors?
Do these ratios have names?
What dimensions do these ratios have?
Do these ratios have meaning in the time domain?
Do they have meaning in the complex frequency domain?

We find the answers by assuming a current and calculating the voltage across each type of element: R, L, and C. Clearly v/i ratios for L and C in the time domain are not possible because integrals or derivatives are involved. However, v/i ratios have significant meaning in the (transformed) complex frequency domain.

$$v_R(t) = Ri(t)$$

$$\mathcal{L}[v_R(t)] = \mathcal{L}[Ri(t)] \Rightarrow V(p) = RI(p) \quad \text{so that} \quad R = \frac{V(p)}{I(p)}$$

(1) If $z_R = \dfrac{V(p)}{I(p)}$ then $z_R = R$

$$v_L(t) = L\frac{di(t)}{dt}$$

$$\mathcal{L}[v_L(t)] = \mathcal{L}\left[L\frac{di(t)}{dt}\right] \Rightarrow V(p) = pLI(p) - Li(0) \quad \text{so that} \quad pL = \frac{V(p)}{I(p)} + \frac{Li(0)}{I(p)}$$

(2) If $z_L = \dfrac{V(p)}{I(p)}$ then $z_L = pL$

$$i_C(t) = C\frac{dv(t)}{dt}$$

$$\mathcal{L}[i_C(t)] = \mathcal{L}\left[C\frac{dv(t)}{dt}\right] \Rightarrow I(p) = pCV(p) - Cv(0) \quad \text{and} \quad pC = \frac{I(p)}{V(p)} + \frac{Cv(0)}{V(p)}$$

(3) If $z_C = \dfrac{V(p)}{I(p)}$ then $z_C = \dfrac{1}{pC}$

Appendix

The generic name for any V(p)/I(p) ratio is impedance z, and the dimension of any V/I ratio is ohms. The generic name for the reciprocal ratio I/V is admittance y, and its dimension is Siemens (it used to be mhos, which we prefer and use). By definition y = 1/z. The generic name for these generic names is Bode's immittance (no symbol). In general z, y, and immittances are functions of the complex frequency p.

A resistor's impedance R is independent of p and consistent with what we have learned about resistive circuit analysis. An inductor's impedance is pL. This is consistent with an inductor (which is just a coil of wire) behaving like a short-circuit in DC circuits, because p=0 in DC circuits. A capacitor's impedance is 1/pC. This is consistent with a capacitor (which is just two insulated metal pieces) behaving like an open-circuit in DC circuits, because 1/p becomes infinite as p approaches zero.

A9.1 Impedance's in Series

Components in series have one path for current. The same current flows in all components, which are in series. Resistors (R), inductors (L), and capacitors (C) are two terminal devices (reference Sections 3.1, 3.2, and 3.3 page 27). R, L, and C devices are connected in series when one terminal of a device with impedance z_1 is connected, via a wire, to one terminal of another device with impedance z_2, and so forth. The connecting wires are assumed to have zero impedance. Suppose we connect the two terminals of a voltage source to the two ends of a series-string of n impedance's (Figure A91). We use Kirchhoff's voltage law to analyze this one-mesh circuit.

(4) $v_1 = i_1 z_1$
$v_2 = i_2 z_2$
$... = ...$
$v_n = i_n z_n$

For impedances in series.
(4) $i = i_1 = i_2 = ... = i_n$
$v = v_1 + v_2 + ... + v_n$
$= i_1 z_1 + i_2 z_2 + ... + i_n z_n$
$= i(z_1 + z_2 + ... + z_n)$
$= iz \quad \text{where} \quad z = z_1 + z_2 + ... + z_n$

Electric Circuits - Analysis and Design

Figure A91 Impedance's in series

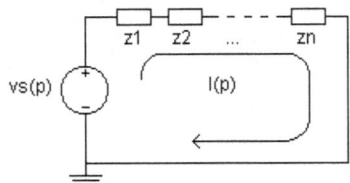

If all the z's are resistors, then the total resistance R equals the sum of the series resistors. If all the z's are inductors, then the total inductance L equals the sum of the series inductors. If all the z's are capacitors, then the reciprocal of the total capacitance C (i.e. 1/C) equals the sum of the reciprocals of the series capacitors. In other words, series combinations of R's, L's, and C's can be reduced to one each R, L, and C.

(5) $z = R_1 + R_2 + ... + R_n$
$R_S = R_1 + R_2 + ... + R_n$

(6) $z = pL_1 + pL_2 + ... + pL_n$
$L_S = L_1 + L_2 + ... + L_n$

(7) $z = \dfrac{1}{pC_1} + \dfrac{1}{pC_2} + ... + \dfrac{1}{pC_n}$
$\dfrac{1}{C_S} = \dfrac{1}{C_1} + \dfrac{1}{C_2} + ... + \dfrac{1}{C_n}$

Figure A92 General branch with series RLC

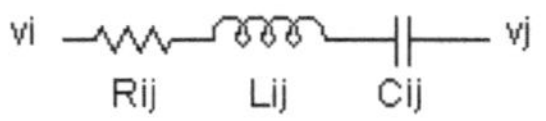

vi —ᴡᴡ—ᵒᵒᵒ—||— vj
   Rij   Lij   Cij

If three z's in series are replaced by one each R, L, and C, then we have the generalized series branch. (Figure A92).

if $z_1 = pL$, $z_2 = R$, $z_3 = \dfrac{1}{pC}$,

(8) $z = z_1 + z_2 + z_3 = pL + R + \dfrac{1}{pC}$

Appendix

A9.2 Admittance's in Parallel

Components in parallel have one current path for each component. The same voltage is across all components in parallel. R, L, and C components are connected in parallel when the two terminals of a component with admittance y_1 (= $1/z_1$) is connected, via two wires, to the two terminals of another component with admittance y_2 (= $1/z_2$), and so forth. As before, the wires are assumed to have zero impedance. Suppose we connect the two terminals of a current source to the two terminals of the parallel set of n components (Figure A93). This creates a one-node-pair circuit which is analyzed as follows:

(9) $i_1 = v_1 y_1$
$i_2 = v_2 y_2$
... = ...
$i_n = v_n y_n$

For admittances in parallel
(9) $v = v_1 = v_2 = ... = v_n$
$i = i_1 + i_2 + ... + i_n$
$= v_1 y_1 + v_2 y_2 + ... + v_n y_n$
$= v(y_1 + y_2 + ... + y_n)$
$= vy$ where $y = y_1 + y_2 + ... + y_n$

Figure A93 Admittance's in parallel

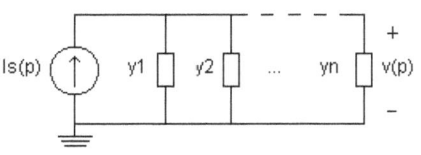

If all the y's are resistors, then the total conductance G equals the sum of the parallel conductance's. G also equals the sum of the reciprocals of the (parallel) resistors. If all the y's are capacitors, then the total capacitance C is the sum of the (parallel) capacitors. If all the y's are inductors, then the reciprocal of the total inductance L (1/L) is the sum of the reciprocals of the (parallel) inductors. In other words, parallel combinations of R's, L's, and C's can be reduced to one each R, L, and C (Figure A94).

231

Electric Circuits - Analysis and Design

$$y = \frac{1}{R_1} + \frac{1}{R_2} + ... + \frac{1}{R_n}$$

(10) $\quad \dfrac{1}{R_P} = \dfrac{1}{R_1} + ... + \dfrac{1}{R_n}$

$$y = \frac{1}{pL_1} + \frac{1}{pL_2} + ... + \frac{1}{pL_n}$$

(11) $\quad \dfrac{1}{L_P} = \dfrac{1}{L_1} + \dfrac{1}{L_2} + ... + \dfrac{1}{L_n}$

$$y = pC_1 + pC_2 + ... + pC_n$$

(12) $\quad C_P = C_1 + C_2 + ... + C_n$

If three y's in parallel are replaced by one each R, L, and C, then we have the generalized parallel branch (Figure A94).

Figure A94 branch with parallel RLC

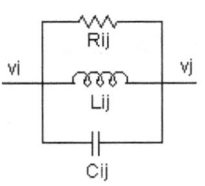

$if\ y_1 = \dfrac{1}{pL},\ y_2 = G,\ y_3 = pC,$

(13) $\quad y = y_1 + y_2 + y_3 = \dfrac{1}{pL} + G + pC$

232

Answers to Most of the Problems

101 I=2mA means 2×10^{-3} coulombs per second are flowing.
102 V=E/C=5/0.2=25 volts

---

211 v_1=11.175V, i_S=0.375A, i_1=0.2375A
212 4.8V
213 1.2A, −111V

---

301 V=IR=0.94V. P=I²R=1.88mW
302 V=Rdq/dt=880mV
303 C=Q/V=15nF
304 i=Cdv/dt=330mA
305 i=Cdv/dt=0.9mA
306 C=Q/V=111fF
307 v=iZ=(2.5/π)A × cos ωt
308 dt=Ldi/V=3.67nS
309 V=Ldi/dt=25KV

401
$$M = k\sqrt{L_1 L_2} = 0.93\sqrt{100\cdot 10^{-3}\cdot 200\cdot 10^{-3}} = 0.93\sqrt{2}\cdot 100\cdot 10^{-3} = 131.5mH$$
$$n = \sqrt{L_2 / L_1} = \sqrt{2}$$

402
$$M = k\sqrt{L_1 L_2} \quad n = \sqrt{L_2 / L_1}$$
$$L_2 = n^2 L_1 = n^2 (M/k)^2 / L_2 \quad \Rightarrow \quad L_2 = nM/k \quad L_1 = L_2 / n^2$$
$$L_2 = 6\cdot 3/0.8 = 22.5mH \quad L_1 = 22.5/36 = 0.625mH$$
$$\text{check } n^2 = 22.5/0.625 = 36 \quad M = 0.8\sqrt{22.5\cdot 0.625} = 3$$

403
$$M = M_{21} = 10^{-8} N_2 \frac{d\Phi_{21}}{di_1} = 10^{-8}\cdot 900\cdot \frac{0.6}{50\cdot 10^{-3}} = 108\cdot 10^{-6} H = 108\mu H$$

$$L_1 = 10^{-8} N_1 \frac{d(\Phi_{21}+\Phi_{11})}{di_1} = 10^{-8}\cdot 300\cdot \frac{0.6+0.2}{50\cdot 10^{-3}} = 48\cdot 10^{-6} H = 48\mu H$$

$$L_2 = \left(\frac{N_2}{N_1}\right)^2 L_1 = \left(\frac{900}{300}\right)^2 48 = 9\cdot 48 = 432\mu H$$

$$k = \frac{M}{\sqrt{L_1 L_2}} = \frac{108}{\sqrt{9\cdot 48\cdot 48}} = 0.75 \quad or \quad k = \frac{\Phi_{21}}{\Phi_1} = \frac{\Phi_{12}}{\Phi_2} = \frac{0.6}{0.6+0.2} = 0.75$$

404

$$v = L\frac{di}{dt} \Rightarrow i = \frac{1}{L}\int v\, dt = \frac{1}{L}\int 5\sin \omega t\, dt = \frac{5}{\omega L}\cos \omega t$$

$$i = \frac{5}{2\pi 10 \cdot 10^6 \cdot 22 \cdot 10^{-6}}\cos \omega t = 3.62\cos \omega t\ mA \quad \omega = 2\pi 10 \cdot 10^6$$

405

$$L_{plus} - L_{minus} = (L_1 + L_2 + 2M) - (L_1 + L_2 - 2M) = 4M$$

406

$$M_{12} = 0.8\sqrt{100 \cdot 60} = 62 mH \quad M_{13} = 0.8\sqrt{100 \cdot 220} = 119 mH$$

$$M_{23} = 0.8\sqrt{60 \cdot 220} = 92 mH$$

$$L = L_1 + L_2 + L_3 - M_{12} + M_{13} - M_{23} = 100 + 60 + 200 - 62 + 119 - 92 = 325 mH$$

407 A transformer has 2 windings n_1, n_2 with turns in the ratio n:1. Resistor R_2 is connected to winding n_2. Find the equation for z_{INPUT} by equating P_1 power in to P_2 power out.

$$v_1 i_1 = v_2 i_2, \quad v_2 = i_2 R_2, \quad v_1 = z_{in} i_1, \quad n_1 = n \cdot n_2 \quad \text{implies}$$

$$z_{in} i_1^2 = R_2 i_2^2 \quad \text{so that} \quad z_{in} = R_2 n^2$$

408 A transformer has 3 windings n_1, n_3, n_2 with turns ratios 1:3:2. Resistors R_3 and R_2 are connected to windings n_3, n_2 respectively. What is the input impedance z_{INPUT} at winding n_1? Hint use power.

$$v_1 i_1 = v_3 i_3 + v_4 i_4$$

$$v_1 \frac{v_1}{z_{in}} = v_3 \frac{v_3}{R_3} + v_4 \frac{v_4}{R_4} = \frac{v_1^2}{(3n_1)^2 R_3} + \frac{v_1^2}{(4n_1)^2 R_3}$$

$$\frac{1}{z_{in}} = \frac{1}{(3n_1)^2 R_3} + \frac{1}{(4n_1)^2 R_3}$$

409

$$R_x = \frac{(1000 + 200)100000}{1000 + 200 + 100000} = 1186\,\Omega$$

$$R_y = (R_2 + z_L)/n^2 = (600 + 5000)/2.5^2 = 896$$

410

$(z_L + R_2)/n^2 = 900/9 = 100\Omega$

$M = k\sqrt{L_1 L_2} = k\sqrt{L_1 \cdot n^2 L_1} = knL_1 = 0.95 \cdot 3 \cdot 50 = 142.5mH$

$M/n = 142.5/3 = 47.5mH$

$L_1 - M/n = 50 - 47.5 = 2.5mH$

411

$kL_1 = 0.95 \times 50 = 47.5mH \qquad (1-k)L_1 = 0.05 \times 50 = 2.5mH$

412

$$f_1 = \frac{1}{2\pi} \cdot \frac{1186 \cdot 896}{1186+896} \cdot \frac{1}{0.99 \cdot 0.1} = \frac{1}{2\pi} \cdot \frac{510.4\Omega}{0.99 \cdot 0.1} = 820.5 Hz$$

in the spice program

$R_1 \| R_2 = 250\Omega$, which reduces f_1 to $\dfrac{250}{510.4} 820.5 = 402 Hz$

413

$$f_H = \frac{1}{2\pi} \cdot \frac{R}{L} = \frac{1}{2\pi} \cdot \frac{(500+500)}{2 \cdot (1-0.99) \cdot 1} = \frac{1}{2\pi} \cdot \frac{1000\Omega}{2 \cdot 0.01} = 7.96 KHz$$

501

$G_1 v_1 = (G_1 + G_2 + G_3) v_2 - G_3 v_3$

$0 = -G_3 v_2 + (G_3 + G_4) v_3$

$$\Delta_G = (G_1 + G_2 + G_3)(G_3 + G_4) - G_3^2 = \left(\frac{1}{22} + \frac{1}{47} + \frac{1}{33}\right)\left(\frac{1}{33} + \frac{1}{33}\right) - \left(\frac{1}{33}\right)^2$$

$\Delta_G = (.04545 + 0.02128 + 0.03030)(0.06061) -$

$\quad = (0.09703)(0.06061) - 0.00092 = 0.00496$

$V_2 = \dfrac{(G_3 + G_4)}{\Delta_G} G_1 V_1 = \dfrac{0.03030 + 0.03030}{0.00496} 0.045455 \cdot 5 = 2.78V$

$V_3 = \dfrac{G_3}{\Delta_G} G_1 V_1 = \dfrac{0.03030}{0.00496} 0.045455 \cdot 5 = 1.39V$

$I_1 = G_1(V_1 - V_2) = 0.045455(5 - 2.78) = 101mA$

$I_2 = G_4 V_3 = 0.03030 \cdot 1.39 = 42mA$

verify $V_2 = R_2(I_1 - I_2) = 47(101 - 42) = 2.773V$

$\qquad V_3 = V_2 - R_3 I_2 = 2.78 - 33(0.042) = 1.39V$

503
$0 = G_1(v_1 - v_S) - i_1$
$v_1 = v_S - R_1 i_1 = 5 - 10^4 \cdot 20 \cdot 10^{-6} = 4.8V \quad \text{and} \quad i_S = i_1$

504
$0 = G_1(v_S - v_L) - i_1$
$v_L = v_S - R_1 i_1 = 9 - 10^2 \cdot 1.2 = -111V \quad \text{and} \quad i_S = i_1$

505
$G_1 v_S = (G_1 + G_2) v_1 - i_2$
$v_1 = \dfrac{G_1 v_S + i_2}{G_1 + G_2} = \dfrac{0.045455 \cdot 12 + 0.2}{0.045455 + 0.02128} = 11.17V$
$i_S = G_1(v_S - v_1) = 0.045455(12 - 11.17) = 37.7 mA$
$i_1 = G_2 V_1 = 0.02128 \cdot 11.17 = 238 mA$

506
(34a) $\quad g_B V_S = [g_B + g_E(1-\alpha) + g_C]V_1 - g_C V_L$
(34b) $\quad 0 = (\alpha g_E - g_C)V_1 + (g_L + g_C)V_L$
$\Delta_G = [g_b + g_e(1-\alpha) + g_c](g_L + g_c) - (-g_c)(-g_c + \alpha g_e)$
$= [g_e(1-\alpha) + g_b + g_c]g_L + [g_e(1-\alpha) + g_b + g_c]g_c + (-g_c g_c + \alpha g_e g_c)$
$= [g_e(1-\alpha) + g_b + g_c]g_L + [g_e + g_b]g_c$

$V_1 = \dfrac{\Delta_{11} g_b}{\Delta_G} V_S \qquad \dfrac{V_1}{V_S} = \dfrac{(g_L + g_c)g_b}{\Delta_G}$

$V_L = \dfrac{-\Delta_{12} g_b}{\Delta_G} V_S \qquad \dfrac{V_L}{V_S} = \dfrac{-(-g_c + \alpha g_e)g_b}{\Delta_G} = \dfrac{g_b(g_c - \alpha g_b)}{\Delta_G}$

$I_S = g_b(V_S - V_1) \quad \text{and} \quad I_S = g_e V_1$

507
(1) $\quad 0 = g_1(v_e - v_S) + g_3(v_e - 0) + i_x$
(2) $\quad 0 = y_2(v_c - 0) - i_x$
(3) $\quad i_x = g_m v_e \quad \text{because} \quad v_e = v_x$

$(1+2) \quad g_1 v_S = (g_1 + g_3) v_e + y_2 v_c \quad \text{sum of 1a and 1b}$
$(sub\ 3\ in\ 2) \quad 0 = \quad -g_m v_e + y_2 v_c$
$\Delta_y = (g_1 + g_3) y_2 + g_m y_2 = (g_1 + g_3 + g_m) y_2$

Answers to Most of the Problems

$$v_e = \frac{g_1 v_S \cdot y_2}{\Delta_y} = \frac{g_1}{g_1 + g_3 + g_m} v_S$$

$$v_c = \frac{g_m g_1 v_S}{\Delta_g} = \frac{g_m g_1 v_S}{(g_1 + g_3 + g_m) p C_2}$$

$$i_2 = p C_2 v_c \quad \text{and} \quad i_1 = g_3 v_e + i_2 = g_3 v_e + p C_2 v_c$$

508

$$g_1 v_1 + g_2 v_3 = (g_1 + g_2) v_2 + i_y \quad (i_y \text{ flows in } v_y, \text{ ignore } i_1, i_2)$$
$$0 = -g_2 v_2 + g_2 v_3 + i_3 \quad (i_3 \text{ flows in } v_3)$$
$$i_x = g_1(v_1 - v_2)$$

Node analysis is a problem here. Try ad hoc.
$$v_2 = v_y = 2 i_x = 2 g_1 (v_1 - v_2)$$

$$v_2 = \frac{2 g_1 v_1}{1 + 2 g_1} = \frac{2 v_1}{R_1 + 2} = \frac{2 \cdot 5}{22 + 2} = 0.417 V$$

Observe the v_2 is set by v_y, which does not depend in v_3.

$$i_3 = g_2 v_2 - g_2 v_3 = g_2 \frac{2 v_1}{R_1 + 2} - g_2 v_3 = \frac{1}{33}(0.417 - 10) = \frac{9.583}{33} = 0.290 A$$

$$i_x = g_1(v_1 - v_2) = \frac{1}{22}(5 - 0.417) = 0.208 A$$

509

$$g_1 v_1 = (g_1 + g_2) v_n - y_2 v_0 \qquad y_2 = 1/p L_2$$
$$0 = -y_2 v_n + (g_S + y_2) v_0 - g_S v_S \qquad v_S = -\mu v_n \text{ and } v_p = 0$$
$$g_1 v_1 = (g_1 + g_2) v_n - y_2 v_0$$
$$0 = -(y_2 - g_S \mu) v_n + (g_S + y_2) v_0$$
$$\Delta_g = (g_1 + y_2)(g_S + y_2) - y_2(y_2 - g_S \mu) = (g_1 + y_2)(g_S) + g_1 y_2 + y_2 g_S \mu$$
$$= \frac{(p L_2 + R_1)(g_S) + 1 + R_1 g_S \mu}{R_1 p L_2}$$

$$\frac{v_0}{v_1} = \frac{(y_2 - g_S \mu) g_1}{\Delta_g} = \frac{R_1 p L_2 (y_2 - g_S \mu) g_1}{(p L_2 + R_1)(g_S) + 1 + R_1 g_S \mu} = -\frac{p L_2}{R_1} \cdot \frac{1 - \dfrac{R_S}{\mu p L_2}}{1 + \dfrac{1}{\mu}\left(1 + \dfrac{p L_2}{R_1} + \dfrac{R_S}{R_1}\right)}$$

Electric Circuits - Analysis and Design

510 Ignore i_1 and i_2. Let R_X=5K, g_M=12mA/v. Use the node voltage method. Solve for V_B, V_E, I_C, I_B, I_E.

$g_1 v_1 = (g_1 + g_x)v_b - g_x v_e$

$0 = -g_x v_b + (g_x + g_2)v_e - g_m v_x \qquad \text{and} \qquad v_x = v_b - v_e$

$g_1 v_1 = (g_1 + g_x)v_b - g_x v_e$

$0 = -(g_x + g_m)v_b + (g_x + g_2 + g_m)v_e$

$\Delta_g = (g_1 + g_x)(g_x + g_2 + g_m) - g_x(g_x + g_m)$

$\Delta_g = g_1 g_x + g_1 g_2 + g_1 g_m + g_x g_2 = \dfrac{R_2 g_x + 1 + g_m R_2 + R_1 g_x}{R_1 R_2}$

$v_e = \dfrac{(g_x + g_m)g_1 v_1}{\Delta_g} = \dfrac{R_2(1 + g_m R_x)v_1}{R_2(1 + g_m R_x) + R_x + R_1} = \dfrac{v_1}{1 + \dfrac{R_x + R_1}{R_2(1 + g_m R_x)}}$

Figure 310 Model   **Figure 311 Model**

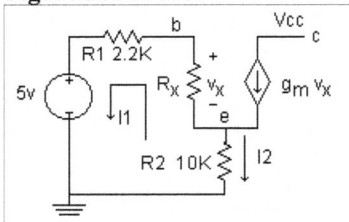 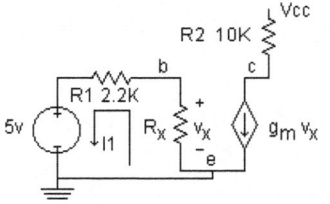

511 Ignore i_1. Let R_X=5K, g_M=12mA/v. Use the node voltage method. Solve for V_B, V_C, I_C, I_B, I_E.

$g_1 v_1 = (g_1 + g_x)v_b + 0 v_c$

$0 = g_m v_x + g_2 v_c \qquad \text{and} \qquad v_x = v_b \qquad \Delta_g = (g_1 + g_x)g_2$

$v_b = \dfrac{g_2 g_1 v_1}{\Delta_g} = \dfrac{g_1 v_1}{(g_x + g_1)} \qquad v_c = -\dfrac{g_m g_1 v_1}{\Delta_g} = -\dfrac{g_m g_1 v_1}{(g_x + g_1)g_2} = -\dfrac{R_x}{R_1 + R_x} g_m R_2 v_1$

512 Ignore i_1 and i_2.

[Circuit: vx=3ix, vy=2ix, v2, v1=5v, i1, ix, R1=39, i2, R2=18]

$0 = g_1 v_2 + i_{xx} - i_{yy}$

$0 = i_{yy} + g_2 v_3 \qquad \text{and} \qquad v_x = 3 g_1 v_2, \; v_y = 2 g_1 v_2$

$v_2 = v_1 + v_x = v_y + v_3$

$$i_{yy} = -g_2 v_3$$
$$i_{xx} = i_{yy} - g_1 v_2 = -g_2 v_3 - g_1 v_2 = -g_2(v_2 - v_y) - g_1 v_2$$
$$= -(g_1 + g_2)v_2 + g_2 v_y = -(g_1 + g_2)v_2 + g_2 2 g_1 v_2$$
$$= (-g_1 - g_2 + 2 g_2 g_1)v_2$$

This is an example of a problem that is not solved conveniently by the node method.

513

write equations in standard form
$$V_S = (R_1 + R_2)I_1 - (R_2)I_2$$
$$0 = -(R_2)I_1 + (R_2 + R_3 + R_4)I_2$$

Evaluate the determinant:
$$\Delta_R = r_{11} r_{22} - r_{12} r_{21}$$
$$= (R_1 + R_2)(R_2 + R_3 + R_4) - R_2^2$$
$$= R_1 R_2 + R_1 R_3 + R_1 R_4 + R_2 R_3 + R_2 R_4$$

apply Cramer's rule to solve the equations:

$$I_1 = \frac{\begin{vmatrix} V_S & r_{12} \\ 0 & r_{22} \end{vmatrix}}{\Delta_R} = \frac{\begin{vmatrix} V_S & -R_2 \\ 0 & R_2 + R_3 + R_4 \end{vmatrix}}{\Delta_R} = \frac{R_2 + R_3 + R_4}{\Delta_R} V_S$$

$$I_2 = \frac{\begin{vmatrix} r_{11} & V_S \\ r_{21} & 0 \end{vmatrix}}{\Delta_R} = \frac{\begin{vmatrix} R_1 + R_2 & V_S \\ -R_2 & 0 \end{vmatrix}}{\Delta_R} = \frac{R_2}{\Delta_R} V_S$$

$$I_1 - I_2 = \frac{R_2 + R_3 + R_4}{\Delta_R} V_S - \frac{R_2}{\Delta_R} V_S = \frac{R_3 + R_4}{\Delta_R} V_S$$

$$V_2 = (I_1 - I_2)R_2 = \frac{(R_3 + R_4)R_2}{\Delta_R} V_S \qquad V_3 = I_2 R_4 = \frac{R_2 R_4}{\Delta_R} V_S$$

Spice result V_1=5V, V_2=2.776V, V_3=1.388V, I_1=101.11mA, I_2=42.054mA

514 V_1=12V, V_2=7.192V, V_3=4.733V, V_4=5V, I_1=4.007mA, I_2=1.343mA, I_3=−0.103mA

515
$$V_S = I_S R_1 + V_1 \qquad I_S = 20E - 6 \qquad R_1 = 10K$$
$$V_1 = V_S - I_S R_1 = 5 - 20 \cdot 10^{-6} \cdot 10^4 = 4.8V$$

516 1.2A, −111V

Electric Circuits - Analysis and Design

517 $V_1=11.171V$, $I_S=0.037681A$, $I_1=0.237681A$
write equations in standard form
$$V_S = (R_1 + R_2)I_1 - (R_2)I_2$$
$$0 = -(R_2)I_1 + (R_2)I_2 + V_1 \quad \text{and} \quad I_2 = 0.2A$$

This is a circuit that is more suitable for solution by the node method.
Hint- ignore the second equation. Substitute numbers in the first equation.

518 $V_1=-4.252V$, $V_2=-3.471V$, $V_3=2.756V$, $I_C=1.360mA$, $I_B=24.201\mu A$, $I_E=-1.385mA$

formulate equations in std form
$$V_S = (R_2 + R_4)I_1 - (R_4)I_2$$
$$0 = -(R_4 - \mu R_2)I_1 + (R_1 + R_3 + R_4)I_2$$

Evaluate the determinant :
$$\Delta_R = r_{11}r_{22} - r_{12}r_{21}$$
$$= (R_2 + R_4)(R_1 + R_3 + R_4) - R_4(R_4 - \mu R_2)$$
$$= (R_2 + R_4)(R_1 + R_3) + R_4^2 - R_4^2 + \mu R_2 R_4$$
$$= R_1 R_2 + R_1 R_4 + R_2 R_3 + R_3 R_4 + \mu R_2 R_4$$

apply Cramer's rule to solve the equations :

$$I_1 = \frac{\begin{vmatrix} V_S & r_{12} \\ 0 & r_{22} \end{vmatrix}}{\Delta_R} = \frac{\begin{vmatrix} V_S & -R_4 \\ 0 & R_1 + R_3 + R_4 \end{vmatrix}}{\Delta_R} = \frac{R_1 + R_3 + R_4}{\Delta_R} V_S$$

$$I_2 = \frac{\begin{vmatrix} r_{11} & V_S \\ r_{21} & 0 \end{vmatrix}}{\Delta_R} = \frac{\begin{vmatrix} R_2 + R_4 & V_S \\ -(R_4 - \mu R_2) & 0 \end{vmatrix}}{\Delta_R} = \frac{R_4 - \mu R_2}{\Delta_R} V_S$$

$$I_1 - I_2 = \frac{R_1 + R_3 + R_4}{\Delta_R} V_S - \frac{R_4 - \mu R_2}{\Delta_R} V_S = \frac{R_1 + R_3 - \mu R_2}{\Delta_R} V_S$$

$$V_1 = (I_1 - I_2)R_4 = \frac{(R_1 + R_3 - \mu R_2)R_4}{\Delta_R} V_S \quad V_2 = I_2 R_3 = \frac{(R_4 - \mu R_2)R_3}{\Delta_R} V_S \; 307$$

formulate equations in std form
$$V_1 = -v_x + R_1 i_1 \quad\quad 0 = -R_1 i_1 - v_y + (R_1 + R_2)i_2 \quad\quad i_x = i_2 - i_1$$
$$V_1 = -3i_x + R_1 i_1 \quad\quad 0 = -R_1 i_1 - 2i_x + (R_1 + R_2)i_2$$
$$V_1 = (3 + R_1)i_1 - 3i_2 \quad\quad 0 = -(R_1 - 2)i_1 + (R_1 + R_2 - 2)i_2$$

substitute numbers
$$5 = (3 + 39)i_1 - 3i_2$$
$$0 = -(39 - 2)i_1 + (39 + 18 - 2)i_2$$

Answers to Most of the Problems

$\Delta = 42 \cdot 55 - 42 \cdot 37 = 756$

$i_1 = \dfrac{5 \cdot 55}{756} = 0.364A \quad i_2 = \dfrac{5 \cdot 37}{756} = 0.245A \quad i_x = 0.245 - 0.364 = -0.119A$

$v_2 = 39 i_x = 4.64V \quad v_x = 3 i_x = -0.36V \quad v_3 = 18 i_2 = 4.41V \quad v_y = 2 i_x = -0.24V$

verify $v_S = -v_x + v_2 = 0.36 + 4.64 = 5 \quad zero = v_2 + v_y - v_3 = 4.64 - .24 - 4.41 = 0$

519

$V_S = (R_1 + R_3) I_1 - (R_3) I_2$

$0 = -(R_3) I_1 + (z_2 + R_3) I_2 + v_{ce}$

and $z_2 = 1/pC_2 \quad I_2 = i_x = g_m v_x = g_m R_3 (I_1 - I_2)$

$(1 + g_m R_3) I_2 = g_m R_3 I_1$

$V_S = (R_1 + R_3) I_1 - (R_3) \dfrac{g_m R_3}{1 + g_m R_3} I_1$

$\dfrac{V_S}{I_1} = R_1 + R_3 \left(1 - \dfrac{g_m R_3}{1 + g_m R_3} \right) = R_1 + R_3 \left(\dfrac{1}{1 + g_m R_3} \right) = \dfrac{R_1 (1 + g_m R_3) + R_3}{1 + g_m R_3}$

$v_c = z_2 I_2 = \dfrac{g_m R_3}{1 + g_m R_3} z_2 I_1 = \dfrac{g_m R_3}{R_1 (1 + g_m R_3) + R_3} \dfrac{V_S}{pC_2}$

$g_m R_3 = 8 mA/V \cdot 0.4 K\Omega = 3.2$

$\dfrac{v_c}{V_S} = \dfrac{3.2}{1(1 + 3.2) + 0.4} 12 = 8.35$

520 $V_1 = 5.000V, V_2 = 0.417V, V_3 = 10.000V, I_1 = 208mA, I_2 = 290mA$

formulate equations in std form

$v_1 = R_1 i_1 + v_y \quad v_3 = v_2 + R_2 i_2 \quad i_x = i_1 \quad v_y = 2 i_x = 2 i_1 = v_2$

$v_1 = (R_1 + 2) i_1 - 0 i_2 \quad v_3 = 2 i_1 + R_2 i_2$

substitute numbers

$5 = (2 + 22) i_1 - 0 i_2$

$10 = 2 i_1 + 33 i_2$

Note: voltage source v_y has zero impedance so i_2 has no effect on v_2.

$\Delta = 24 \cdot 33 + 0 \cdot 2 = 792$

$i_1 = \dfrac{5 \cdot 33}{24 \cdot 33} = 0.2083 A \quad i_2 = \dfrac{24 \cdot 10 - 2 \cdot 5}{24 \cdot 33} = 0.290 A$

$i_x = i_1 = 0.2083 A \quad v_2 = v_y = 2 i_x = 0.417 V$

verify $v_1 = 22 i_x + v_2 = 4.583 + 0.417 = 5$

$v_3 = 33 i_2 + v_y = 9.57 + .417 = 9.987 \approx 10$

241

521
There is one mesh current flowing clockwise around the "outer" mesh.

$v_1 = (R_1 + z_2 + R_S)i_1 - \mu v_p \quad (v_n = 0, \; z_2 = pL_2)$

$v_p = v_1 - i_1 R_1 \quad \rightarrow \quad v_1 = (R_1 + z_2 + R_S)i_1 - \mu(v_1 - i_1 R_1)$

$(1+\mu)v_1 = [(1+\mu)R_1 + z_2 + R_S]i_1$

$i_1 = \dfrac{(1+\mu)}{(1+\mu)R_1 + z_2 + R_S} v_1 \qquad \text{for } \mu \gg 1 \quad i_1 = \dfrac{v_1}{R_1}$

$v_0 = v_1 - (R_1 + z_2)i_1 = v_1 - \dfrac{(1+\mu)(R_1 + z_2)}{(1+\mu)R_1 + z_2 + R_S} v_1$

$\dfrac{v_0}{v_1} = -\dfrac{R_S - \mu z_2}{(1+\mu)R_1 + z_2 + R_S} = -\dfrac{pL_2}{R_1} \cdot \dfrac{\left(1 - \dfrac{R_S}{\mu pL_2}\right)}{1 + \dfrac{1}{\mu}\left(1 + \dfrac{pL_2}{R_1} + \dfrac{R_S}{R_1}\right)}$

522
V_{cc} is AC ground.

$V_S = (R_2 + R_x + R_1)I_1 - (R_2)I_2$
$I_2 = g_m v_x = -g_m R_x I_1$

$V_S = (R_2 + R_x + R_1)I_1 + (R_2)g_m R_x I_1$
$V_S = [(1 + g_m R_x)R_2 + R_x + R_1]I_1$
$\dfrac{I_1}{V_S} = \dfrac{1}{(1 + g_m R_x)R_2 + R_x + R_1}$

$v_e = (I_1 - I_2)R_2 = (1 + g_m R_x)R_2 I_1$

$\dfrac{v_e}{V_S} = \dfrac{(1+g_m R_x)R_2}{(1+g_m R_x)R_2 + R_x + R_1} = \dfrac{1}{1 + \dfrac{R_x + R_1}{(1+g_m R_x)R_2}} \approx 1$

$\dfrac{v_e}{V_S} = \dfrac{1}{1 + \dfrac{5 + 2.2}{(1 + 12mA/V \cdot 5K\Omega)10}} = \dfrac{1}{1 + \dfrac{7.2}{610}} = 0.988$

Figure 510 Model

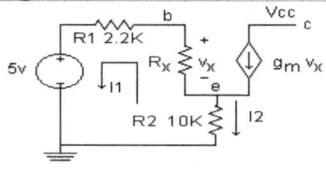

Figure 511 Model

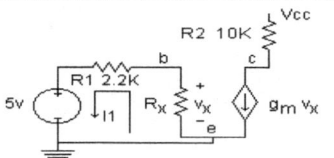

523
V_{cc} is AC ground.

$V_S = (R_x + R_1)I_1$
$I_2 = g_m v_x = -g_m R_x I_1$

$V_c = R_2 I_2 = -R_2 g_m R_x I_1 = -R_2 g_m R_x \dfrac{V_S}{R_x + R_1} = -g_m R_2 V_S \dfrac{R_x}{R_x + R_1}$

$\dfrac{V_c}{V_S} = -g_m R_2 \dfrac{R_x}{R_x + R_1} = -12 mA/V \cdot 10^4 \Omega \cdot \dfrac{5}{5 + 2.2} = -83.33$

Answers to Most of the Problems

601

(3a) $i_s = (g_1 + g_2)v_1 - g_2 v_2$

(3b) $0 = -g_2 v_1 + (g_2 + g_3 + g_4)v_2$

$\Delta_g = (g_1 + g_2)(g_2 + g_3 + g_4) - g_2(g_2)$

$\Delta_g = g_1 g_2 + g_1 g_3 + g_1 g_4 + g_2 g_3 + g_2 g_4$

Apply Cramer's rule to solve the equations.
apply Cramer's rule to solve the equations:

$$v_1 = \frac{\begin{vmatrix} i_s & -g_2 \\ 0 & g_2 + g_3 + g_4 \end{vmatrix}}{\Delta_g} = \frac{g_2 + g_3 + g_4}{\Delta_g} i_s$$

$$v_2 = \frac{\begin{vmatrix} g_1 + g_2 & i_s \\ -g_2 & 0 \end{vmatrix}}{\Delta_g} = \frac{g_2}{\Delta_g} i_s$$

Insert numerical values to get
$i_1 = 5.82\text{mA}$, $v_1 = 41.85\text{V}$, $i_2 = 3.44\text{mA}$, $v_2 = 9.28\text{V}$

---

701 From the definition of the Laplace transform calculate
a) $F(p) = \mathcal{L}[f(t)]$ for $f(t) = 3t$
b) $F(p) = \mathcal{L}[f(t)]$ for $f(t) = 7t e^{-3t}$
c) $F(p) = \mathcal{L}[f(t)]$ for $f(t) = \cosh \omega t$
d) $F(p) = \mathcal{L}[f(t)]$ for $f(t) = t + e^{-at}$

702 Solve for i(t) when initial conditions equal zero.

$$L\frac{di}{dt} + Ri + \frac{1}{C}\int_0^t i(t)dt = V_m \sin \omega t$$

Figure P703

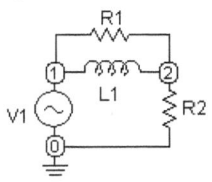

Figure P704

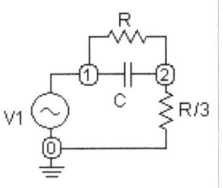

Problem 703 Reference Figure P703. Write Spice program FigP703.ckt. Plot the v_2 transient response.

704 Reference Figure P704. Write Spice program FigP704.ckt. Plot the v_2 transient response.

Figure P705 Band pass

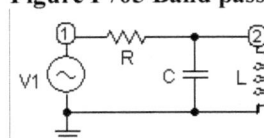

705 Reference Figure P705. Write Spice program Figp705.ckt. Plot the transient response when Q=5.

706

$$Step3: mesh1 \quad V_S = \left[R + pL + \frac{1}{pC}\right]I_1 - \frac{1}{pC}I_2$$

$$mesh2 \quad 0 = -\frac{1}{pC}I_1 + \left[R + pL + \frac{1}{pC}\right]I_2$$

$$\Delta_Z = \left[R + pL + \frac{1}{pC}\right]^2 - \left[\frac{1}{pC}\right]^2 = [R + pL]\left[R + pL + \frac{2}{pC}\right]$$

$$V_0 = I_2 R = \frac{R}{pC}\frac{V_S}{\Delta_Z} = \frac{V_S}{[p^2 LC + pCR + 2]\left[1 + \frac{pL}{R}\right]}$$

707 See Spice program PROB709.ckt

708

$$R = 2\sqrt{\frac{L}{C}} \quad \rightarrow \quad RC = 2\sqrt{LC}$$

$$Y(p) = \frac{1}{R + pL} + pC = \frac{(R + pL)(pC) + 1}{R + pL}$$

$$V_0(p) = \frac{I_S(p)}{Y(p)} = \frac{I_m}{p}\frac{R + pL}{(R + pL)(pC) + 1} = \frac{I_m R}{p} \cdot \frac{1 + pLG}{p^2 LC + RCp + 1}$$

$$= \frac{I_m R}{p} \cdot \frac{1 + pLG}{p^2 LC + R2\sqrt{LC}\, p + 1} = \frac{I_m R}{p} \cdot \frac{1 + pLG}{\left(p\sqrt{LC} + 1\right)^2}$$

709 $V_0(p) = I_S(p)Z(p) = \frac{(p^2 LC + pLG + 1)}{(pCR + 1)} \times \frac{I_m R}{p}$

710

$$V_1(p) = \frac{I_S(p)}{Y(p)} = \frac{I_m}{p + a}\frac{1}{G_1 + pC_1} = \frac{I_m}{C_1}\frac{1}{p + a}\frac{1}{p + b} \text{ where } b = \frac{1}{R_1 C_1}$$

$$V_1(p) = \frac{I_m}{C_1} = \frac{I_m}{C_1}\frac{1}{a - b}\left[\frac{1}{p + b} - \frac{1}{p + a}\right]$$

Answers to Most of the Problems

711

$$I(p) = \frac{V_1(p)}{Z(p)} = \frac{V_m}{L_1} \frac{1}{p+a} \cdot \frac{1}{p+b} \quad \text{where } a = -j\omega \text{ and } b = \frac{R_1}{L_1}$$

$$I(p) = \frac{V_m}{L_1} \frac{1}{a-b}\left[\frac{1}{p+b} - \frac{1}{p+a}\right] \quad i(t) = \frac{V_m}{R_1 + j\omega L_1}\left[e^{-at} - e^{-j\omega t}\right]$$

712

$$V_3(p) = \frac{V_S(p)}{Z} R = \frac{V_m R}{pL + R + \frac{1}{pC}} \cdot \frac{\omega}{p^2 + \omega^2} = \frac{p2\sqrt{LC}}{(p\sqrt{LC}+1)^2} \cdot \frac{\omega}{p^2 + \omega^2}$$

and $R = 2\sqrt{\dfrac{L}{C}}$

713

$v_S(t)/V_m = u(t) - u(t-T)$

$v_S(t)/2V_m = -1/2 + [u(t) - u(t-T/2)] + [u(t-T) - u(t-3T/2)] + \ldots$

$v_S(t)/V_m = t[u(t) - u(t-T)] + t[u(t-T) - u(t-2T)] + t[u(t-2) - u(t-3T)] + \ldots$

In all of the following problems you can also do a partial fraction expansion followed by the inverse transform.

714 $V_3(p) = \dfrac{V_1(p)}{Z} R_2 = \dfrac{R_2}{R_1 + R_2} \cdot \dfrac{p + \dfrac{R_1}{L_1}}{p + \dfrac{R_1 R_2}{L_1(R_1 + R_2)}} \cdot \dfrac{V_m}{p}$

715 $V_2(p) = \dfrac{V_1(p)}{Z} Z_1 = \dfrac{R_1}{L_2} \cdot \dfrac{1}{p + \dfrac{(L_1 + L_2)R_1}{L_1 L_2}} \cdot \dfrac{V_m}{p}$

716 $V_3(p) = \dfrac{V_S(p)}{Z} pL = \dfrac{V_m pL}{pL + R + \dfrac{1}{pC}} \cdot \dfrac{1}{p} = V_m \dfrac{p}{p^2 + p\dfrac{R}{L} + \dfrac{1}{LC}}$

717 $V_3(p) = \dfrac{V_S(p)}{R} \dfrac{1}{Y} = \dfrac{V_m}{Rp} \cdot \dfrac{1}{pC + G + \dfrac{1}{pL}} = \dfrac{V_m}{R} \dfrac{L}{p^2 LC + pGL + 1}$

803

Find the current $I = I_m e^{j\theta}$ when
$V_s = V_m$ volts, $\omega = 2\pi 10^6$, $R = 100\Omega$, $C = 500\,pF$, $L = 5\mu H$

$$I = \frac{V_m}{100 + j2\pi 10^6 \cdot 5 \cdot 10^{-6} + \dfrac{1}{j2\pi 10^6 \cdot 500 \cdot 10^{-12}}} = \frac{V_m}{100 + j10\pi - j\dfrac{10^3}{\pi}}$$

$$I = \frac{V_m}{100 + j31.4 - j318.3} = \frac{V_m}{100 - j286.9} = \frac{V_m}{303.9} e^{j70.7°} = 0.00329 V_m e^{j70.7°}$$

804

$\dfrac{1}{\omega_0 C} = R \qquad \omega_0 L = R \qquad \omega_0 L = \dfrac{1}{\omega_0 C}$

805

$Q_p = \dfrac{R_{parallel}}{\omega_0 C} = \omega_0 L \cdot R_{parallel} \qquad Q_S = \dfrac{\omega_0 L}{R_{series}} = \dfrac{1}{\omega_0 C \cdot R_{series}}$

Useful expressions

$\omega_0 L = R \qquad \dfrac{1}{\omega_0 C} = R \qquad \dfrac{1}{\omega_0 C} = \omega_0 L \qquad \lambda = \dfrac{p}{\omega_0}$

806 to 811 You can use mesh or node methods to solve for T(p). Here is

806.

$V_1 = (R_1 + pL_1)I_1 \qquad V_2 = R_1 I_1 \qquad \text{and} \qquad \omega_0 L_1 = R_1$

$T = \dfrac{V_2}{V_1} = \dfrac{R_1}{R_1 + pL_1} \qquad \lim_{p \to 0} T = 1 \qquad \lim_{p \to j\omega_0} T = \dfrac{1}{1 + j1} \qquad \lim_{p \to \infty} T = \dfrac{R_1}{pL_1}$

812 to 814 You can use mesh or node methods to solve for T(p). Here is

812.

$0 = -(g_1 + pC_1)V_1 + (g_2 + g_1 + pC_1)V_2$

$\dfrac{V_2}{V_1} = \dfrac{g_1 + pC_1}{g_2 + g_1 + pC_1} = \dfrac{g_1}{g_2 + g_1} \cdot \dfrac{1 + pC_1 R_1}{1 + pC_1 \dfrac{R_1 R_2}{R_1 + R_2}} = \dfrac{R_2}{R_1 + R_2} \cdot \dfrac{1 + \lambda}{1 + k\lambda} = k\dfrac{1 + \lambda}{1 + k\lambda}$

where $\lambda = \dfrac{p}{\omega_0} \qquad C_1 R_1 = \dfrac{1}{\omega_0} \qquad k = \dfrac{R_2}{R_1 + R_2}$

Answers to Most of the Problems

902

1) $v(t) = V_m \cos(\omega t + d) = Re\left[V_m e^{j(\omega t+d)}\right] = Re\left[V_m e^{jd} e^{j\omega t}\right] \Rightarrow F = V_m e^{jd}$

2) $i(t) = I_m \cos(\omega t - a) = Re\left[I_m e^{j(\omega t-a)}\right] = Re\left[I_m e^{-ja} e^{j\omega t}\right] \Rightarrow F = I_m e^{-ja}$

3) $v(t) = 1.8\cos(2\pi \cdot 10^6 t + \pi/4) = Re\left[1.8 e^{j(2\pi \cdot 10^6 t + \pi/4)}\right] = Re\left[1.8 e^{j\pi/4} e^{j2\pi \cdot 10^6 t}\right]$

$\Rightarrow F = 1.8 e^{j\pi/4}$

904

```
FigP9021.ckt   series RL circuit

V1 1 0 AC 1
L1 1 2 100m
R1 2 0 1000
.AC DEC 200 10 100000
.TEMP 27
.PLOT AC VDB(1) VDB(2) -40,10
.end
```

905

```
FigP9031.ckt   series RL circuit

I1 1 0 AC 1
L1 1 0 100m
R1 1 0 1000
.AC DEC 200 10 100000
.TEMP 27
.PLOT AC VDB(1) 0,75
.end
```

906

$v_2(t) = 5\left(1 - e^{-t/\tau}\right)$ and so $V_2(p) = \dfrac{5}{p} - \dfrac{5}{p + 1/\tau} = \dfrac{5}{p} \cdot \dfrac{1}{p + 1/\tau}$

circuit analysis $V_2(p) = V_1(p) \dfrac{1}{(L/R)p + 1}$

if $V_1(p) = \dfrac{5}{p}$ and $\dfrac{1}{(L/R)p + 1} = \dfrac{1}{p\tau + 1}$ then $v_1(t) = 5u(t)$ and $\tau = \dfrac{L}{R}$

$L = R\tau = \dfrac{1000}{2000} = 0.5H$

Electric Circuits - Analysis and Design

907

$$V_1(p) = I(p)Z(p) = \frac{I_m}{p} \cdot \frac{pLR}{R+pL} = I_m R \cdot \frac{1}{p+(R/L)} \quad \text{so that}$$

$$v_1(t) = I_m R e^{-Rt/L}, \quad \tau = \frac{L}{R} = \frac{2 \cdot 10^{-6}}{5 \cdot 10^3} = 400\,ps,$$

$$i_1(0) = I_m = 0.02A, \quad v_1(0) = I_m R = 20mA \cdot 5K\Omega = 100V$$

908

$$KVL \quad Ri + L\frac{di}{dt} = 0 \quad \Rightarrow \quad (R+pL)I(p) - Li(0) = 0$$

$$I(p) = Li(0)\frac{1}{R+pL} = i(0)\frac{1}{p+(R/L)} \quad \text{so that} \quad i(t) = i(0)e^{-Rt/L}$$

$$p = \int_0^\infty i^2 R\,dt = Ri^2(0)\int_0^\infty e^{-2Rt/L}dt = Ri^2(0)\frac{L}{2R}e^{-2Rt/L}\bigg|_0^\infty = \frac{1}{2}Li^2(0)$$

909

$$v(t) = (I_m R)e^{-\frac{Rt}{L}} = 12e^{-10^6 t} \quad \Rightarrow \quad v(0) = I_m R = 12\,volts \text{ and } \tau = 10^{-6} = 1\mu s = \frac{L}{R}$$

$$i(0) = I_m = 7mA \quad R = \frac{12}{I_m} = \frac{12}{0.007} = \frac{12}{7}K\Omega \quad L = R\tau = \frac{12}{7}10^{-3}H \quad \omega_0 = \frac{R}{L} = 10^6$$

910

$$v(t) = (I_m R)e^{-\frac{Rt}{L}} = 5e^{-4 \cdot 10^9 t} \quad \Rightarrow \quad v(0) = I_m R = 12\,volts \text{ and } \tau = \frac{1}{4 \cdot 10^9} = 0.25ns = \frac{L}{R}$$

$$i(0) = I_m = \frac{v(0)}{R} = \frac{5}{1.2K} = \frac{5}{1.2}mA \quad L = R\tau = 1.2 \cdot 10^3 \frac{10^{-9}}{4} = 0.3\mu H \quad \omega_0 = \frac{R}{L} = \frac{1}{\tau} = 4 \cdot 10^9$$

911
```
FigP911.ckt series RL circuit
V5 5 0 PULSE(0 1 100p 000p 000p 500p 2000p)
R5 5 6 1000
L5 6 0 100n

V3 3 0 PULSE(0 1 105p 000p 000p 200p 1000p)
R2 3 4 1000
L2 4 0 100n

V1 1 0 PULSE(0 1 110p 000p 000p 100p 1000p)
R1 1 2 1000
L1 2 0 100n
.TRAN 1e-011 1e-009 0
.TEMP 27
.PLOT TRAN V(2) V(4) V(6)  -1,1
.end
```

Answers to Most of the Problems

912

```
FigP9023.ckt series RL circuit

V5 5 0 PULSE(0 1 100p 000p 000p 500p 2000p)
L5 5 6 100n
R5 6 0 1000

V3 3 0 PULSE(0 1 105p 000p 000p 200p 1000p)
L2 3 4 100n
R2 4 0 1000
V1 1 0 PULSE(0 1 110p 000p 000p 100p 1000p)
L1 1 2 100n
R1 2 0 1000
.TRAN 1e-011 1e-009 0
.TEMP 27
.PLOT TRAN V(2) V(4) V(6) -0,1
.end
```

913

(14) $i(t) = \dfrac{V_m}{R}(1 - e^{-\dfrac{t}{\tau}}) = 0.02(1 - e^{-100t})$

(15) $v_2(t) = V_m e^{-\dfrac{Rt}{L}} = 3e^{-t/\tau}$ and $v_1(t) = 10u(t)$

$\tau = \dfrac{1}{100} = 10^{-2} = \dfrac{L}{R}$ $\dfrac{V_m}{R} = 0.02$ so $R = \dfrac{3}{0.02} = 150\Omega$

$L = R\tau = 150 \cdot 10^{-2} = 1.5H$

914

```
FigP10051.ckt   series RC circuit   sine
V1 1 0 AC 1
C1 1 2 100p
R1 2 0 1000
.AC DEC 200 10000 1e+009
.TEMP 27
.PLOT AC VDB(1) VDB(2) -40,10
.end
```

915

```
FigP10061.ckt   parallel RC circuit   sine
I1 1 0 AC 1m
R1 1 0 1000
C1 1 0 100p
.AC DEC 200 10000 1e+009
.TEMP 27
.PLOT AC VDB(1) -40,10
.end
```

$$KVL \quad v_I(t) = \frac{1}{C}\int_0^\infty i\,dt + Ri = 0 \quad \Rightarrow \quad V_1(p) = \left(\frac{1}{pC} + R\right)I(p)$$

$$I(p) = V_1(p)\frac{pC}{1+pCR} = \frac{V_m}{p} \cdot \frac{pC}{1+pCR} \quad \text{so that} \quad i(t) = \frac{V_m}{R}e^{-t/RC} = 0.003e^{-2\cdot 10^6 t}$$

$$\tau = RC = \frac{10^{-6}}{2} \quad \frac{V_m}{R} = 0.003 \text{ so that } R = \frac{9}{0.003} = 3K\Omega \quad C = \frac{\tau}{R} = \frac{1}{6}10^{-9}F \quad v_2(t) = Ri(t)$$

917

```
FigP9052.ckt series RC circuit step
V1 1 0 Pulse(0 0.9 1n 0n 0n 2n 10n)
C1 1 2 100f
R1 2 0 1000
.TRAN 1e-012 5e-009 0
.TEMP 27
.PLOT TRAN V(1) V(2) -1,1
.end
```

918

```
FigP9062.ckt parallel RC circuit step
I1 1 0 Pulse(0 0.9m 1n 0n 0n 2n 10n)
C1 1 0 100f
R1 1 0 1000
.TRAN 1e-012 5e-009 0
.TEMP 27
.PLOT TRAN V(1) -1,1
.end
```

919

```
FigP9043.ckt series RC circuit pulse
V1 1 0 PULSE(0 0.9 100p 000p 000p 500p 1000p)
R1 1 2 1000    ; 1000 ohms
C1 2 0 100f    ; 100fF
V3 3 0 PULSE(0 0.9 100p 000p 000p 200p 1000p)
R3 3 4 1000    ; 1000 ohms
C3 4 0 100f    ; 100fF

V5 5 0 PULSE(0 0.9 100p 000p 000p 100p 1000p)
R5 5 6 1000    ; 1000 ohms
C5 6 0 100f    ; 100fF
.TRAN 1e-011 1e-009 0
.TEMP 27
.PLOT TRAN V(2) V(4) V(6) -1,1
.end
```

920
```
FigP9053.ckt series RC circuit pulse
V1 1 0 PULSE(0 0.9 100p 000p 000p 500p 1000p)
C1 1 2 100f    ; 100fF
R1 2 0 1000   ; 1000 ohms

V3 3 0 PULSE(0 0.9 100p 000p 000p 200p 1000p)
C3 3 4 100f    ; 100fF
R3 4 0 1000   ; 1000 ohms

V5 5 0 PULSE(0 0.9 100p 000p 000p 100p 1000p)
C5 5 6 100f    ; 100fF
R5 6 0 1000   ; 1000 ohms

.TRAN 1e-011 1e-009 0
.TEMP 27
.PLOT TRAN V(2) V(4) V(6) -1,1
.end
```

921

(41) $i(t) = \dfrac{V_m}{R} e^{-t/RC} = 0.001 e^{-10^8 t}$ (42) $v_C(t) = V_m\left(1 - e^{-t/RC}\right)$

$\tau = RC = 10^{-8} = 10 ns \qquad \dfrac{V_m}{R} = 0.001 \text{ so that } R = \dfrac{5}{0.001} = 5 K\Omega$

$C = \dfrac{\tau}{R} = \dfrac{10 \cdot 10^{-9}}{5 \cdot 10^3} = 2 pF$

922

(41) $i(t) = \dfrac{V_m}{R} e^{-t/RC} = 0.002 e^{-10^6 t}$ (43) $v_R(t) = V_m e^{-t/RC}$

$\tau = RC = 10^{-6} = 10 \mu s \qquad \dfrac{V_m}{R} = 0.002 \text{ so that } R = \dfrac{8}{0.002} = 4 K\Omega$

$C = \dfrac{\tau}{R} = \dfrac{10^{-6}}{4 \cdot 10^3} = 0.25 nF$

923
```
FigP9062.ckt  parallel RC circuit
V3 3 0 PULSE(0 0.9 100p 500p 000p 1000p 2000p)
I1 1 0 PULSE(0 -0.9m 100p 500p 000p 1000p 2000p)
R1 1 0 1000   ; 1000 ohms
C1 1 0 100f   ; 100fF
.TRAN 1e-011 1e-009 0
.TEMP 27
.PLOT TRAN V(1) V(3) 0,1
.end
```

1001 Reference Figure 100311

$$\frac{v_2(p)}{v_1(p)} = \frac{1}{1+\lambda} = \frac{1}{1+\dfrac{p}{\omega_0}} = \frac{\omega_0}{p+\omega_0} = \frac{\dfrac{R}{L}}{p+\dfrac{R}{L}} = \frac{R}{R+pL} \qquad \omega_0 = \frac{R}{L}$$

1002 Reference Figures 100211 and 100311

$$\frac{v_2(p)}{v_1(p)} = \frac{\lambda}{1+\lambda} = \frac{p}{p+\omega_0} = \frac{pL}{pL+R} = pL \cdot \frac{1}{pL+R}$$

1003 Reference Figures 100211 and 100311

$$\frac{v_2(p)}{i_1(p) \cdot R} = \frac{\lambda}{1+\lambda} = \frac{p}{p+\omega_0} = \frac{pL}{pL+R} = pL \cdot \frac{1}{pL+R}$$

1004 Reference Figure 100311

$$\frac{v_2(p)}{v_1(p)} = \frac{1}{1+\lambda} = \frac{1}{1+\dfrac{p}{\omega_0}} = \frac{1}{1+pRC} = \frac{G}{pC+G}$$

1005 Reference Figures 100211 and 100311

$$\frac{v_2(p)}{v_1(p)} = \frac{\lambda}{1+\lambda} = \frac{p}{p+\omega_0} = \frac{pC}{pC+G} = pC \cdot \frac{1}{pC+G}$$

1007 Reference Figures 100211 and 100511

$$\frac{v_2(p)}{v_1(p)} = \frac{\lambda^2}{\lambda^2+1} = \frac{p^2}{p^2+\omega_0^2} = \frac{p^2}{p^2+\dfrac{1}{LC}} = p \cdot p \cdot \frac{1}{p^2+\dfrac{1}{LC}}$$

1009 Reference Figures 100211 and 100511

$$T(p) = \frac{v_2}{v_1} = \frac{1}{Q}\frac{\lambda}{1+\dfrac{1}{Q}\lambda+\lambda^2} = \frac{\omega_0}{Q}\frac{p}{\omega_0^2+\dfrac{\omega_0}{Q}p+p^2} = \frac{R}{L}\frac{p}{\dfrac{1}{LC}+\dfrac{R}{L}p+p^2}$$

$$= R\frac{1}{\dfrac{1}{pC}+R+pL}$$

1012 Reference Figures 100311 and 100411

$$T(p) = \frac{v_2}{i_1} = k\frac{1+\lambda}{1+k\lambda} \qquad \omega_0 = \frac{1}{R_1C_1}$$

$$\frac{v_2}{i_1} = k\frac{\omega_0+p}{\omega_0+kp} = \frac{\omega_0+p}{\dfrac{\omega_0}{k}+p}$$

Ten Experiments

The experiments support the text. You teach yourself how to use electronic instruments, tools, how to make measurements, and learn about parts. You teach yourself how to build the circuits shown in the figures on a solderless breadboard, and measure their performance.

Electrical engineering book learning is necessary but not sufficient, because an EE has to know

1 how to make measurements in order to evaluate a design.
2 what parts are available as well as their properties.
3 what parts look like, and how to read any part's label.
4 the equivalent circuit of a part given a frequency range,
5 that all parts have parasitic components attached to them.
6 what parasitic components are added when a part is placed in a circuit.
7 how to use the tools of the trade
8 and so on.

Here you design and build circuits from the get go, while referring to the text and other writings to learn what you need to know to implement the design. The idea is that you seek answers on an ongoing basis to the many questions that arise as you try to design and build circuits.

You take time outs to seek answers to the questions by reading the text, doing the problems, and perhaps doing a search on the Internet. In this way theory and practice merge.

Parts

A comprehensive view of what parts are available is found in the Product Index of every electronics parts distributor. The Electric Circuits experiments only use a small subset of each type of part such as resistors R, capacitors C, inductors L, transformers, and potentiometers R.

Associated with every part is a *data sheet*, which presents the part's characteristics in many formats: text descriptions, drawings, maximum ratings, thermal characteristics, tables of electrical characteristics, graphs of typical characteristics, available electrical values, package dimensions, pin assignments, test circuits and application notes.

Electric Circuits - Analysis and Design

A *data sheet* in effect tells you what a part is about.

Rarely will you find in a *data sheet* the equivalent circuit of a part given a frequency range, nor any information about the part's parasitic components attached to them.. This information is usually found in *technical articles* or *application notes* issued by the manufacturer.

Every manufacturer has a web site from which you can download data sheets, application notes, white papers and so forth.

Each experiment specifies the parts used in the experiment's circuits.

Instruments

What do you need to measure or observe, while evaluating circuit performance? As a minimum you need to measure or observe ohms, DC voltages, DC currents, steady state AC signal voltages, and transient state signal voltages.

You need an oscilloscope so that you can "see" the DC voltages and AC signal voltages at circuit nodes. Two channels allow you to compare what is going on at two nodes, such as an input and its corresponding output. For our purposes here a two channel oscilloscope, with 1MHz bandwidth or better, is satisfactory as well as cost effective.

You need a signal generator, specifically a function generator, to generate the signals driving the circuit under test. Function generator, because you will need sinewaves, square waves, triangular waves, and pulses. A function generator with maximum frequency 1MHz or better will do.

A very important feature of any signal generator is that a signal amplitude does NOT change when frequency is changed. This must be verified as experiments are executed.

You need a DC multimeter, which measures wide ranges of volts, ohms, and amperes. Analog or digital meter? Your choice. We use both types.

To us a Power Supply is a *generator* of zero frequency DC voltages. In this text's experiments you only need ±5V.

Electric Circuits Experiments   Experiments, Parts, Instruments, Tools

Tools

Side cutters and long nose pliers cut and form wire. A wire stripper strips insulation from the ends of wire without damaging the wire. A lead forming tool accurately forms wire leads of parts for insertion into the breadboard holes. Tweezers facilitate picking up and placing small parts. Clip leads connect terminals as needed.

1   reactance chart (do a Google search for "reactance chart")
1   breadboard
1   side cutters, 4"
1   long nose pliers, 4"
1   wire stripper
1   lead forming tool
1   tweezers, fine point
?   clip leads

Color Code
An effective way to learn the color code is to sort a pile of resistors by value. *Another way is to use the program colorcode.exe.*

1st band   1st digit
2nd band   2nd digit
3rd band   number of zeros
4th band   tolerance, gold 5%, silver 10%

digit	0	1	2	3	4	5	6	7	8	9
color	black	brown	red	orange	yellow	green	blue	violet	gray	white

10% values	100	120	150	180	220	270	330	390	470	560	680	820	1000
5% values	100	110	120	130	150	160	180	200	220	240	270	300	330
	360	390	430	470	510	560	620	680	750	820	910	1000	

Examples
22   red red black
220K   red, red, yellow
1.2K   brown, red, red
47K   yellow, violet, orange
910   white, brown, brown
8.2 meg   gray, red, green

Electric Circuits - Analysis and Design

Contents

The Solderless Breadboard and the Power Supply 257
Experiment 1 Circuit with Resistors 260
Experiment 2 Kirchhoff's Laws.. 265
Experiment 3 Bridged-T Attenuator 267
Experiment 4 AC Voltmeter... 270
Experiment 5 Capacitance.. 274
Experiment 6 Inductance 277
Experiment 7 Transformers.. 279
Experiment 8 Circuit Analysis 282
Experiment 9 Frequency Response 285
Experiment 10 Transient Response 291
Solved... 297

SAFETY FIRST!!!!!!!

SAFETY FIRST!!!!!!! Electricity is silent so be very careful. We remove all metal objects from our hands and wrists such as rings and watches. If you do not remove them, then know you are taking an unnecessary risk.

Furthermore you do the experiments at your own risk, because there is no way we can supervise your work.

Word to the wise - NEVER GRAB anything, because if its "hot' you will have difficulty letting go.

The AC line voltage is extremely dangerous.

The Solderless Breadboard and the Power Supply

Solderless breadboards are designed to connect parts together without using solder. A power supply energizes the circuits on the solderless breadboard.

You build a circuit by "plugging in" parts. For example a resistor has 2 leads, which when suitably bent can be inserted into 2 pin holes.

A part's lead (28 to 20 AWG, 0.0126 to 0.0320 inches diameter) is inserted into a hole whose spring loaded metal insert "grabs" the lead. The metal inserts are not visible. *The metal inserts in horizontal rows of 5 holes are shorted together.* Consequently leads inserted in the same 5-hole-row are shorted together. The leads placed in a 5 hole row are the equivalent of leads soldered together.

Figure 101 Part of a Solderless breadboard showing fields of holes.

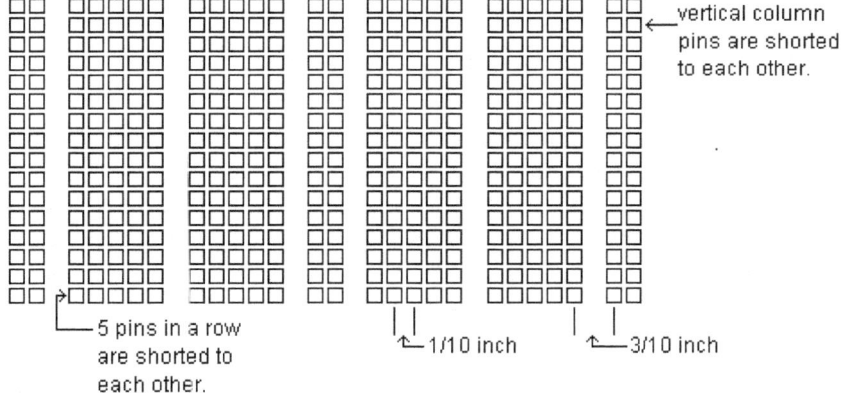

A board is an assembly of *two types of hole patterns* mounted on a metal plate (Figure 101). One type of hole pattern has two 5×59 arrays of holes separated by a narrow gutter. The columns of holes parallel and adjacent to the channel are 0.3 inches apart, because 0.3 inches is the IC DIP package minimum pin row spacing. Each array of holes is a column of 59 five hole rows. All holes are on a 0.1 inch grid so that vertical and horizontal separation is 0.1 inch. The five holes in each row are shorted together. The rows are not shorted together.

IC pins are inserted into the board so that the IC straddles the gutter and each IC pin plugs into one hole of one row. Then the four other holes in each "pin" row are available to receive leads, which are automatically connected to the IC pin. In this way a circuit is wired from node to node. Larger ICs have 0.4-inch and 0.6-inch pin row spacing. These are inserted in the same way that the narrower 0.3 inch ICs are inserted; however, the covered up row holes are not available for point-to-point wiring.

Power and Ground The other type of hole pattern is formatted to distribute power and ground. The pattern has two columns of 50 holes. The 50 holes *in each column* are shorted together. The two columns are not shorted together so that one column can distribute 5 volts, for example, and the other 0 volts (ground). Another pattern across the top of the board has two rows of 40 holes. The 40 holes *in each row* are shorted together.

Above the rows of 40 holes the solderless breadboard has binding posts whose insulated "knobs" unscrew to reveal a hole in the post in which a wire is inserted. The other end of the wire is plugged into a row of 40 holes. In turn jumpers connect the 40 hole rows to the 50 hole columns. (The black post is shorted to the metal base plate, and the red ones are insulated from the base plate.)

Verifying the solderless breadboard "shorts"

Select the R × 10 or higher ohm range, because the R × 1 range drains the 1.5 volt battery (use this range only when you have to).

Connect a pin to each lead of your ohmmeter. Place one pin in the first hole of a 5 hole row. Place the other pin in each of the other 4 holes in the row. Verify that the resistance is essentially zeros ohms (the short).

Repeat for the columns of holes.

Measure the resistance between rows, which should be infinite (open circuit). Repeat for row to columns, etc. Check out all possibilities to KNOW how the holes are wired.

Power Supply You can select any $\pm 5V$ power supply. We use a linear open frame $\pm 5V$ power supply. We have to add a power cord. COVER THE AC TERMINALS WITH TAPE. Plug the power cord into a plug strip outlet. The plug strip on/off switch becomes the power supply on/off switch. Think of a power supply as a constantly recharged battery.

The source impedance R_S of the "battery" is estimated as follows. If a 5V output drops by 2%, or 0.1V, when a 0.25 ampere load current is drawn, then R_S=0.1V/0.25A=0.4 ohms. However if plus is shorted to minus, then potentially I=5/0.4=12.5 amperes, which may or may not flow. This is why unprotected parts are destroyed (and perhaps the power supply).

We have to be very careful to avoid shorting plus to minus.

> *Shut off the power supply OR Disconnect the voltage leads AT THE SUPPLY* when making changes on the solderless breadboard.

Connecting the power supply to the solderless breadboard

Connect the binding posts to the solderless breadboard. The solderless breadboard has "binding" posts whose insulated "knobs" unscrew to reveal a hole in the post. The black post is shorted to the metal base plate, and the red ones are insulated from the base plate. Unscrew the black post and insert a black wire in the post hole. Tighten the black knob to secure the wire. Insert the other end of the wire into a hole in a column of holes that you want to be grounded. Repeat with a red wire from a red post to what becomes the "power supply voltage, the B+" column of holes (e.g. +5V). Repeat for B− (e.g. −5V).

Connect the power supply to the binding posts First turn off the AC power to the supply. Use clip leads or wires to connect the power supply voltage terminals to the solderless breadboard binding posts.

Note: you do not have to use the binding posts.

AFTER you have built a circuit, turn on the AC power to the supply. Turn the power off *before* you make circuit changes.

Use the multimeter to verify the $\pm 5V$ voltages.

Electric Circuits - Analysis and Design

Experiment 1 Circuits with Resistors

This is about analysis of circuits with resistors. We show that resistors (in fact any combination of components) can be wired in series, in parallel, and in series-parallel. Then we show to how to find solutions for voltage v and current i in these circuits.

Solutions to "resistor" problems apply the concepts of current and voltage, Kirchhoff's Laws (the connection constraints), and Ohm's Law (the resistor vi constraint).

Resistors in Series *The purpose here is to show that input voltage is divided up by the resistors, and to emphasize the fact the same current flows in components that are in series.*

Kirchhoff's voltage law KVL	$0 = -v_S + v_1 + v_2 + v_3$
Resistor vi constraint - Ohm's Law	$v = iR$
Using the vi constraint	$v_S = iR_1 + iR_2 + iR_3 = i(R_1 + R_2 + R_3)$
Resistors in series add	$R_{series} = R_1 + R_2 + R_3$

Construct the circuit in Figure 201. Let V_s=5V, R_1=1K, R_2=2.2K. Measure V_1 (3.44V). Calculate the current. Measure the current (1.56mA). See Figure 2011. Reminder: when evaluating results take into account the fact that the resistors have a ±5% tolerance.

Figure 201

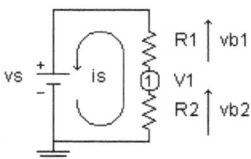

Figure 202

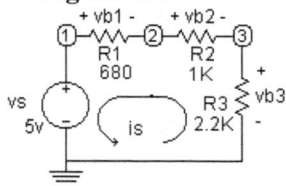

Figure 2011 Circuit for Figure 201

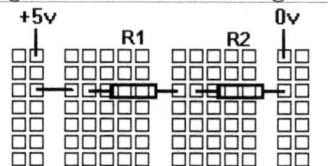

Figure 2021 Circuit for Figure 202

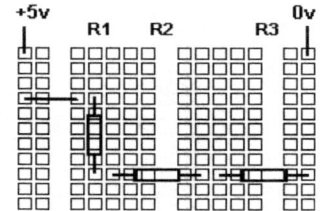

260

Construct the circuit in Figure 202. Measure vb_1 (0.88V), vb_2 (1.29V), vb_3 (2.84V). Calculate the current. Measure the current (1.29mA). See Figure 2021.

Resistors in Parallel The purpose here is to show that current in any parallel component is independent of current in the other components.

Kirchhoff's current law KCL $\qquad 0 = -i_S + i_1 + i_2 + i_3$

Resistor vi constraint - Ohm's Law $\qquad v = iR$

Using the vi constraint $\qquad i_S = \dfrac{v}{R_1} + \dfrac{v}{R_2} + \dfrac{v}{R_3} = v\left(\dfrac{1}{R_1} + \dfrac{1}{R_2} + \dfrac{1}{R_3}\right)$

Resistors in parallel $\qquad \dfrac{1}{R_{parallel}} = \dfrac{1}{R_1} + \dfrac{1}{R_2} + \dfrac{1}{R_3}$

2 Resistors in parallel $\qquad \dfrac{1}{R_p} = \dfrac{1}{R_1} + \dfrac{1}{R_2} = \dfrac{R_1 + R_2}{R_1 R_2}$

Solving for R_2 $\qquad R_p = \dfrac{R_1 R_2}{R_1 + R_2} \quad \Rightarrow \quad R_2 = \dfrac{R_1 R_p}{R_1 - R_p}$

Construct the circuit in Figure 203 Let V_s=5V, R_1=R_2=R_3=10K. Add one resistor at a time and measure the current (0.5mA, 1mA, 1.5mA). See Figure 2031. Each resistor draws 0.5mA - right?

Figure 203

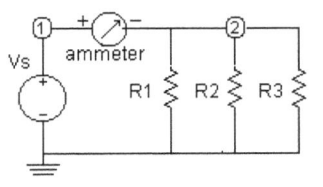

Figure 2031 Circuit for Figure 203

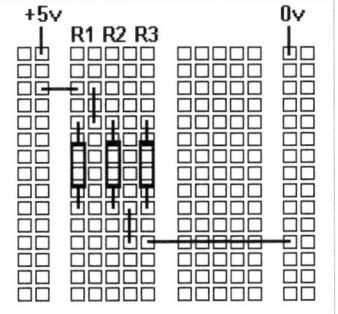

Electric Circuits - Analysis and Design

Resistors in Series-Parallel Combinations   *The purpose here is to assemble, and evaluate a series parallel circuit as a 2 mesh circuit..*

Select $R_1 = 2.2K\Omega$, $R_2 = 1K\Omega$, $R_3 = 2.2K\Omega$, and $R_4 = 10K\Omega$. Use an ohmmeter to measure and record their actual values. See Figure 2071.

Figure 207 Two Mesh Circuit

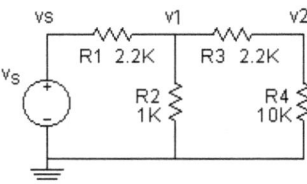

Assemble the four resistors as shown in Figure 2071, and let V_s be a 5v power supply. Measure and record the power supply voltage. Measure the voltages across each resistor (V_{R1}=3.52V, V_{R2}=1.48V, V_{R3}=0.27V, V_{R4}=1.21V).

A formal analysis produces input to output *transfer function* v_2/v_S that shows $v_2/v_S = 1.212/5$ for the values in Figure 207. Compare to measured v_2/v_S.

Figure 2071 Circuit for Figure 207

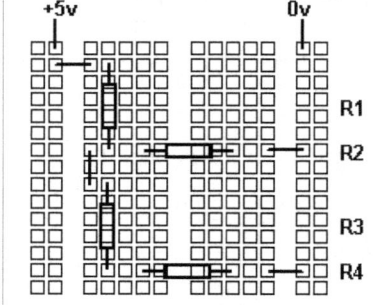

Electric Circuits Experiments   Experiment 1 Circuits with Resistors

Potentiometers (a.k.a. "pots") *The purpose here is to show that a potentiometer, a "pot", is a variable divider of voltage.*

Construct the circuit in Figure 204. Let R=1K. Connect the voltmeter to measure V_1. Rotate the 1K pot arm full ccw. Measure V_1 (0V). Rotate pot cw slowly, and watch voltmeter reading increase from 0V to almost 2.5V. See Figure 2041. Why almost?

Change the circuit to Figure 205. Rotate the 1K pot arm full ccw. Measure V1 (0V). Now rotate the pot arm full cw. Measure V1 (5V). What changed?

A common application of a pot is to vary the gain, transmission, of a voltage amplifier. In such a case the 5v is replaced by a signal voltage (Figure 205).

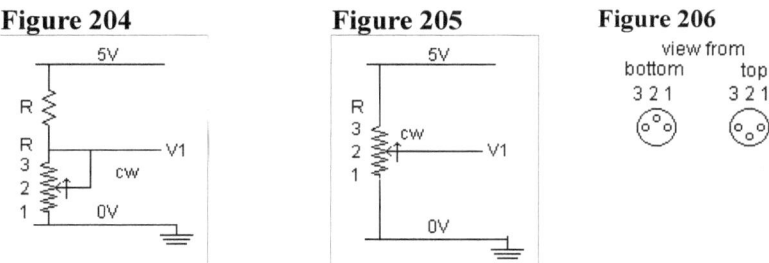

Figure 204   **Figure 205**   **Figure 206**

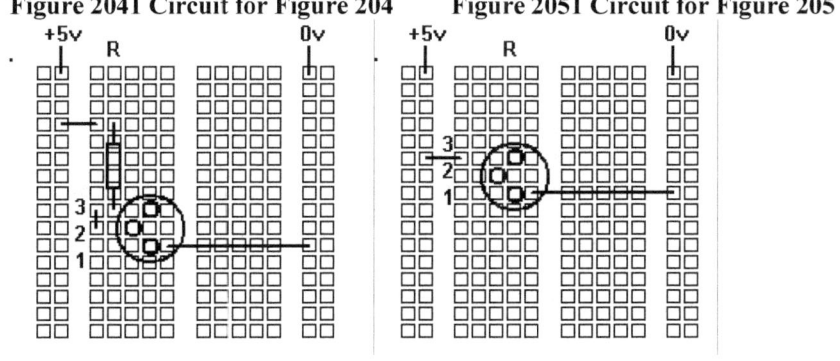

Figure 2041 Circuit for Figure 204   **Figure 2051 Circuit for Figure 205**

263

Electric Circuits - Analysis and Design

Multimeter Voltage Input Impedance (Resistance) The purpose here is to emphasize the fact every instrument has an input impedance that becomes part of the circuit.

Wire a circuit (Figure 2101) on the breadboard

A multimeter has 20K input resistance per scale volt. Select the 10v scale, which has an input resistance of 200K. Connect the multimeter to nodes 2 and ground. Measure voltages to show that V_2=3.33V, V_1=5V.

The 100K and 200K resistors "divide" the 5V into 3 parts: 1 part (1.67V) across the 100K and 2 parts (3.3V) across the 200K (the voltmeter). Do you agree? (Figure 2101)

Figure 210 Resistor   **Figure 2101 Circuit for Figure 210**

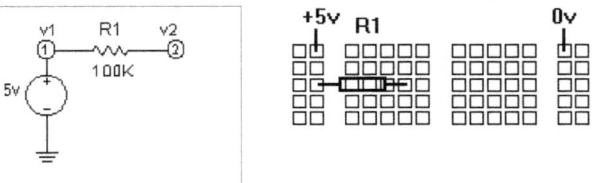

Note: If your multimeter is different, then change the above to suit.

Multimeter Current Input Impedance (Resistance)

The purpose here is to know that an ammeter has a very low impedance. In fact the ammeter has about 50mV across its terminals when current has full scale value. You decide if 50mV is negligible.

Wire a circuit (Figure 2111) on the breadboard *Before* connecting the +5V power supply select the 1mA scale. Now connect the +5V power supply. Does the meter read 0.5mA? Why? (Figure 211). Show that the power consumed by the resistor is 2.5mW. Calculate it two ways.

Figure 211 Current   **Figure 2111 Circuit for Figure 211**

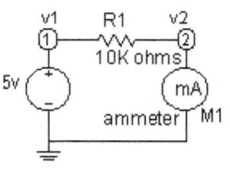 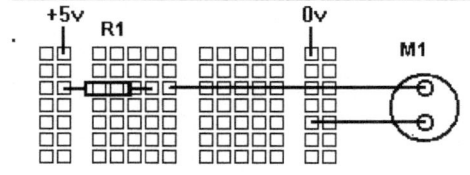

Experiment 2 Kirchhoff and Superposition Laws

This is about Kirchhoff's voltage and current laws constraining circuits as well as the Superposition Law.

Kirchhoff's Laws

Kirchhoff's laws are *connection constraints*, because they relate voltages and currents to how components are wired into a circuit.

Kirchhoff's Current Law (KCL) *The algebraic sum of all currents entering and leaving one node is zero.*

Kirchhoff's Voltage Law (KVL) *The algebraic sum of the voltage differences around every closed circuit contour is zero.*

Let us make some measurements to check out KCL and KVL.

Figure 207 Two Mesh Circuit

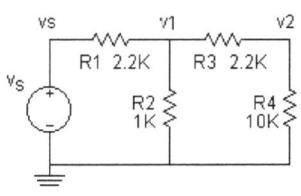

Build the circuit in Figure 2071. Connect a 5v power supply as shown. Measure the voltage across each resistor. Mark the positive ends. There are three loops in the circuit. Show that the algebraic sum of the voltages in each loop is zero.

Figure 2071 Circuit for 207

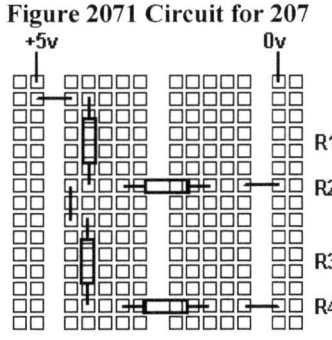

Measure the currents flowing in the circuit.

First turn off the power supply. Then remove a wire jumper in series with a resistor. Select the multimeter's 10mA scale. Connect the multimeter leads to the wire jumper's holes so that an ammeter has replaced the wire jumper. Pay attention to + and − leads.

Now turn on the power supply. Record the current in the resistor. Repeat the process for the other three wire jumpers. Show that the sum of the currents at each node is zero.

Electric Circuits - Analysis and Design

Superposition Law:

Build the circuit shown in Figure 200.

Figure 200

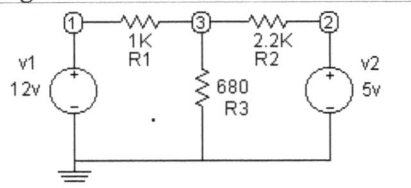

Make three measurements:
1 With source V1 shorted (we mean remove V1 and then ground node 1), measure current through R2 and voltage at node 3 due to source V2 only.

Restore V1.

2 With source V2 shorted (we mean remove V2 and then ground node 2), measure current through R2 and voltage at node 3 due to source V1 only.

Restore V2.

3 With V1 and V2 sources connected measure current through R2 and voltage at node 3.

Show that sums of the currents and voltages from measurements 1 and 2 equal the currents and voltages from measurement #3. This is a result of *superposition.*

Superposition of the voltages and currents produced by two sources, V1 and V2, connected to the circuit applies here, because the circuit is *linear*.

Experiment 3 Bridged-T Attenuator

Design, Assemble, and Test a Bridged-T Attenuator.

Figure 208 Bridged-T Attenuator

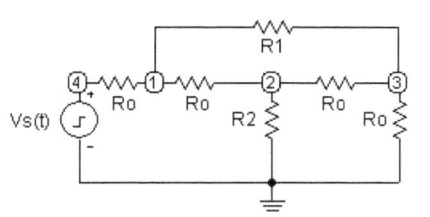

How they work: Bridged-T Attenuators are designed to have an input resistance equal to the load resistance R_0. R_0 is referred to as the *characteristic resistance* of the attenuator. A typical value for R_0 is 600 ohms. Note that $V_1 = 0.5 V_s$ with the Bridged-T *in* or *out* of the circuit.

Convert the R_0, R_0, R_1 node 1,2,3 delta (Figure 208) to an R_a, R_b, R_c node 1,2,3 wye (Figure 209).

Figure 209

$$z_a = \frac{z_2 z_3}{z_3 + z_1 + z_2} = \frac{R_0 R_1}{R_1 + R_0 + R_0} = R_a$$

$$z_b = \frac{z_1 z_3}{z_3 + z_1 + z_2} = \frac{R_0 R_1}{R_1 + R_0 + R_0} = R_b$$

$$z_c = \frac{z_1 z_2}{z_3 + z_1 + z_2} = \frac{R_0 R_0}{R_1 + R_0 + R_0} = R_c$$

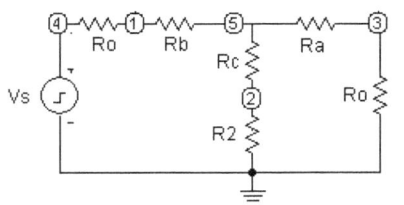

Delta to Wye Transformation

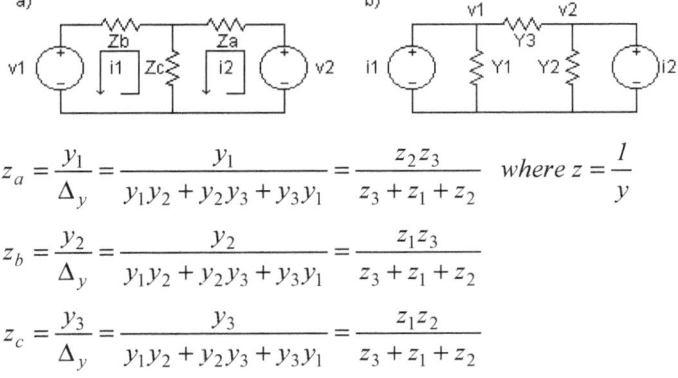

$$z_a = \frac{y_1}{\Delta_y} = \frac{y_1}{y_1 y_2 + y_2 y_3 + y_3 y_1} = \frac{z_2 z_3}{z_3 + z_1 + z_2} \quad \text{where } z = \frac{1}{y}$$

$$z_b = \frac{y_2}{\Delta_y} = \frac{y_2}{y_1 y_2 + y_2 y_3 + y_3 y_1} = \frac{z_1 z_3}{z_3 + z_1 + z_2}$$

$$z_c = \frac{y_3}{\Delta_y} = \frac{y_3}{y_1 y_2 + y_2 y_3 + y_3 y_1} = \frac{z_1 z_2}{z_3 + z_1 + z_2}$$

Figure 209

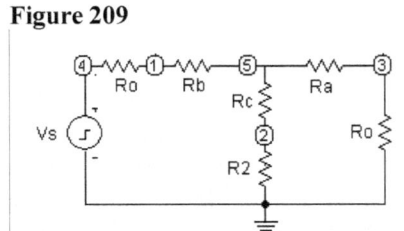

Show that R_0 is the input impedance of a Bridged-T terminated with R_0.

See *Solved*.

Use $R_1 R_2 = R_0^2$ to show that the resistance to the right of R_b is

$$R_p = \frac{(R_a + R_0)(R_c + R_2)}{(R_a + R_0) + (R_c + R_2)} = R_0 \frac{2R_0}{R_1 + 2R_0} \quad \text{and} \quad R_b + R_p = R_0$$

Using the above results find the expression for V_3/V_1.

1. The resistance to the right of node 1 is R_0.
   Show that $\dfrac{V_1}{V_S} = \dfrac{1}{2}$

2. The resistance to the right of node 5 is R_p. If V_1 is known then
   Show that $\dfrac{V_5}{V_1} = \dfrac{R_p}{R_b + R_p} = \dfrac{R_p}{R_0} = \dfrac{2R_o}{R_1 + 2R_o}$

3. The resistance to the right of node 3 is R_0. If V_5 is known then
   Show that $\dfrac{V_3}{V_5} = \dfrac{R_o}{R_o + R_a} = \dfrac{R_1 + 2R_o}{2(R_1 + R_o)}$

4. The ratio V_3/V_s is the product of the three prior ratios.
   Show that $\dfrac{V_3}{V_S} = \dfrac{V_1}{V_S} \dfrac{V_5}{V_1} \dfrac{V_3}{V_5} = \dfrac{R_0}{2(R_0 + R_1)}$

5. Find the expression for R_1/R_0 as a function of n when V_3/V_S equals $1/n$.
   Show that $\dfrac{R_1}{R_0} = \dfrac{n-2}{2}$

Electric Circuits Experiments   Experiment 3 Bridged-T Attenuator

Design ($R_0 =$) 600Ω Bridge-T attentuators with insertion loss of 6dB, 20dB, and 100dB. Assemble a 20dB Bridge-T and measure its attenuation. inserting a Bridge-T between the source and load R_0 resistors increases the attenuation

Insertion loss in dB = $-20 \log n$.
E.g. n=10 produces a 20 dB insertion loss and $R_1 / R_0 = 4$.

Why would a more practical 40dB attenuator consist of two 20dB sections inserted between the source and load R_0 resistors instead of one 40dB T-pad?

Build an attenuator. Let $r_0 = 680Ω$, $r_1 = 2200Ω$, $r_2 = 210Ω$. Calculate and measure the attenuation. Compare results.

Figure 208 Bridged-T attenuator

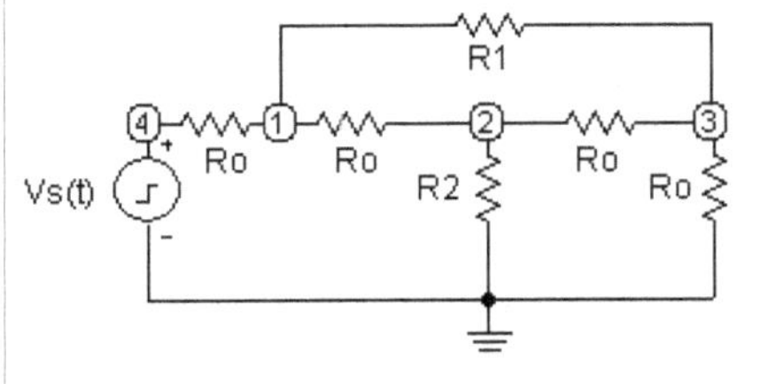

Figure 2081 Circuit for Figure 208

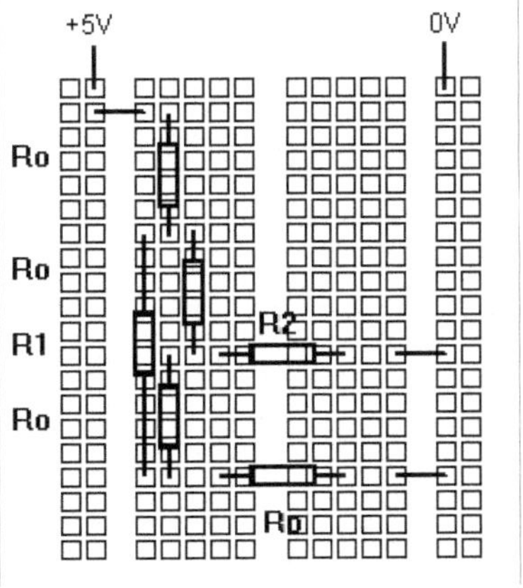

Electric Circuits - Analysis and Design

Experiment 4 AC Voltmeter

Build the AC Voltmeter circuit (Figure 401) at the right side of the breadboard. You will use it in experiments. The meter is a 1mA full scale DC current meter. We used the BiMOS quad op amp **TLC074CN**. Any comparable MOS op amp will do. The diodes are 1N914. Any low current diodes will do. R_1 is 10^6 ohms, R_2 is 1K ohms.

Make sure all connections to the breadboard are zero ohms. This BiMOS op amp has a parasitic 20pF capacitor at its inputs ($10^6 \Omega$ at 8KHz, $10^4 \Omega$ at 800KHz).

Connect a sinewave signal generator as V_{in}. Adjust the signal's frequency to about 1KHz. Adjust the signal's amplitude until the meter reads 1mA. The peak amplitude should be 1.57 volts.

Use an oscilloscope to verify that the signal at pins 5,6,7,2,3 is an undistorted sinewave. To check meter bandwidth increase the frequency until the meter reading drops to 0.89mA (−1dB). When we did that the frequency was 1MHz.

$for\ sinewave\ I_{max} \sin \omega t \quad I_{average} = \frac{2}{\pi} I_{max} \Rightarrow I_{max} = \frac{\pi}{2} I_{average}$

$V_{2\max} = I_{\max} R_2 = \frac{\pi}{2} I_{average} R_2 = \frac{\pi}{2} 1mA \times 1K\Omega = 1.57 volts$

$V_{2\max} = V_{3\max} = V_{7\max} = V_{5\max} = V_{in\max}$

$V_{in\max} = 1.57 volts\ for\ full\ scale\ 1mA\ meter\ dc\ current$

Figure 401 AC Voltmeter 1MHz

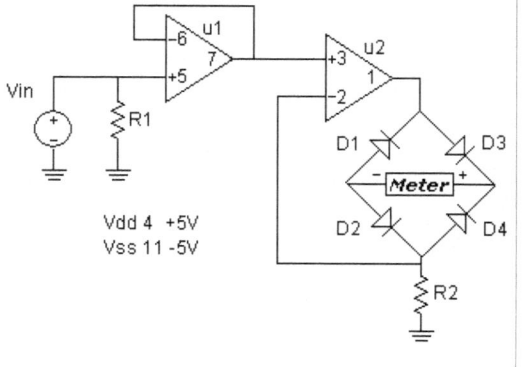

Figure 402 TI TLC074CN

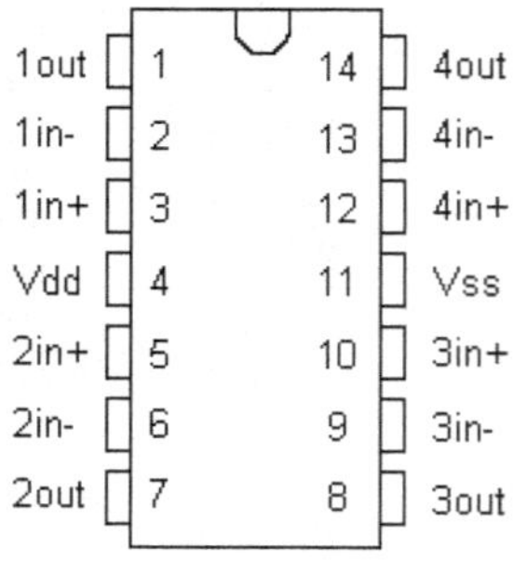

Electric Circuits Experiments   Experiment 4 AC Voltmeter

There is a lot going on here. This is an opportunity to teach yourself about the measuring instruments.

Connect meter + to D3-D4 junction, meter − to D1-D2 junction

Figure 4011 AC Voltmeter Circuit

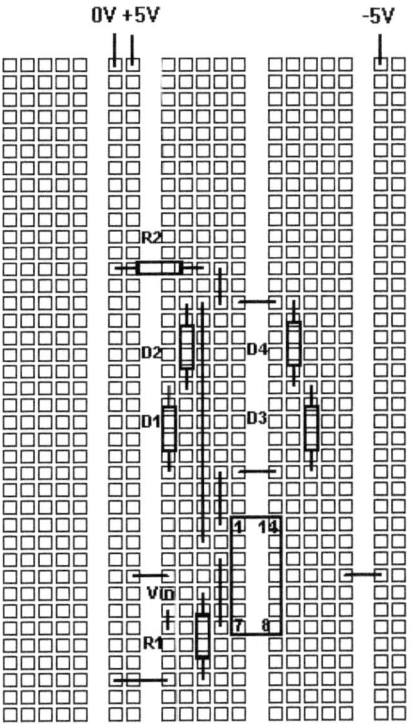

Electric Circuits - Analysis and Design

Voltage Feedback Operational Amplifier (VFA)

We start with the ideal op amp voltage transfer function (Figure 902). The op amp output ranges from 0 to V_{DD} or 0 to $-V_{SS}$ as the differential input varies from 0 to $+V_{DD}/\mu$ or from 0 to $-V_{SS}/\mu$, when the op amp gain is μ. As the differential input increases the amplifier output increases linearly until it reaches the maximum value that is referred to as saturation (Figure 902). Any further input increase results in the output remaining at the maximum value, because the output cannot exceed the supply voltages. The transition from linear operation to saturation is abrupt by design. By definition the ideal amplifier has 'infinite' gain, and zero input currents.

Figure 901 Typical Voltage Op Amp in the original 8-pin DIP package

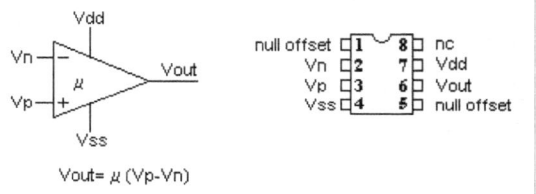

Figure 902 Voltage Op Amp ideal model and transfer function

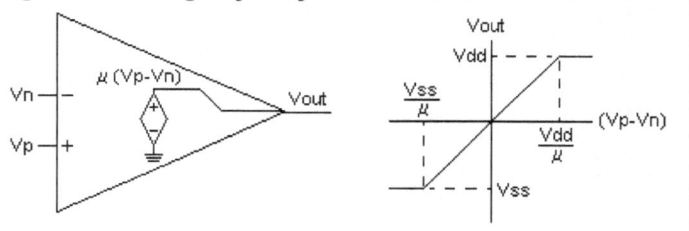

The ideal op amp has zero input current, the input impedance is infinite, the output impedance is zero, and the voltage gain is μ. The linear transfer function saturates when the output is close to the power supply voltages. The gain while in saturation is zero. For example with ±5 volt supplies, and a voltage gain of 400,000 a differential input voltage of 5/400000 = 12.5 microvolts saturates the amplifier. In the linear region V_P and V_N differ by less than 12.5 microvolts that is why they are considered equal in circuit analysis as a practical matter.

The voltage feedback inverting amplifier (Figure 903) is a widely used and general configuration. A simplified analysis proceeds as follows.

V_N equals zero volts, because the gain is infinite so that V_N equals V_P, which is connected to ground. Then the currents i_1 and i_2 flowing through z_1 and z_2 are equal in magnitude, because ideal op amp input current is zero. Therefore $V_1/z_1 = -V_3/z_2$ ($V_2=0$), and the voltage gain $V_3/V_1 = -z_2/z_1$. Here is a formal analysis.

Figure 903 Inverting Op Amp

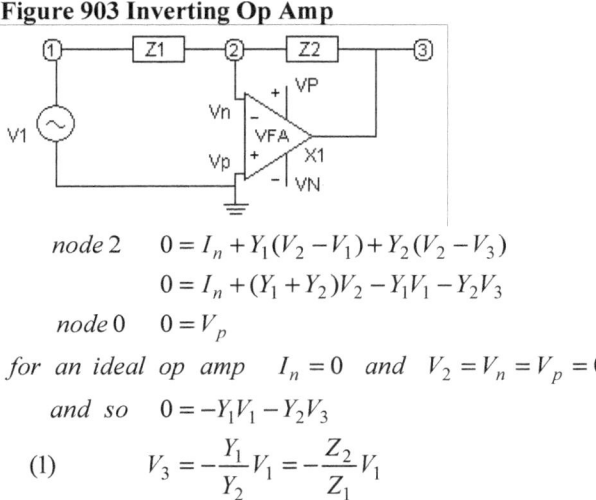

$$\text{node 2} \quad 0 = I_n + Y_1(V_2 - V_1) + Y_2(V_2 - V_3)$$
$$0 = I_n + (Y_1 + Y_2)V_2 - Y_1 V_1 - Y_2 V_3$$
$$\text{node 0} \quad 0 = V_p$$

for an ideal op amp $\quad I_n = 0 \quad$ and $\quad V_2 = V_n = V_p = 0$

and so $\quad 0 = -Y_1 V_1 - Y_2 V_3$

(1) $\quad V_3 = -\dfrac{Y_1}{Y_2} V_1 = -\dfrac{Z_2}{Z_1} V_1$

Most circuits using a voltage op amp are structures with feedback; there is a signal path from the output back to one of the op amp's inputs. (Feedback theory is a complex subject presented in Chapter 8.) A feedback amplifier loop is a signal path from output to input, through the amplifier, and back to the output. The relevant point is that when the gain around a feedback loop, the loop gain, is high the circuit properties are independent of the amplifier's parameters.

Emphasis If one input terminal is connected to ground (or voltage V) then an excellent approximation is that the other input terminal is also at ground potential (or voltage V).

> The ideal op amp output impedance is zero, and so the ideal op amp output is an ideal voltage source.

Design Practice A path for direct current to both op amp input terminals must exist, because real amplifiers require bias currents and have leakage currents.

Electric Circuits - Analysis and Design

Experiment 5 Capacitance

We intend to show that capacitors C *in parallel* add. Whereas reciprocals of capacitors 1/C *in series* add.

We will use the corner frequency idea in the measurements. The idea is explained in *Electric Circuit Analysis and Design Section 10.1* (page 168). Corner frequency is the radian frequency ω_0 where $R_1 = X_C = 1/\omega_0 C_1$. The transmission T(p) is −1dB at $\omega_0/2$, −3dB at ω_0, and −7dB at $2\omega_0$ (*Section 10.5 Table 1001*, page 178). For example

$$T(p) = \frac{v_2}{v_1} = \frac{\frac{1}{pC_1}}{R_1 + \frac{1}{pC_1}} = \frac{1}{1+pC_1 R_1} = \frac{1}{1+j\omega C_1 R_1} = \frac{1}{1+j\frac{\omega}{\omega_0}} \quad \left(R_1 = \frac{1}{\omega_0 C_1}\right)$$

$T(0) = 1 = 0 \; dB \qquad T(\infty) = 0$

We connect oscilloscope *inputs* to circuit nodes whose properties we want to observe and measure. We connect function generator *outputs* to circuit nodes we want to drive with known signals. The connections change the circuit, because the oscilloscope has an impedance Z_L at its inputs (Figure 302a). The impedance is a parallel $R_L C_L$ circuit where $R_L = 10^6$ ohms and $C_L = 20pf$. The function generator has a 50 ohm source impedance R_S.

Therefore, whenever you connect anything to a circuit you want to know what you are adding to the circuit. The instrumentation industry has tried to make these input and output impedances negligible when added to most circuits. Always check.

NOTE Measuring signal voltages over a frequency range requires a wide band AC meter with a dB scale. This is an expensive instrument not widely available today. We can build a meter on the breadboard (experiment 4). Or, we can avoid the need for a wide band AC meter by using a function generator with a Bode plotting capability or by viewing signals on an oscilloscope display.

Figure 302 Connected RC Circuit

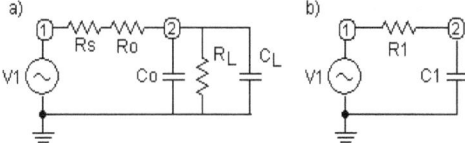

Wire the circuit (Figure 302b) on the breadboard

V_1 is a sinewave signal source. R_1=2.2K ± 5%. Select C_1=0.01µF ± 5% so that f_0 = 7.23 KHz ± 10%. Verify using a reactance chart.

Measure any frequency f=1/T by measuring period T on the oscilloscope. Set the V_1 frequency to 100Hz where the transmission $T(f)=v_2/v_1=1=0$dB. Adjust the signal amplitude until the AC voltmeter reads 1. Now increase the frequency until the AC voltmeter reads 0.707=−3dB where it may turn out that f=f_0 is different from 7.23KHz. Why? Answer - tolerances.

Add C_2=0.01µF in parallel. $T(f_0)$ should now equal 0.44= −7dB. Why?

Decrease the frequency until T(f)=0.707=−3dB again. This f, f_{0new}, should be about 0.5×7.23KHz = 3.62KHz demonstrating that $C_1+C_2=2C_1$ (parallel capacitors add).

Remove C_2 in parallel, and insert it in series with C_1. Use a wire to short out C_2, and increase the frequency until T(f)=0.707=−3dB (f should equal 7.23KHz again, i.e. the original f_0). Remove the short and now T(f) is about 0.89=−1dB. Why?

Increase the frequency until T(f)=0.707=−3dB. This f, f_{0new}, should be 2×7.23KHz=14.46KHz so that $(1/C_1)+(1/C_2)=2/C_1$ (reciprocals of series capacitors add).

Figure 3021 Circuit for Figure 302b

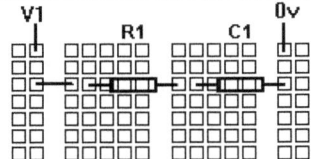

Electric Circuits - Analysis and Design

```
Fig3011.ckt    capacitors RC circuits

*asymptote source voltage
V1 1 0   AC 1 0     ;sin(0 1 50K 0 0)   ; volts
* names/nodes/values
R1 1 2 2.2K
C1 2 0 0.02E-6

R11   1 12 2.2K
C11  12   0 0.01E-6

R12   1 22 2.2K
C12  22   0 0.005E-6

*.PLOT AC VP(2) -100,0
.AC DEC 200 100 100000
.TEMP 27
.PLOT AC VDB(2) VDB(12) VDB(22) -4,1
.PRINT AC VP(2)
.PRINT AC VDB(2)
.end
```

These plots do NOT prove capacitors in parallel add, nor that capacitors in series divide. Proof requires experiments evaluating physical circuits.

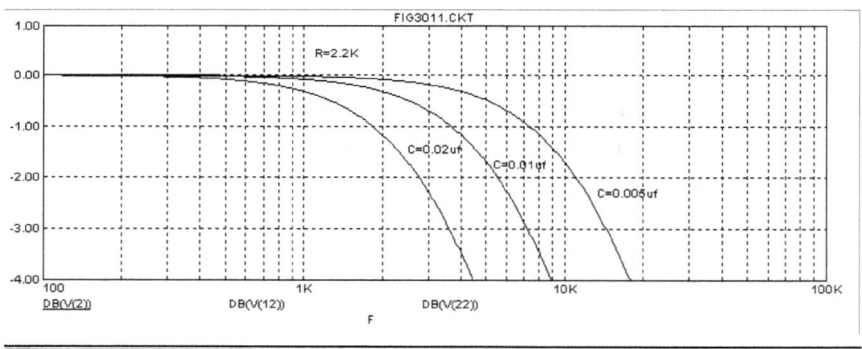

Experiment 6 Inductance

We intend to show that inductors L *in series* add. Whereas reciprocals of inductors 1/L in *parallel* add. And, that an inductor's resonant frequency is due to parasitic capacitance,

We will use the corner frequency idea in the measurements. The idea is explained in *Electric Circuit Design Section 10.1* (page 168). Corner frequency is the radian frequency ω_0 where $R_1 = X_L = \omega_0 L_1$. The transmission T(p) is −1dB at $\omega_0/2$, −3dB at ω_0, and −7dB at $2\omega_0$ (*Section 10.5 Table 1001,* page 178). For example

$$T(p) = \frac{v_2}{v_1} = \frac{pL_1}{R_1 + pL_1} = \frac{\frac{pL_1}{R_1}}{1 + \frac{pL_1}{R_1}} = \frac{j\frac{\omega L_1}{R_1}}{1 + j\frac{\omega L_1}{R_1}} = \frac{j\frac{\omega}{\omega_0}}{1 + j\frac{\omega}{\omega_0}} \quad (R_1 = \omega_0 L_1)$$

$$T(0) = 0 \qquad T(\infty) = 1$$

We connect oscilloscope *inputs* to circuit nodes whose properties we want to observe and measure. We connect function generator *outputs* to circuit nodes we want to drive with known signals. The connections change the circuit, because the oscilloscope has an impedance Z_L at its inputs (Figure 302a). The impedance is a parallel $R_L C_L$ circuit where $R_L = 10^6$ ohms and $C_L = 20$pf. The function generator has a 50 ohm source impedance R_S.

Therefore, whenever you connect anything to a circuit you want to know what you are adding to the circuit. The instrumentation industry has tried to make these input and output impedances negligible when added to most circuits. Always check.

NOTE *Measuring signal voltages over a frequency range requires a wide band AC meter with a dB scale. This is an expensive instrument not widely available today. We can build a meter on the breadboard (experiment 4). Or, we can avoid the need for a wide band AC meter by using a function generator with a Bode plotting capability or by viewing signals on an oscilloscope display.*

Electric Circuits - Analysis and Design

Wire the circuit (Figure 902) in the breadboard

V_1 is a sinewave signal source. R_1=1K ± 5%. Select L_1=1000µH ± 5% so that f_0 = 159 KHz ± 10% . Verify using a reactance chart.

Set the frequency to 10KHz where the transmission $T(f)=v_2/v_1=1=0dB$. Adjust the signal amplitude until the AC meter reads 1=0dB. Now increase the frequency until $T(f)=0.707=-3dB$ where it turns out that f is different from 159KHz. Why?

Add L_2=1000µH in parallel. $T(f)$ should now be 0.89=−1dB. Why?

Increase the frequency until $T(f)=0.707=-3dB$ again. This f, f_{0new}, should be 2×159KHz = 318KHz demonstrating that $1/L_1+1/L_2=2/L_1$ (reciprocals of parallel inductors add).

Remove L_2 in parallel, and insert it in series with L_1. Use a wire to short out L_2, and increase the frequency until $T(f)=0.707=-3dB$ (f should equal 159KHz again i.e. the original f_0). Remove the short and now $T(f)$ is about 0.44=−7dB. Why?

Decrease the frequency until $T(f)=0.707= -3dB$. This f, f_{0new}, should be 1/2×159KHz = 80KHz demonstrating that $(L_1+L_2)=2L_1$ (series inductors add).

Figure 902   Figure 9021 Circuit for Figure 902

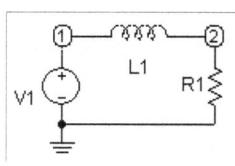
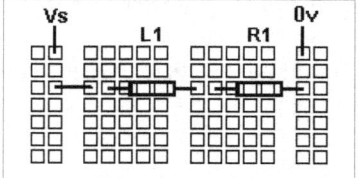

Experiment 7 Transformer

The basic transformer is an assembly of two or more windings (coils of insulated wire) on a core. Each winding has self inductance L and mutual inductance M with every other winding.

Mutual inductance (M) is the circuit property that arises when a changing current in one wire induces a voltage in *another* wire.

A alternating current in a wire produces an alternating magnetic field that spreads out in space. The field induces a voltage in any wire (or any conductive material) the field passes across. A changing current in one wire that induces a changing voltage in a second wire extends the concept of inductance. This extension is referred to as mutual inductance (M) because it arises from a magnetic field *flux* coupling *both* wires. The difference is that the induced voltage is across one inductance (wire) while the changing current flows in another inductance (wire). Mutual inductance (M) has the same type of vi constraint as inductance. *M is represented in Spice by the coupling coefficient k* (Electric Circuits Section 4.1.3, page 46).

There is no distinct mutual inductor you can touch. Mutual inductance is a property of an assembly of two or more inductors that may just be adjacent lengths of wire.

Figure 403

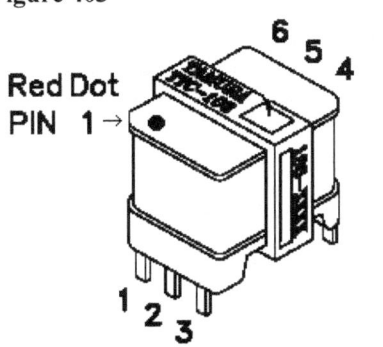

Electric Circuits - Analysis and Design

Tamura TTC-108 Transformer Specifications

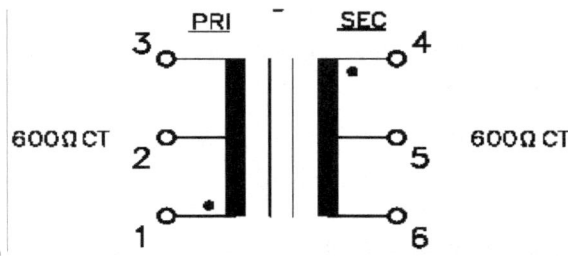

TELECOMMUNICATION DRY COUPLING TRANSFORMER DESIGNED TO OPERATE AT A MAX LEVEL OF +7dBm AND TO REFLECT A PRIMARY SOURCE IMPEDANCE OF APPROXIMATELY 600ΩCT WITH 600ΩCT LOAD ON SECONDARY

A. Electrical Specifications (@ 25° C)
1. Pri Source Impedance; 600Ω CT
2. Sec Load Impedance; 600Ω CT
3. Operating Level; −45 dBm to +7 dBm
4. Insertion Loss;
   1.4 dB MAX @ 1 KHz, 0 dBm
5. Frequency Response;
   ±0.5 dB 300 Hz to 3.5 KHz @ 0 dBm
6. Primary Impedance;
   600 Ω +15%, −5% @ 300 Hz to 3.5 KHz, 0dBm
   600 Ω +10%, −5% @ 500 Hz to 2.5 KHz, 0dBm
7. Longitudinal Balance;
   60 dB MIN @ 200 Hz to 1 KHz
   40 dBm MIN @ 4 KHz
8. DC Resistance;
   (1−3) = 44 Ω ±20%
   (4−6) = 56 Ω ±20%
9. Turns Ratio; (1−3) : (4−6) = 1 : 1.00 ±2%
10. Dielectric Strength;
    1500 Vrms 1 minute @ Pri to Sec, and Pri to Core
    1000 Vrms 1 minute @ Sec to Core
11. Total Harmonic Distortion;
    0.5% MAX @ 300 Hz to 3.5 KHz, 0 dBm

Measure the transformer equivalent circuit components of a transformer

Figure 409 Transformer Equivalent Circuit

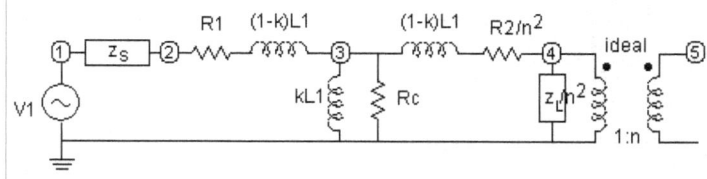

1 measure windings ohms

Measure to show that (*transformer* pins 1 to 3) R_1=56 Ω and (pins 4 to 6) R_2=77 Ω (see R_1, R_2 in Figure 409). Note deviation from specification 8.

2 measure turns ratio n Ground *transformer* pins 3 and 6 (Figure 413, omit R_1). Connect signal generator V_1 to transformer pin 1. Connect AC voltmeter input to transformer pin 4. Set frequency to 1KHz. Measure 1KHz AC signal at transformer pins 1, 4.

3 measure kL_1
Add $R_1=470\Omega$ to signal generator 50Ω R_S to make R_{total} about 580Ω including R_1, R_2 winding ohms (Figure 409). Connect 470Ω shown as R_1 in Figure 413. Connect 680Ω resistor to pins 4, 6. This is R_L. Omit C_L. Connect AC voltmeter input to pin 1.

Set frequency to about 1KHz. Adjust signal generator voltage so that AC voltmeter reads 1=0dB at transformer pin 1. Reduce frequency gradually until AC voltmeter reads 0.707=−3dB. Record the frequency. Calculate kL_1. (Section 4.4.1 page 51). kL_1 is about 1H.

Figure 410 Low Frequency Model   **Figure 413 Test Setup**

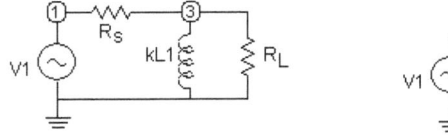

4 measure (1−k)L1
Signal generator $R_S=50\Omega$. Add 470 Ω to make $R_{total}=581\Omega$ as before. Set frequency to about 1KHz. Connect AC voltmeter input to transformer pin 4 (Figure 413). Adjust signal generator voltage so that AC voltmeter reads 1=0dB.

Increase frequency gradually until AC voltmeter reads 0.707=−3dB. Record the frequency. Calculate $(1-k)L_1$. (Section 4.4.2, page 53).

Calculate k Given kL_1 and $(1-k)L_1$, calculate k.

Figure 411 Transformer High Frequency Model

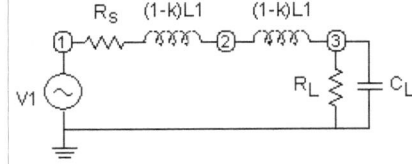

Experiment 8 Circuit Analysis

The Bridge T (Figure 501) has many applications, which are implemented by selecting components for z_1 and $z_2 = z_0^2/z_1$.

Figure 501 Bridge T

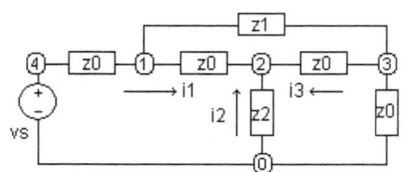

Write Spice AC programs that show frequency response of the four bridge T circuits you build (see solutions in *Solved*)

Bridge T node analysis
Write the 3 node equations (see solutions in *Solved*).
Show that the node equation determinant Δ is

(2) $\Delta = 8y_0^3 + 4y_0^2(y_1 + y_2) = \dfrac{4y_0^2}{y_1}(y_0 + y_1)^2$

Show that the transfer function is

(3) $\dfrac{v_3}{v_s} = \dfrac{z_0}{2(z_0 + z_1)}$

Bridge T mesh analysis
Write the 3 mesh equations.
Show that the mesh equation determinant Δ is

(2) $\Delta = 8z_0^3 + 4z_0^2 z_1 + 4z_0^2 z_2 = \dfrac{4z_0^2}{z_1}(z_0 + z_1)^2$

Show that the transfer function is

(3) $\dfrac{v_3}{v_s} = \dfrac{z_0}{2(z_0 + z_1)}$

Attenuator Design and build a −20dB attenuator (Figure 5011). Let $r_0 = 680\Omega$. Show that $r_1 = 4r_0 = 2720\Omega$ (use 2.7K), and $r_2 = r_0/4 = 170\Omega$ (use 180Ω). Measure the attenuation.

(4a) $z_0 = r_0,\ z_1 = r_1,\ z_2 = \dfrac{r_0^2}{r_1}$

(4b) $loss\ in\ dB = 20\log 2(1 + r_1/r_0)$

Low Pass Filter Design and build a low pass filter (**Figure 5011**). Select the z_1, z_2 components to be L=1000µH, and C=1000pF. This sets the −3dB corner frequency at 159KHz. Given L and C r_0=1000Ω. Measure the frequency response. Write a Spice program and plot the frequency response. Compare to your measurements.

(5) $\quad \dfrac{v_3}{v_s} = \dfrac{z_0}{2(z_0 + z_1)} = \dfrac{r_0}{2(r_0 + pL)} = \dfrac{1}{2(1 + pL/r_0)}$

High Pass Filter Design and build a high pass filter (**Figure 5011**). Select z_1, z_2 components to be C=1000pF, L=1000µH, and r_0=1KΩ. This sets the −3dB corner frequency at 159KHz. Measure the frequency response. Write a Spice program and plot the frequency response. Compare to your measurements.

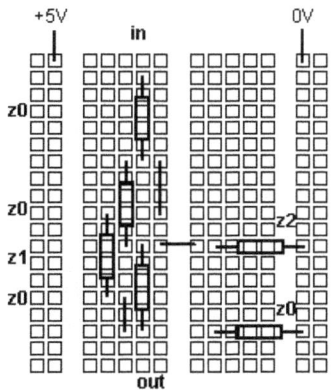

Figure 5011 Attenuator or Low Pass or High Pass Bridge T Filter

(6) $\quad \dfrac{v_3}{v_s} = \dfrac{z_0}{2(z_0 + z_1)} = \dfrac{pCr_0}{2(1 + pCr_0)}$

Band Pass Filter Design and build a band pass filter (**Figure 5012**). Select the resonant frequency to be at 159KHz. Show that the z_1 components are L and C in series, and that the z_2 components are L and C in parallel where C=1000pF and L=1000µH, and r_0=1KΩ. Measure the frequency response. Write a Spice program and plot the frequency response. Compare to your measurements.

(7) $\quad \dfrac{v_3}{v_s} = \dfrac{z_0}{2(z_0 + z_1)} = \dfrac{1}{2} \dfrac{r_0}{r_0 + pL + \dfrac{1}{pC}} = \dfrac{1}{2} \times \dfrac{pCr_0}{1 + pCr_0 + p^2 LC}$

Band Stop Filter Design and build a band stop filter (**Figure 5013**). Select the resonant frequency to be at 159KHz, and r_0=1KΩ. Show that the z_1 components are L and C in parallel, and that the z_2 components are L and C in series where C=1000pF and L=1000µH. Measure the frequency response. Write a Spice program and plot the frequency response. Compare to your measurements.

Electric Circuits - Analysis and Design

(8) $\dfrac{v_3}{v_s} = \dfrac{z_0}{2(z_0+z_1)} = \dfrac{1}{2}\dfrac{r_0}{r_0 + \dfrac{1}{pC+\dfrac{1}{pL}}} = \dfrac{1}{2}\times\dfrac{p^2LC+1}{1+pL/r_0+p^2LC}$

Resonant Frequency

(9a) $\; 0 = 1 + pCr_0 + p^2LC = 1 + \dfrac{1}{Q}\dfrac{p}{\omega_0} + \dfrac{p^2}{\omega_0^2} = 1 + \dfrac{1}{Q}\lambda + \lambda^2$

(9b) $\; \lambda_1,\lambda_2 = -1/2Q \pm \sqrt{(1/2Q)^2 - 1} = -1/2Q \pm j\sqrt{1-(1/2Q)^2}$

(9c) $\; \dfrac{\omega_{max}^2}{\omega_0^2} = 1 - (1/2Q)^2$

(9d) $\; \omega_{max}^2 = \omega_0^2 - \dfrac{\omega_0^2}{4Q^2} = \omega_0^2 - \Delta\omega^2 = \dfrac{1}{LC} - \dfrac{r_0^2}{4L^2}$

Figure 5012 Band Pass Bridge T Filter z_1 series LC, z_2 parallel LC

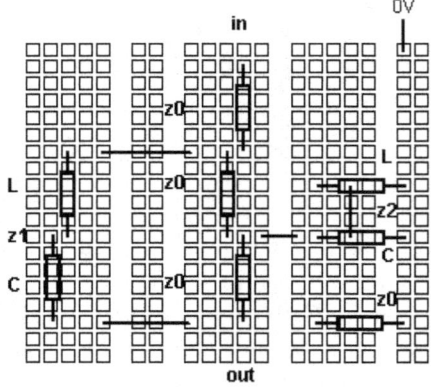

Figure 5013 Band Stop Bridge T Filter z_1 parallel LC, z_2 series LC

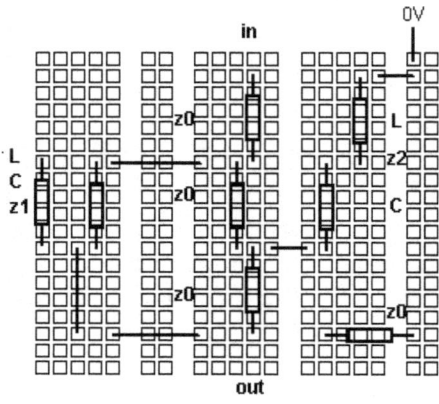

Experiment 9 Frequency Response

The purpose of these experiments is to study frequency response of RC, RL, and RLC circuits, *because frequency response is a measure of how sinewaves are transmitted through a circuit* from input(s) to output(s).

We show how R, L, C vi constraints insert these components into KVL and KCL circuit equations. We discuss their steady state behavior, how they modify waveforms, and how to plot the frequency response.

In all experiments the sources $v_S(t)$ and $i_S(t)$ have the same amplitude at all frequencies. The source amplitude does not vary as frequency varies.

Capacitor C - RC Circuits

Figure 601   Figure 602

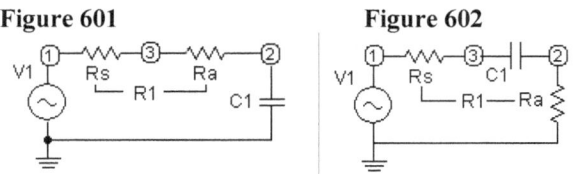

In the time domain the equation for both RC circuits is

(1) $\quad v_1(t) = R_1 i_1(t) + \dfrac{1}{C_1} \displaystyle\int_0^t i_1(x)\, dx$

In the complex frequency domain (Section 7.9 page 125) the transforms equations are (Laplace Transforms of vi Constraints page 113).

(2) $\quad v_1(p) = R_1 i_1(p) + \dfrac{1}{pC_1} i_1(p) = \left(R_1 + \dfrac{1}{pC_1} \right) i_1(p) \;\Rightarrow\; i_1(p) = \dfrac{v_1(p)}{\left(R_1 + \dfrac{1}{pC_1} \right)}$

(3a) $\quad T_{2-601}(p) = \dfrac{v_2(p)}{v_1(p)} = \dfrac{1}{(1 + pC_1 R_1)}$ $\qquad$ (3b) $\quad T_{2-602}(p) = \dfrac{v_2(p)}{v_1(p)} = \dfrac{pC_1 R_1}{(1 + pC_1 R_1)}$

The $v_1(t)$ waveform *is* NOT modified by the differential equation if $V_1 = v_1(t)$ is a sinewave.

601 is a low pass filter, where low frequency input signals are passed to the output with zero or low attenuation. And high frequency input signals are severely attenuated, because the impedance of the capacitor decreases to zero at high frequency. Transformed equation 3a is the input to output voltage transfer function when $p = j\omega$.

Experimental procedure for Low Pass Filter - Measure $T_{2\text{-}601}$. Signal generator V_1 has a 50Ω source resistance. Connect 950Ω resistor R_a to nodes 3 and 2 to make $R_1 = 1000$Ω (Figure 601). In effect V_1 is connected to node 1, and R_1 (50 + 950) is connected to circuit 601 nodes 1 and 2. Connect 0.01μF capacitor C_1 to nodes 2 and ground. This completes circuit 601.

Connect the AC voltmeter input to circuit 601 node 2.

Measure the signal at node 2. Set frequency to about 1KHz. Monitor V_2. Adjust signal generator voltage so that AC voltmeter reads 1 (0dB).

Increase frequency gradually until AC voltmeter falls to 0.707 (−3dB). Record the frequency. This is the corner frequency (page 168).

Increase frequency gradually until AC voltmeter falls to 0.447 (−7dB). Record the frequency. This is 2 times the corner frequency (page 178).

602 is a high pass filter, where high frequency input signals are passed to the output with zero or low attenuation. And low frequency input signals are severely attenuated, because the impedance of the capacitor increases to infinity at low frequency. Transformed equation 3b is the input to output voltage transfer function when $p = j\omega$.

Experimental procedure for High Pass Filter - Measure $T_{2\text{-}602}$. Signal generator V_1 has a 50Ω source resistance. Connect 950Ω resistor R_a to nodes 2 and ground to make $R_1 = 1000$Ω (Figure 602). In effect V_1 is connected to node 1, and R_1 (50 + 950) is connected to circuit 601 as shown. Connect 0.01μF capacitor C_1 to nodes 2 and 3. This completes circuit 602.

Connect the AC voltmeter input to circuit 602 nodes 2 and ground.

Measure the signal at node 2. Set frequency to about 50KHz. Monitor V_2. Adjust signal generator voltage so that AC voltmeter reads 1 (0dB).

Decrease frequency gradually until the AC voltmeter falls to 0.707 (−3dB). Record the frequency. This is the corner frequency (page 168).

Decrease frequency gradually until voltage falls to 0.447 (−7dB). Record the frequency. This is 1/2 the corner frequency (page 154).

Inductor L - RL Circuits

Figure 603

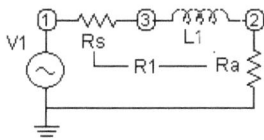

Figure 604

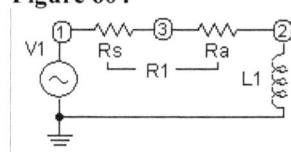

In the time domain the equation for both RL circuits is

(4) $\quad v_1(t) = R_1 i_1(t) + L_1 \dfrac{di_1(t)}{dt}$

In the complex frequency domain (Section 7.9 page 125) the equations are as follows (see Laplace Transforms of vi Constraints page 113).

(5) $\quad v_1(p) = R_1 i_1(p) + p L_1 i_1(p) = (R_1 + p L_1) i_1(p) \quad \Rightarrow \quad i_1(p) = \dfrac{v_1(p)}{(R_1 + p L_1)}$

(6a) $\quad T_{2-603}(p) = \dfrac{v_2(p)}{v_1(p)} = \dfrac{R_1}{(R_1 + p L_1)} \qquad$ (6b) $\quad T_{2-604}(p) = \dfrac{v_2(p)}{v_1(p)} = \dfrac{p L_1}{(R_1 + p L_1)}$

The $v_1(t)$ waveform *is* NOT modified by the differential equation if $V_1 = v_1(t)$ is a sinewave.

603 is a low pass filter, where low frequency input signals are passed to the output with zero or low attenuation. And high frequency input signals are severely attenuated, because the impedance of the inductor increases to infinity at high frequency. Transformed equation 6a is the input to output voltage transfer function when $p = j\omega$.

Experimental procedure for Low Pass Filter - Measure $T_{2\text{-}603}$. Signal generator V_1 has a 50Ω source resistance. Connect 150Ω resistor R_a to nodes 2 and ground to make $R_1 = 200Ω$ (Figure 603). In effect V_1 is connected to node 1 and R_1 (50 + 150) is connected to circuit 601 as shown. Connect 1000μH inductor L_1 to nodes 2 and 3. This completes circuit 603. Connect the AC voltmeter input to circuit 603 node 2.

Measure the signal at node 2. Set frequency to about 1KHz. Monitor V_2. Adjust signal generator voltage so that AC voltmeter reads 1 (0dB).

Increase frequency gradually until the AC voltmeter falls to 0.707 volts (−3dB). Record the frequency. This is the corner frequency (page 168).

Increase frequency gradually until voltage falls to 0.447 volts (−7dB). Record the frequency. This is 2 times the corner frequency (page 178).

604 is a high pass filter, where high frequency input signals are passed to the output with zero or low attenuation. And low frequency input signals are severely attenuated, because the impedance of the inductor decreases to zero at low frequency. Transformed equation 6b is the input to output voltage transfer function when $p = j\omega$.

Experimental procedure for High Pass Filter - Measure $T_{2\text{-}604}$. Signal generator V_1 has a 50Ω source resistance. Connect 150Ω resistor R_a to nodes 2 and 3 to make $R_1 = 200Ω$ (Figure 602). In effect V_1 is connected to node 1, and R_1 (50+150) is connected to circuit 604 as shown. Connect 1000μH inductor L_1 to nodes 2 and ground. This completes circuit 604. Connect the AC voltmeter input to circuit 604 node 2.

Measure the signal at node 2. Set frequency to about 80KHz. Monitor V_2. Adjust signal generator voltage so that AC voltmeter reads 1 (0dB).

Decrease frequency gradually until AC voltmeter falls to 0.707 (−3dB). Record the frequency. This is the corner frequency (page 168).

Decrease frequency gradually until voltage falls to 0.447 (−7dB). Record the frequency. This is 1/2 the corner frequency (page 178).

RLC Circuit

Figure 606

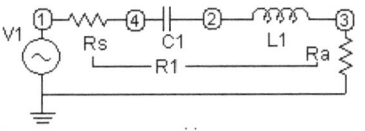

Read Section 6.4 page 100.

In the time domain the equation is as follows (Laplace Transforms of vi Constraints page 113).

(7) $\quad v_1(t) = R_1 i_1(t) + L_1 \dfrac{di_1(t)}{dt} + \dfrac{1}{C_1} \int_0^t i_1(x) dx$

In the complex frequency domain (Section 7.9 page 125) the equations are as follows

(8) $\quad v_1(p) = R_1 i_1(p) + pL_1 i_1(p) + \dfrac{1}{pC_1} i_1(p) = \left(R_1 + pL_1 + \dfrac{1}{pC_1}\right) i_1(p)$

$\Rightarrow \quad i_1(p) = \dfrac{v_1(p)}{\left(R_1 + pL_1 + \dfrac{1}{pC_1}\right)}$

(9) $\quad T_{3-606}(p) = \dfrac{v_3(p)}{v_1(p)} = \dfrac{R_1}{\left(R_1 + pL_1 + \dfrac{1}{pC_1}\right)} = \dfrac{pC_1 R_1}{\left(p^2 C_1 L_1 + pC_1 R_1 + 1\right)}$

(10) *band center* $\quad \omega_0 = \dfrac{1}{\sqrt{C_1 L_1}}$

606 is a band pass filter, where resonant frequency input signals are passed to the output with zero or low attenuation. High as well as low frequency input signals are severely attenuated, because the impedance of the inductor and capacitor decrease to zero at low and high frequencies respectively. Transformed equation 9 is the input to output voltage transfer function when p = jω.

Figure 606

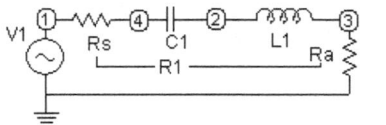

Experimental procedure for Band Pass Filter - Measure $T_{3\text{-}606}$. Signal generator V_1 has a 50Ω source resistance R_s. Connect 150Ω resistor R_a to nodes 3 and ground to make R_1= 200Ω (Figure 606). In effect V_1 is connected to node 1, and R_1 (50 + 150) is connected to circuit 606 as shown. Connect 2 parallel 1000pF capacitors C_1 to nodes 4 and 2. Connect 1000μH inductor L_1 to nodes 2 and 3. This completes circuit 606.

Connect the AC voltmeter input to circuit 606 node 3.

Monitor the signal at node 3. Set frequency to about 500KHz.

Decrease frequency gradually until voltage rises to a maximum. Adjust signal generator voltage so that AC voltmeter reads 1 (0dB). Record the resonant frequency f_0 (Figure 60611 page 103).

Decrease frequency gradually until AC voltmeter falls to 0.707 (−3dB). Record the frequency f_1.

Increase frequency gradually past the center frequency until AC voltmeter again falls to 0.707 (−3dB). Record the frequency f_2.

The −3dB bandwidth is f_2-f_1. Calculate the Q (page 100).

Calculate f_0 (equation 10 above). The reactance chart shows it is about 112.5KHz.

Calculate the reactance X_C at f_0. The reactance chart shows it is about 700 ohms. Now calculate the Q, which should be about 700/200.

Experiment 10 Transient Response

The purpose of these experiments is to study transient response of RC, RL, and RLC circuits, *because transient response shows how a waveform transmitted through a circuit is modified by the circuit.* Digital waveform modification from input(s) to output(s) is a particular concern. We show how RLC vi constraints insert these components into KVL and KCL circuit equations. Then we discuss their transient behavior, how they modify waveforms, and how to plot the transient response.

Capacitor C - RC Circuits

Figure 601   **Figure 602**

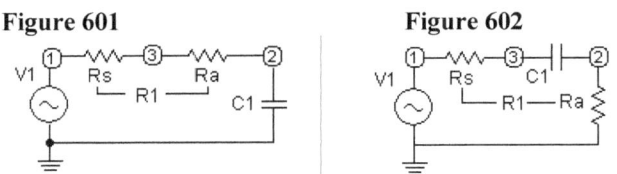

In the time domain the equation for both circuits is

(1) $\quad v_1(t) = R_1 i_1(t) + \dfrac{1}{C_1} \displaystyle\int_0^t i_1(x)\,dx$

Show that the responses to a step function $V_m u(t)$ (pages 107, 114, 115) are

(2a) $\quad v_{2\text{-}601}(t) = V_M \left(1 - e^{-\frac{t}{R_1 C_1}} \right)$
(2b) $\quad v_{2\text{-}602}(t) = V_M e^{-\frac{t}{R_1 C_1}}$

The input signal step function u(t) has zero *rise time* (pages 107, 114). The output signal rise time is increased to 2.2 time constants τ where $\tau = R_1 C_1$.

Since circuit 601 is a low pass filter the response to an input signal step function u(t) is the output $v_{2\text{-}601}$ capacitor voltage, which starts at zero volts and rises to V_m volts. Whereas the output $v_{2\text{-}602}$ of high pass filter circuit 602 is the resistor voltage, which starts at V_m volts and falls to zero volts. The basic reason for these behaviors is that since the initial response to u(t) is an infinite *dv/dt* trying to produce an infinite current so that initially a capacitor acts like a short circuit (page 115).

Electric Circuits - Analysis and Design

Capacitor C - RC Circuits

Figure 601   **Figure 602**

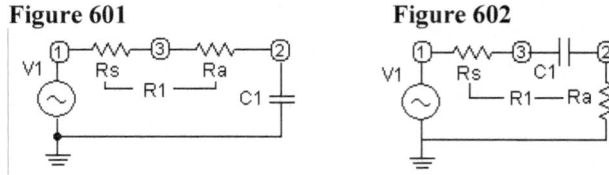

Experimental procedure for Low Pass Filter - Measure $T_{2\text{-}601}$. Signal generator V_1 has a 50Ω source resistance. Connect 950Ω resistor R_a to nodes 3 and 2 to make $R_1 = 1000Ω$ (Figure 601). V_1 is connected to node 1. Connect 0.01μF capacitor C_1 to nodes 2 and ground. This completes circuit 601. Time constant $\tau = 1000 \times 0.01\text{E}{-}6 = 10$ microseconds.

Note: a transient goes to 0 in $5\tau = 50\mu s$.

Set oscilloscope channel 1 to 1V/cm, channel 2 to 1V/cm.
Connect channel 1 to input node 1 and channel 2 to output node 2

Set 0V level at center screen. Select time base to be *10μs/cm so that full screen 10 cm sweep takes 100μs.*

On the Function Generator Set up a periodic pulse waveform with a 100μs period (frequency 10KHz), and 50μs pulse width. Adjust output pulse voltage to range from 0V to 5V.

Observe that output V_2 has a $1-e^{-x}$ waveform.

Experimental procedure for High Pass Filter - Measure $T_{2\text{-}602}$. build circuit Figure 602 and use the same procedure as above modified accordingly. Observe that output V_2 has an e^{-x} waveform.

Inductor L - RL Circuits

Figure 603　　　　　　　**Figure 604**

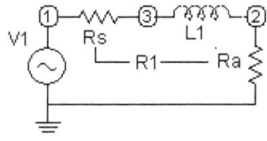 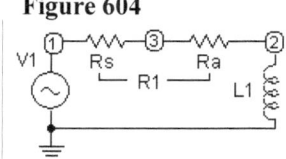

In the time domain the equation for both RL circuits is

(3) $\quad v_1(t) = R_1 i_1(t) + L_1 \dfrac{di_1(t)}{dt}$

Show that the responses to a step function $V_m u(t)$ (pages 107, 111, 114) are

(4a) $\quad v_{2\text{-}603}(t) = V_M \left(1 - e^{-\frac{R_1}{L_1}t} \right) \qquad$ (4b) $\quad v_{2\text{-}604}(t) = V_M e^{-\frac{R_1}{L_1}t}$

The input signal step function u(t) has zero *rise time* (page 107). The output signal rise time increased to 2.2 time constants τ where $\tau = L_1/R_1$ (page 112).

Since circuit 603 is a low pass filter the response to an input signal step function u(t) is the output $v_{2\text{-}603}$ the resistor voltage, which starts at zero volts and rises to V_m volts. And the output $v_{2\text{-}604}$ of high pass filter circuit 604 is the inductor voltage, which starts at V_m volts and falls to zero volts. The basic reason for these behaviors is that since the initial response to u(t) is an infinite *di/dt* wanting to produce an infinite voltage an inductor acts like an open circuit (page 115).

Electric Circuits - Analysis and Design

Inductor L - RL Circuits

Figure 603 **Figure 604**

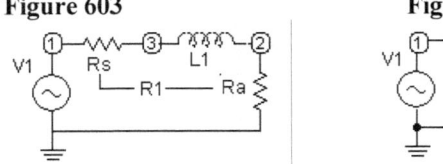

Experimental procedure for Low Pass Filter - Measure T_{2-603}. Signal generator V_1 has a 50Ω source resistance. Connect 950Ω resistor R_a to nodes 2 and ground to make $R_1 = 1000Ω$ (Figure 603). V_1 is connected to node 1, and R_1 (50 + 950) is connected to circuit 603 as shown. Connect 1000µH inductor L_1 to nodes 2 and 3. This completes circuit 603. Time constant $\tau = L_1/R_1 = 1000E-6/1000 = 1\mu s$. Note that a transient goes to 0 in $5\tau = 5\mu s$.

Set oscilloscope channel 1 to 1V/cm, channel 2 to 1V/cm.
Connect channel 1 to input node 1 and channel 2 to output node 2

Set 0v level at center screen. Select time base to be *1µs/cm so that full screen 10 cm sweep takes 10µs.*

On the Function Generator Set up a periodic pulse waveform with a 10µs period (frequency 100KHz), and 5µs pulse width. Adjust output pulse voltage to range from 0V to 5V.

Observe that node 2 has an $1-e^{-x}$ waveform.

Experimental procedure for High Pass Filter - Measure T_{2-604}. Use the same procedure as above modified accordingly. Observe that node 2 has an e^{-x} waveform.

RLC Circuit

Figure 606

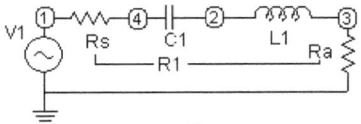

Read Section 6.4.

In the time domain the equation is (see 7.6 page 118).

(5) $\quad v_1(t) = R_1 i_1(t) + L_1 \dfrac{di_1(t)}{dt} + \dfrac{1}{C_1} \int_0^t i_1(x)dx$

(6) $\quad I(p) = \dfrac{1}{pL + R + \dfrac{1}{pC}} \cdot \dfrac{V_m}{p} = \dfrac{1}{p^2 + p\dfrac{R}{L} + \dfrac{1}{LC}} \cdot \dfrac{V_m}{L} = \dfrac{V_m}{L} \dfrac{1}{(p+p_1)(p+p_2)}$

(7) $\quad \text{where } p_1, p_2 = -\alpha \mp \beta = -\dfrac{R}{2L}\left[1 \pm \sqrt{1 - \dfrac{4L}{R^2 C}}\right] = -\dfrac{\omega_0}{2Q_s}\left[1 \pm \sqrt{1 - 4Q_s^2}\right]$

Show that the responses to a step function $V_m u(t)$ are

(8) $\quad v_1(p) = R_1 i_1(p) + pL_1 i_1(p) + \dfrac{1}{pC_1} i_1(p) = \left(R_1 + pL_1 + \dfrac{1}{pC_1}\right) i_1(p)$

$\Rightarrow i_1(p) = \dfrac{v_1(p)}{\left(R_1 + pL_1 + \dfrac{1}{pC_1}\right)}$

(9) $\quad T_{606}(p) = \dfrac{v_3(p)}{v_1(p)} = \dfrac{R_1}{\left(R_1 + pL_1 + \dfrac{1}{pC_1}\right)} = \dfrac{pC_1 R_1}{\left(p^2 C_1 L_1 + pC_1 R_1 + 1\right)}$

The $v_1(t)$ waveform *is* modified by the integral and differential operations.

The input signal step function $u(t)$ has zero *rise time* (page 107).

Figure 606

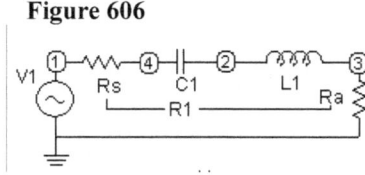

Experimental procedure for Band Pass Filter - Measure $T_{3\text{-}606}$. Signal generator V_1 has a 50Ω source resistance R_s. Connect 150Ω resistor R_a to nodes 3 and ground to make R_1= 200Ω (Figure 606). In effect V_1 is connected to node 1, and R_1 (50 + 150) is connected to circuit 606 as shown. Connect 2000pF capacitor C_1 to nodes 4 and 2. Connect 1000μH inductor L_1 to nodes 2 and 3. This completes circuit 606.

Calculate f_0 (equation 10). The reactance chart shows it is about 112KHz (period T=8.9μs).

(10) $\quad band\ center\ \omega_0 = \dfrac{1}{\sqrt{C_1 L_1}}$

Connect an Oscilloscope channel input to circuit 606 node 3. Set 0v level at screen bottom. Select the time base to be *10μs/cm so that full screen 10 cm sweep takes 100μs.*

Set oscilloscope channel 1 to 1V/cm, channel 2 to 0.5V/cm.
Connect channel 1 to input node 1 and channel 2 to output node 3

On the Function Generator Set up a periodic pulse waveform with a 100μs period (frequency 10KHz), and 50μs pulse width. Adjust output pulse voltage to range from 0V to 5V. Connect to node 1.

Observe the "ringing" (a damped sinewave) at node 3.

Select the time base to be 2μs/cm so that full screen 10 cm sweep takes *20μs.*

Measure the time for the first damped sinewave period (about 8.9μs).

Electric Circuits Experiments   Solved

Solved

Resistors in Series-Parallel Combinations *The purpose here is to assemble, analyze, and evaluate this series parallel circuit as a 2 mesh circuit..*

Figure 207 Two Mesh Circuit

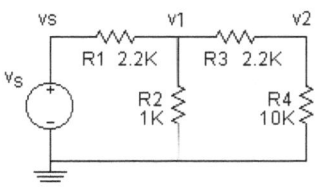

We will solve for currents i and voltages v first using KCL, a node method, and then KVL, a mesh method. In Chapter 5 these methods, restricted here to resistor only circuits, are presented as general methods for any circuits.

The KCL solution - the current connection constraints. The currents are the currents flowing in branches between nodes where i_1 flows through R_1, etc.

node v_1 $0 = i_1 + i_2 + i_3$
node v_2 $0 = i_3 + i_4$

Apply Ohm's law where $i = gv$
$i_1 = g_1(v_1 - v_s)$ $i_2 = g_2(v_1 - 0)$ $i_3 = g_3(v_1 - v_2)$ $i_4 = g_4(v_2 - 0)$

Use the vi constraint to insert resistors into the node equations
node v_1 $0 = g_1(v_1 - v_s) + g_2(v_1 - 0) + g_3(v_1 - v_2)$
node v_2 $0 = g_3(v_2 - v_1) + g_4(v_2 - 0)$

Collect terms
node v_1 $g_1 v_s = (g_1 + g_2 + g_3)v_1 - g_3 v_2$
node v_2 $0 = -g_3 v_1 + (g_3 + g_4)v_2$

calculate the determinant

$\Delta = (g_1 + g_2 + g_3)(g_3 + g_4) - g_3^2 = (g_1 + g_2)g_3 + (g_1 + g_2 + g_3)g_4$
$\Delta = g_1 g_3 + g_2 g_3 + g_1 g_4 + g_2 g_4 + g_3 g_4$
$R_1 R_3 \Delta = 1 + R_1 g_2 + R_3 g_4 + R_1 R_3 g_2 g_4 + R_1 g_4$

Use Cramer's rule

$$\frac{v_2}{v_s} = \frac{1}{v_s} \frac{\begin{vmatrix} g_1 + g_2 + g_3 & g_1 v_s \\ -g_3 & 0 \end{vmatrix}}{\Delta} = \frac{g_1 g_3}{\Delta} = \frac{1}{1 + R_1 g_2 + R_1 g_4 + R_3 g_4 + R_1 R_3 g_2 g_4}$$

Now the KVL solution - the voltage connection constraints

297

Electric Circuits - Analysis and Design
Figure 207 Two Mesh Circuit

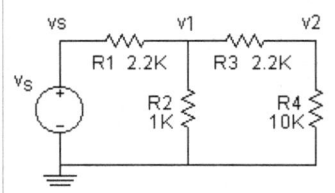

for convenience let $v_{pq} = v_p - v_q > 0$ where p and q are node numbers
mesh 1 $v_s = v_{s1} + v_{10}$
mesh 2 $0 = v_{12} + v_{20} - v_{10}$

apply Ohm's Law where $v = iR$ and clockwise $i_{mesh1} = i_1$, $i_{mesh2} = i_2$
$v_{s1} = i_1 R_1$ $v_{10} = (i_1 - i_2)R_2$ $v_{12} = i_2 R_3$ $v_{20} = i_2 R_4$

Use the vi constraint to insert resistors into the node equations
mesh 1 $v_s = i_1 R_1 + (i_1 - i_2)R_2$
mesh 2 $0 = i_2 R_3 + i_2 R_4 - (i_1 - i_2)R_2$

collect terms
mesh 1 $v_s = (R_1 + R_2)i_1 - R_2 i_2$
mesh 2 $0 = -i_1 R_2 + (R_2 + R_3 + R_4)i_2$
calculate the determinant
$$\Delta = (R_1 + R_2)(R_2 + R_3 + R_4) - R_2^2 = R_1 R_2 + R_1 R_3 + R_2 R_3 + R_1 R_4 + R_2 R_4$$

Use Cramer's rule

$$\frac{v_2}{v_S} = \frac{i_2 R_4}{v_s} = \frac{R_4}{v_s} \frac{\begin{vmatrix} R_1 + R_2 & v_s \\ -R_2 & 0 \end{vmatrix}}{\Delta} = \frac{R_2 R_4}{\Delta} = \frac{1}{1 + R_1 g_2 + R_1 g_4 + R_3 g_4 + R_1 R_3 g_2 g_4}$$

Bridge T node analysis
(1a) node 1 $y_0 v_s = (2y_0 + y_1)v_1$ $- y_0 v_2$ $- y_1 v_3$
(1b) node 2 $0 =$ $- y_0 v_1 + (2y_0 + y_2)v_2$ $- y_0 v_3$
(1c) node 3 $0 =$ $- y_1 v_1$ $- y_0 v_2 + (2y_0 + y_1)v_3$

Take care, because errors are very easy to make while calculating the determinant Δ. Minus signs can be a trap. Cramer's rule a_{jk} are coefficients

Electric Circuits Experiments   Solved

of node voltages or mesh currents. The a_{jk} include any minus signs. (see Cramer's rule).

(2) $\Delta = (2y_0 + y_1)\left[(2y_0 + y_2)(2y_0 + y_1) - y_0^2\right]$
$\qquad -(-y_0)[-y_0(2y_0 + y_1) - y_0 y_1]$
$\qquad +(-y_1)\left[y_0^2 - (-y_1)(2y_0 + y_2)\right]$

$\Delta = 8y_0^3 + 4y_0^2(y_1 + y_2) = \dfrac{4y_0^2}{y_1}(y_0 + y_1)^2$

(3) $v_3 = \dfrac{y_0 v_s \Delta_{13}}{\Delta} = \dfrac{y_0 v_s \left[y_0^2 + y_1(2y_0 + y_2)\right]}{\Delta}$

$\qquad = \dfrac{y_0 v_s \left[y_0^2 + 2y_0 y_1 + y_1 y_2\right]}{8y_0^3 + 4y_0^2(y_1 + y_2)}$ and $y_1 y_2 = y_0^2$

$\dfrac{v_3}{v_s} = \dfrac{y_0^2 + 2y_0 y_1 + y_0^2}{8y_0^2 + 4y_0(y_1 + y_2)} = \dfrac{2y_0(y_0 + y_1)}{4y_0(2y_0 + y_1 + y_0^2/y_1)}$

$\qquad = \dfrac{y_1(y_0 + y_1)}{2(2y_0 y_1 + y_1^2 + y_0^2)} = \dfrac{y_1}{2(y_0 + y_1)} \dfrac{z_0 z_1}{z_0 z_1} = \dfrac{z_0}{2(z_0 + z_1)}$

Attenuator

(4a) $z_0 = r_0, \; z_1 = r_1, \; z_2 = \dfrac{r_0^2}{r_1}$

(4b) $\dfrac{v_3}{v_s} = \dfrac{z_0}{2(z_0 + z_1)} = \dfrac{r_0}{2(r_0 + r_1)} = \dfrac{1}{2(1 + r_1/r_0)}$

(4c) loss in dB $= 20 \log 2(1 + r_1/r_0)$

Low Pass Filter

(5a) $z_0 = r_0, \; z_1 = pL, \; z_2 = \dfrac{r_0^2}{z_1} = \dfrac{r_0^2}{pL} = \dfrac{1}{pC} \;\rightarrow\; C = \dfrac{L}{r_0^2} \quad r_0^2 = \dfrac{L}{C} = z_1 z_2$

(5b) $\dfrac{v_3}{v_s} = \dfrac{z_0}{2(z_0 + z_1)} = \dfrac{r_0}{2(r_0 + pL)} = \dfrac{1}{2(1 + pL/r_0)}$

High Pass Filter

(6a) $z_0 = r_0, \; z_1 = \dfrac{1}{pC} \quad z_2 = \dfrac{r_0^2}{z_1} = pCr_0^2 = pL \;\rightarrow\; L = Cr_0^2 \quad r_0^2 = \dfrac{L}{C} = z_1 z_2$

(6b) $\dfrac{v_3}{v_s} = \dfrac{z_0}{2(z_0 + z_1)} = \dfrac{pCr_0}{2(1 + pCr_0)}$

Electric Circuits - Analysis and Design

Band Pass Filter

(7a) $\quad z_0 = r_0, \quad z_1 = pL + \dfrac{1}{pC} = \dfrac{1+p^2LC}{pC},$

$$y_2 = \dfrac{1}{z_2} = \dfrac{z_1}{r_0^2} = \dfrac{1+p^2LC}{pCr_0^2} = \dfrac{1+p^2LC}{pL}$$

$$pCr_0^2 = pL \;\rightarrow\; z_1 z_2 = \dfrac{L}{C} = r_0^2 = z_0^2$$

(7b) $\quad \dfrac{v_3}{v_s} = \dfrac{z_0}{2(z_0+z_1)} = \dfrac{1}{2}\dfrac{r_0}{r_0+pL+\dfrac{1}{pC}} = \dfrac{1}{2}\times\dfrac{pCr_0}{1+pCr_0+p^2LC}$

Band Stop Filter

(8a) $\quad z_0 = r_0, \quad y_1 = pC + \dfrac{1}{pL} = \dfrac{1+p^2LC}{pL}, \quad z_2 = pL + \dfrac{1}{pC} = \dfrac{1+p^2LC}{pC},$

$$z_0^2 = z_1 z_2 = \dfrac{L}{C}$$

(8b) $\quad \dfrac{v_3}{v_s} = \dfrac{z_0}{2(z_0+z_1)} = \dfrac{1}{2}\dfrac{r_0}{r_0+\dfrac{1}{pC+\dfrac{1}{pL}}} = \dfrac{1}{2}\times\dfrac{p^2LC+1}{1+pL/r_0+p^2LC}$

Bridge T mesh analysis

(1a) mesh 1 $\quad v_s = (2z_0+z_2)i_1 \quad -z_2 i_2 \quad\quad -z_0 i_3$
(1b) mesh 2 $\quad 0 = \quad -z_2 i_1 + (2z_0+z_2)i_2 \quad -z_0 i_3$
(1c) mesh 3 $\quad 0 = \quad -z_0 i_1 \quad\quad -z_0 i_2 + (2z_0+z_1)i_3$

(2) $\Delta = (2z_0+z_2)\left[(2z_0+z_2)(2z_0+z_1)-z_0^2\right]$
$\quad\quad -(-z_2)\left[-z_2(2z_0+z_1)-z_0^2\right]$
$\quad\quad +(-z_0)\left[z_2 z_0 - (-z_0)(2z_0+z_2)\right]$

$$\Delta = 8z_0^3 + 4z_0^2 z_1 + 4z_0^2 z_2 = \dfrac{4z_0^2}{z_1}(z_0+z_1)^2$$

(3) $i_2 = \dfrac{v_s \Delta_{12}}{\Delta} = \dfrac{v_s\left[-z_2(2z_0+z_1)-z_0^2\right]}{\Delta} = \dfrac{v_s\left[-2z_0 z_2 - z_1 z_2 - z_0^2\right]}{\Delta}$

$\quad = \dfrac{-2z_0 v_s [z_2+z_0]}{\Delta} = \dfrac{-2z_0 v_s \left[z_0^2/z_1+z_0\right]}{\Delta} = \dfrac{-2z_0^2 v_s [z_0+z_1]}{z_1 \Delta}$

$\dfrac{v_3}{v_s} = \dfrac{-i_2 z_0}{v_s} = \dfrac{z_0 \times 2z_0^2 [z_0+z_1]}{z_1 \dfrac{4z_0^2}{z_1}(z_0+z_1)^2} = \dfrac{1}{2}\times\dfrac{z_0}{(z_0+z_1)}$

Bridge T Low Pass Filter

```
Fig5012.ckt    bridge T lpf

*asymptote source voltage
Vs   4 0   AC 1 0sin(0 1 50K 0 0); volts
* names/nodes/values
R1 4 1 680
R2 1 2 680
R3 2 3 680
R4 3 0 680
L1 1 3 470E-6
C2 2 0 1000p
*for high pass filter exchange L1 and C2
*.PLOT AC VP(2) -100,0
.AC DEC 200 10 1e+007
.TEMP 27
.PLOT AC VDB(2) -40,10
.PRINT AC VP(2)
.PRINT AC VDB(2)
.end
```

Figure 50121 Low Pass Filter

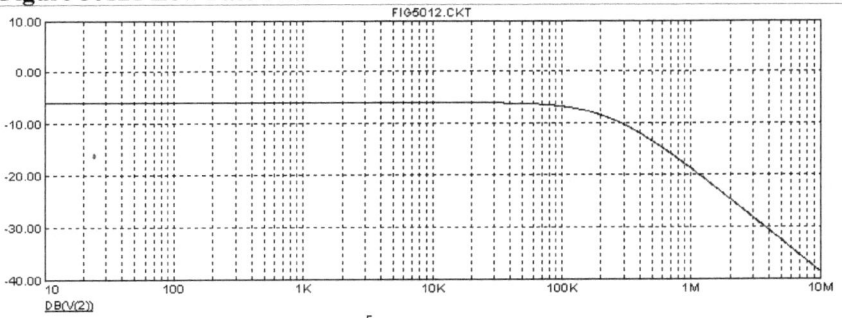

Figure 50131 High Pass Filter

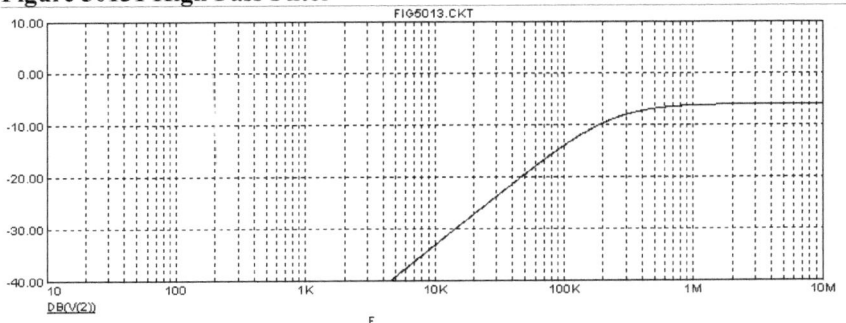

Electric Circuits - Analysis and Design

```
Fig5014.ckt   bridge T band pass filter

*asymptote source voltage
Vs   4 0   AC 1 0 sin(0 1 50K 0 0); volts
* names/nodes/values
R1 4 1 680
R2 1 2 680
R3 2 3 680
R4 3 0 680
L1 1 5 470E-6
C1 5 3 1000p
L2 2 0 470E-6
C2 2 0 1000p
*for band stop filter exchange series and parallel LC
*.PLOT AC VP(2) -100,0
.AC DEC 200 10 1e+007
.TEMP 27
.PLOT AC VDB(2) -40,10
.PRINT AC VP(2)
.PRINT AC VDB(2)
.end
```

Figure 50141 Band Pass Filter

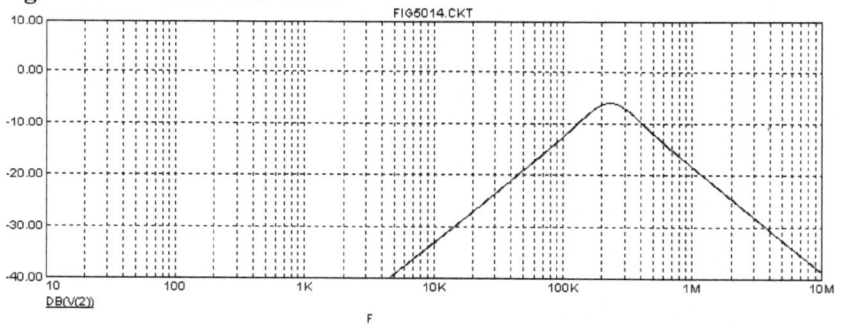

Figure 50151 Band Stop Filter

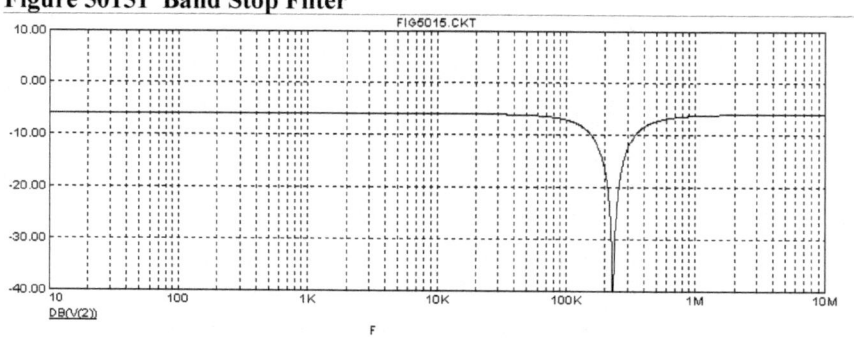

INDEX

ac analysis, traditional 128
ac voltage and current, sine 130
admittance 228
admittance in parallel 231
analog multimeter design 21
answers to some problems 233
attenuator circuit equations 74
attenuator circuit solution 77

Bode method 166
    Bode plots 180
    corner frequency 168
    factor K 170
    factors p and 1/p 171
    factors $p+\omega_0$ and $1/p+\omega_0$. 173
    factors complex 176
    logarithms 167
    Table 1001 neg real pole .. 178
    Table 1002 complex pole . 179
branch currents 14
branch with current source 71
bridge circuit 84

capacitor C 28
    impedance 35
    in a circuit 32
    models 33
    parallel 34
    series 33
    Spice 30
    theory 28
    vi constraint 28
charge 4
circuit analysis 59
    node method 64
    mesh method 69
    practice 81
circuit analysis process 59
circuit components 27

circuit variables
    charge 4
    current 6
    electric field 1
    electrons 4
    power 10
    voltage 8
    vi constraints 59
circuits with resistors 14
    resistors in series 14
    resistors in parallel 16
    resistors in series-parallel .. 18
    voltage division 20
    current division 20
    multimeter design 21
coefficient of coupling k 46
complex frequency plane 125
complex numbers 85, 203
connection constraint 11
corner frequency 168
Cramer's rule 76
current 6
current divider 20

decibels 167
delta-wye 84, 217

easy method, transforms 210
electrostatic field theory 1
electric field 1
electrons 4
equivalent circuits 219
Euler's relation 123
Experiments 253
exponential function 114

303

Electric Circuits - Analysis and Design

fields of force 5
forest and the trees 13
frequency response 87
    resistor 88
    RC low pass filter 90
    RC phase equalizer 95
    RLC resonant circuit 100
full wave rectifier 57

graphs, network 214

ideal transformers 48
immittance 228
impedance 228
impedance in the steady state 132
impedance in series 229
induced voltage 43
inductor L 36
    in a circuit 38
    impedance 41
    models 40
    parallel 40
    series 40
    Spice 39
    theory 37
    vi constraint 38

Kirchhoff's KCL 11
Kirchhoff's KVL 12

ladder circuit 83
Laplace transform 106
    transforms simplify f(t) 107
    transforms simplify ops 109
    transforms-vi constraints .. 113
logarithmic scales 184
logarithms 167
low pass filter circuit equations .. 75
low pass filter circuit solution 78

Maxwell's mesh method 69
    circuit mesh equations 69
    branch with i source 71
meshes M 215
mesh currents 216
mesh method 69
model (see device)
mutual inductance M 43
    dot convention 45
    vi constraint 44
    coefficient k 46

network graphs 214
nodes 213
node method 64
    circuit node equations 64
    branch with v source 66
node pairs 216

Ohm's law 7
Oliver Heaviside 206
one mesh circuits 141
    RL steady state 142
    RL transient state 145
    RC steady state 152
    RC transient state 155

parallel resistors 16
partial fractions 200
phasor AC analysis 133
phasor method 127
poles and zeros 125
power 10

Q 33,40,79,100,126,176,189

Index

RC steady state 152
RC transient state 155
    differentiator 162
    pulses 158
    ramps 160
    square wave fidelity 164
    steps 155
reactance chart method 184
    impedance plots 185
    transfer function plots 190
real and complex frequencies 123
reference Laplace Text 106
reference circuit ... 11,12,64,69,74,77
resistors, circuits 14
resistor R 27
resistance to the right 83
RL steady state 142
RL transient state 145
    pulses 148
    ramps 150
    steps 145
RLC steady state impedances ... 132
rms value 25

Series RC differentiator 162
SI units xvi
sign of current 14
signal sources 226
sinusoidal alternating v and i 86
Spice, AC, DC, TRAN 192
Spice, DC 194
Spice, AC 196
Spice, TRAN 198
Square wave fidelity 164
steady state 62,121,127
subcircuits S 212

Table 1001 neg real pole 178
Table 1002 complex pole pair ... 179
T network 74
total inductance of coils 47
topology 212
traditional AC analysis 128

transformers 42
    ideal transformers 48
    equivalent circuit 49
    frequency response 50
    full wave rectifier 57
    high frequency corner 53
    high & low corner 56
    low frequency corner 51
transforms, easy method 210
transforms simplify functions ... 107
    exponential 107
    damped sin and cos 108
    ramp 108
    step 107
transforms simplify operations .. 109
    constant 109
    first derivative 109
    integral 110
    time delay 110
transient response 105
    Laplace Transform 106
    RL circuit 111
    RC circuit 115
    RLC circuit 118
transient and steady state 121
trees 215

vi constraint (see devices)
vi constraints 59
voltage divider 20

ω phasor examples 135
ω phasor method 127
ω phasor transforms 131
ω phasors and AC analysis 133
wye-delta 84,217

www.ingramcontent.com/pod-product-compliance
Lightning Source LLC
Chambersburg PA
CBHW051626170526
45167CB00001B/74